CURRENT RESEARCH IN NUCLEAR REACTOR TECHNOLOGY IN BRAZIL AND WORLDWIDE

Edited by **Amir Zacarias Mesquita**

Current Research in Nuclear Reactor Technology in Brazil and Worldwide
http://dx.doi.org/10.5772/56032
Edited by Amir Zacarias Mesquita

Contributors

Nikolay Klassen, Yusuke Kuno, Hugo Dalle, João Roberto Mattos, Marcio Dias, Fernando Lameiras, Wilmar Barbosa Ferraz, Rafael Pais, Ana Maria Santos, Hugo Cesar Rezende, André Augusto Campangnole Dos Santos, Moysés Alberto Navarro, Amir Zacarias Mesquita, Elizabete Jordão, Daniel Palma, Aquilino Martinez, Alessandro Gonçalves, Fábio Branco Vaz De Oliveira, Delvonei Alves Andrade, Juliana Pacheco Duarte, Paulo Frutuoso e Melo, Jose Rivero Oliva, Georgy Levanovich Khorasanov, Cristian Ghezzi, Walter Cravero, Nestor Edgardo Sanchez Fornillo, Maria Moreira, Antonio Cesar Guimarães, Igor Leonardovich Pioro, Motoo Fumizawa

Published by InTech

Janeza Trdine 9, 51000 Rijeka, Croatia

Publishing Process Manager Danijela Duric
Technical Editor InTech DTP team
Cover InTech Design team

Additional hard copies can be obtained from orders@intechopen.com

Current Research in Nuclear Reactor Technology in Brazil and Worldwide, Edited by Amir Zacarias Mesquita
p. cm.
ISBN 978-953-51-0967-9

For EU product safety concerns:
IN TECH d.o.o., Prolaz Marije Krucifikse Kozulić 3, 51000 Rijeka, Croatia,
info@intechopen.com or visit our website at intechopen.com.

Meet the editor

Amir Zacarias Mesquita is Doctor of Science in Chemical Engineering - State University of Campinas/Brazil (2005), Master of Science and Nuclear Technologies - Federal University of Minas Gerais/Brazil (1981) and graduated in Electrical Engineering - Federal University of Minas Gerais/Brazil (1978). He is Titular Researcher at Nuclear Technology Development Center (CDTN), a research institute of the Brazilian Nuclear Energy Commission (CNEN), Scholarship of Productivity and Technological Development of Innovative Extension at Brazilian Council for Scientific and Technological Development (CNPq), and Professor of Postgraduate Program in Science and Radiation Technology, Minerals and Materials at CDTN. Dr. Mesquita performs experimental studies on thermal-hydraulic and neutronic of nuclear reactor. He is author of more than one hundred technical papers published in international journals and conference proceedings. He is expert of the International Atomic Energy Agency (IAEA), and the National Centre of Science and Technology Evaluation of Kazakhstan. He is member of the Brazilian Association for Nuclear Engineering (ABEN), Brazilian Society of Mechanical Sciences and Engineering (ABCM), and Editorial Board Member of several journals.

Contents

Preface

Nuclear reactor technology play a number of significant roles in improving the quality of our environment while at the same time has the potential to generate virtually limitless energy with no greenhouse gas emissions during operations. New generations of power plants, safer than the old ones, are in various stages of design and construction. In addition, basic research and nuclear technology applications in chemistry, physics, biology, agriculture, health and engineering have been showing their importance in the innovation of nuclear technology applications with sustainability.Today, there are about 440 nuclear power reactors in operation in 30 countries, including several developing nations. They provide about 15% of the world's electricity. Many more nuclear power stations are under construction or planned. The reliability, safety and economic performance of nuclear power relative to coal or oil have been demonstrated in many countries.

The aim of this book is to disseminate state-of-the-art research and advances in the area of nuclear reactors technologyof authors from Brazil and around the world.It will also serve as a landmark source to the nuclear community, non-nuclear scientists, regulatory authorities,researchers, engineers, politicians, journalists, decision makers and students (our hope for the future). It can be used as a basis for them to critically assess the potential of nuclear techniques to benefit human development, to contribute to the needs of our society, and to help in solving some particular questions.

The book was divided in two parts: the first shows some Brazilian nuclear studies, and the second part shows the investigation from authors across the globe. Topics discussed in the first part of this compilation include: experimental investigation and computational validation of thermal stratification in PWR reactors piping systems, new methods in doppler broadening function calculation for nuclear reactors fuel temperature, isothermal phase transformation of uranium-zirconium-niobium alloys for advanced nuclear fuel, reactivity Monte Carloburnup simulations of enriched gadolinium burnable poison for PWR fuel, utilization of thermal analysis technique for study of uranium-molybdenum fuel alloy, probabilistic safety assessment applied to research reactors, and a review on thestate-of-the art and current trends of next generation reactors.

In the second part of the book include the follow topics: thermal hydraulics study for a ultra high temperature reactor with packed sphere fuels, benefits in using lead-208 coolant for fast reactors and accelerator driven systems, nuclear power as a basis for future electricity production in the world: Generation III and IV reactors, nanostructural materials and shaped solids for essential improvement of energetic effectiveness and safety of nuclear reactors and radioactive wastes, multilateral nuclear approach to nuclear fuel cycles, and a cold analysis of the Fukushima accident.

Finally, I would like to thank all the researchers who attended the call and submitted their works, and also the support of Intech for this opportunity to disseminate our research.

Amir Zacarias Mesquita, ScD.
Researcher and Professor of
Nuclear Technology Development Center (CDTN)
Brazilian Nuclear Energy Commission (CNEN)
Belo Horizonte – Brazil

Section 1

Nuclear Reactors Technology Research in Brazil

Chapter 1

Experimental Investigation and Computational Validation of Thermal Stratification in Piping Systems of PWR Reactors

Hugo Cesar Rezende,
André Augusto Campagnole dos Santos,
Moysés Alberto Navarro,
Amir Zacarias Mesquita and Elizabete Jordão

Additional information is available at the end of the chapter

http://dx.doi.org/10.5772/52614

1. Introduction

One phase thermally stratified flow occurs in horizontal piping where two different layers of the same liquid flow separately without appreciable mixing due to the low velocities and difference in density (and temperature). This condition results in a varying temperature distribution in the pipe wall and in an excessive differential expansion between the upper and lower parts of the pipe walls. This phenomenon can induce thermal fatigue in the piping system threatening its integrity. In some safety related piping systems of pressurized water reactors (PWR) plants, temperature differences of about 200 °C can be found in a narrow band around the hot and cold water interface. To assess potential piping damage due to thermal stratification, it is necessary to determine the transient temperature distributions in the pipe wall (Häfner, 2004) (Schuler and Herter, 2004).

Aiming to improve the knowledge on thermally stratified flow and increase life management and safety programs in PWR nuclear reactors, experimental and numerical programs have been set up at Nuclear Technology Development Center, a researcher institute of the Brazilian Nuclear Energy Commission (CDTN/CNEN) (Rezende, 2012), (Rezende et al. 2012). The Thermal Stratification Experimental Facility (ITET) was built to allow the study of the phenomenon as broadly as possible. The first test section was designed to simulate the steam generator injection nozzle and has the objective of studying the flow configurations and understanding the evolution of the thermal stratification process. The driving parameter considered to characterize flow under stratified regime due to difference in specific masses is the Froude number. Different Froude numbers, from 0.019 to 0.436, were obtained in different testes by setting injection cold water flow rates and hot water initial temperatures.

The use of Computational Fluid Dynamics (CFD) in nuclear reactor safety analyses is growing due to considerable advancements made in software and hardware technology. However, it is still necessary to establish quality and trust in the predictive capabilities of CFD methodologies. A validation work requires comparisons of CFD results against experimental measurements with high resolution in space and time. Recently, some research laboratories have been implementing experimental programs aiming to assist this demand.

The organization of the XVII ENFIR – Seventeenth Meeting on Nuclear Reactor Physics and Thermal Hydraulics propose the Special Theme on Thermal Hydraulics for CFD Codes as a contribution for the validation of CFD methodologies (Rezende el al. 2011a). The experimental results of thermal stratification developed at the Thermal Hydraulics Laboratory of CDTN/CNEN were used for comparisons with CFD results. This Chapter shows the results of the validation done by the Brazilian researcher (Rezende, 2012). The purposes of this special theme are: CDF simulations of a transient with a coolant thermally stratified single phase flow in the steam generation injection nozzle simulating experimental facility; performing comparisons of different CFD simulations and comparisons of CFD simulations with experimental results. Two sets of experimental data are proposed for the numerical simulation.

Numerical simulation was performed with the commercial finite volume Computational Fluid Dynamic code CFX. A vertical symmetry plane along the pipe was adopted to reduce the geometry in one half, reducing mesh size and minimizing processing time. The RANS two equations RNG *k*- turbulence model with scalable wall function and the full buoyancy model were used in the simulation. In order to properly evaluate the numerical model a Verification and Validation (V&V) process was performed according to an ASME standard. Numerical uncertainties due to mesh and time step were evaluated. The performed validation process showed the importance of proper quantitative evaluation of numerical results. In past studies a qualitative evaluation of the results would be considered sufficient and the present model would be (as it has been) considered very good for the prediction and study of thermal stratification. However, with the present V&V study it was possible to identify objectively the strengths and weaknesses of the model.

Results show the influence of Froude number on the hot and cold water interface position, temperature gradients and thermal striping occurrence. Results are presented in terms of wall temperature, internal temperature, vertical probe temperature, temperature contours and velocity fields.

2. The thermal stratification experimental facility at CDTN

The Thermal Stratification Experimental Facility (ITET) wasbuilt in the Thermal-hydraulic Laboratory at Nuclear Technology Development Center (CDTN) (Fig. 1) to allow the study of the phenomenon as broadly as possible. The first test section was designed to simulate the steam generator injection nozzle. Figure2shows a drawing of this test section that consists of a stainless steel tube (AISI 304 L), 141.3 mm in outside diameter and9.5 mm thick.

It was made of two pieces of this tube connected each other by a 90° curve, a vertical and a horizontal piece respectively 500 mm and 2000 mm length. A flanged extension of the tube was placed inside a pressure vessel, which simulates the steam generator. Thermocouples were placed in four Measuring Stations along the length of the test section tube. Measuring Stations I, II and III, located in the horizontal length of the tube were instrumented with thermocouples, measuring both fluid and wall temperature at several positions of each Measuring Station. Measuring Station A, positioned in the vertical length of the tube, was instrumented with three thermocouples just to determine the moment when the injected cold water reaches its position.

Figure 1. Thermal-hydraulic Laboratory at Nuclear Technology Development Center (CDTN)

Experiment	Flow rate [kg/s]	Pressure [Pa]	Initial system temperature [°C]	Cold water injection temperature [°C]
2146	0.76	2.11×10^6	32	220
2212	1.12	2.14×10^6	28	221

Table 1. Input data for the proposed experiments

Before the beginning of each test the whole system is filled with cold water. Then it is pressurized and heated by steam supplied by a boiler. A temperature equalization pump ensures that the entire system is heated in a homogeneous way. After the heating process, the equalization pump is turned off and both the steam supply and equalization lines are isolated by closing valves V3, V5 and V6. The test itself begins then by injecting cold water from the lower end of the vertical tube after opening valve V4. The cold water flow rate was previously adjusted at a value planned in the test matrix. This flow rate and the system pressure are maintained stable through a set of safe (V1) and relieve (V2) valves at the upper side of the pressure vessel, which controls upstream pressure. The water flows from the injection nozzle simulator pipe to the steam generator simulator vessel through 11 holes at the upper side of the extension tube placed inside the vessel. These holes are 12 mm in diameter and they are displaced 42 mm from each other. The center of the first hole is 20 mm from the end of the tube.

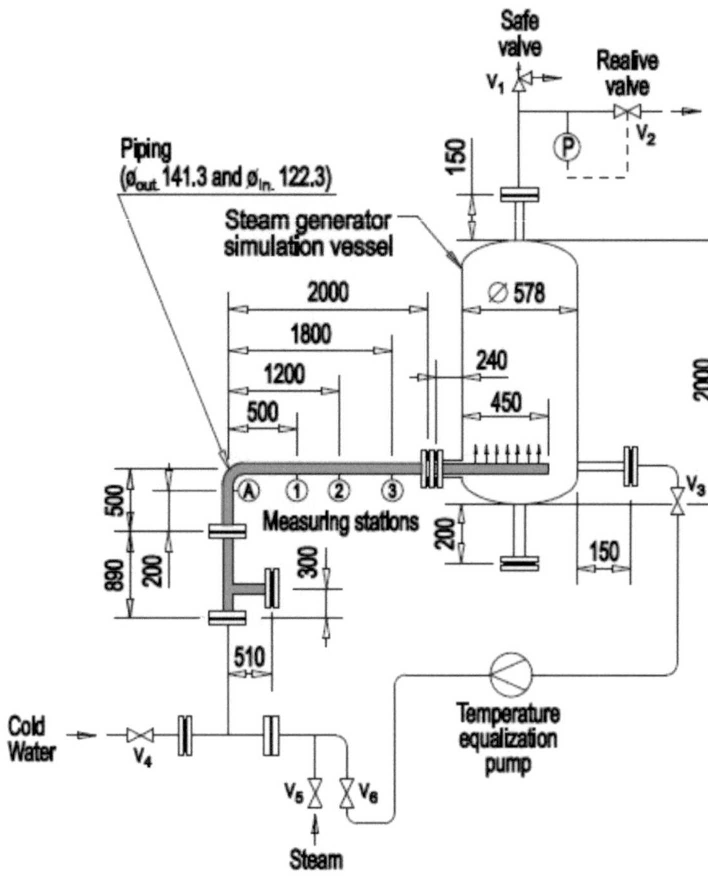

Figure 2. Position of the Measuring Stations A, I, II and III in the steam generator injection nozzle simulating test section

2.1. The instrumentation

Measurement Stations I, II and III, positioned along the longitudinal length of the tube simulating the steam generator injection nozzle, as shown in Figure 2, were used for temperature measurements. Figures 3, 4 and 5 show the thermocouples distribution in Measuring Stations I, II and III, respectively. To measure fluid temperature on Measurement Station I a set of 12 thermocouples was angularly distributed along the tube's internal wall (3 mm from the wall), shown in Figure 3 by circle symbols. These internal thermocouples were named clockwise starting from the highest vertical position as T1I01, T1I02, ..., T1I11 and T1I12. To measure the tube's wall temperature another set of 12 thermocouples was brazed on the outside wall at the same angular position as the internal thermocouples, displayed by triangle symbols in Figure 3. These external thermocouples were named clockwise starting from the highest vertical position as T1E01, T1E02, ..., T1E11 and T1E12. Finally, a removable probe was placed along the cross section's vertical diameter, containing a set of 9 fluid thermocouples placed at the same vertical position of each of the internal thermocouples, shown by square symbols in Figure 3. These probe thermocouples were named from the highest to the lowest vertical position as T1S01, T1S02,.....T1S08 and T1S09.

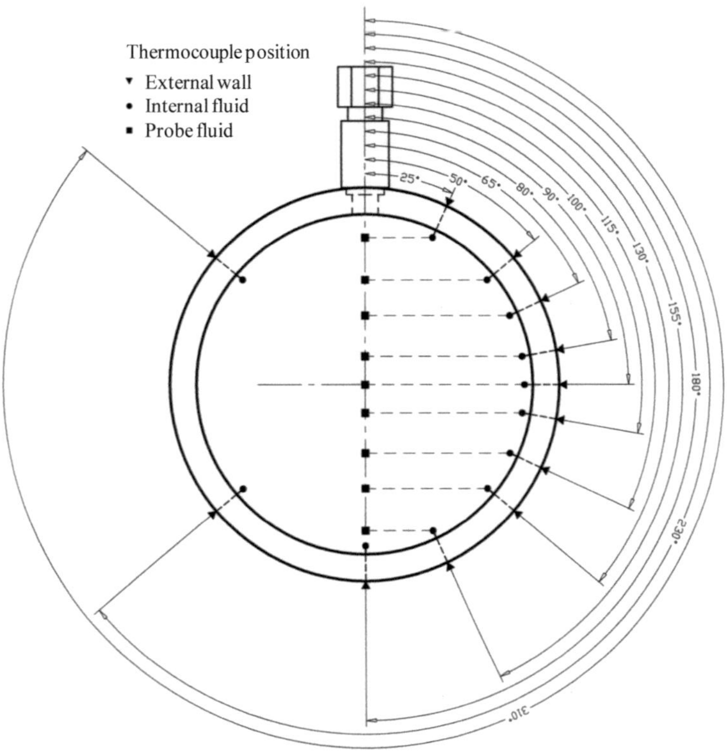

Figure 3. Positions of the thermocouples at Measurement Station I

Figure 4 shows the thermocouple distribution on Measurement Station II. A set of 19 thermocouples was angularly distributed along the tube's internal wall (3 mm from the wall) to measure fluid temperature, shown in Fig. 4 by circle symbols. Close to the angular position of 90° a set of 5 internal thermocouples was positioned in close proximity, displaced 2 mm from each other, to capture fluctuations of the cold-hot water interface. In the opposite side 2 internal thermocouples were positioned in the same manner to capture asymmetrical behaviors of the interface. These internal thermocouples were named clockwise starting from the highest vertical position as T2I01, T2I02, ..., T2I18 and T2I19. Another set of 14 thermocouples was brazed on the outside wall at the same angular position as the internal thermocouples (only 1 external thermocouple was positioned at the angular positions of 90° and 270°), shown in Fig. 4 by triangle symbols. These external thermocouples were named clockwise starting from the highest vertical position as T2E01, T2E02, …, T2E13 and T2E14. Finally, a removable probe was placed along the cross section's vertical diameter, containing a set of 10 fluid thermocouples placed at the same vertical position of each of the internal thermocouples, as shown in Fig. 4 by square symbols. These probe thermocouples were named from the highest to the lowest vertical position as T2S01, T2S02,...., T2S09 and T2S10.

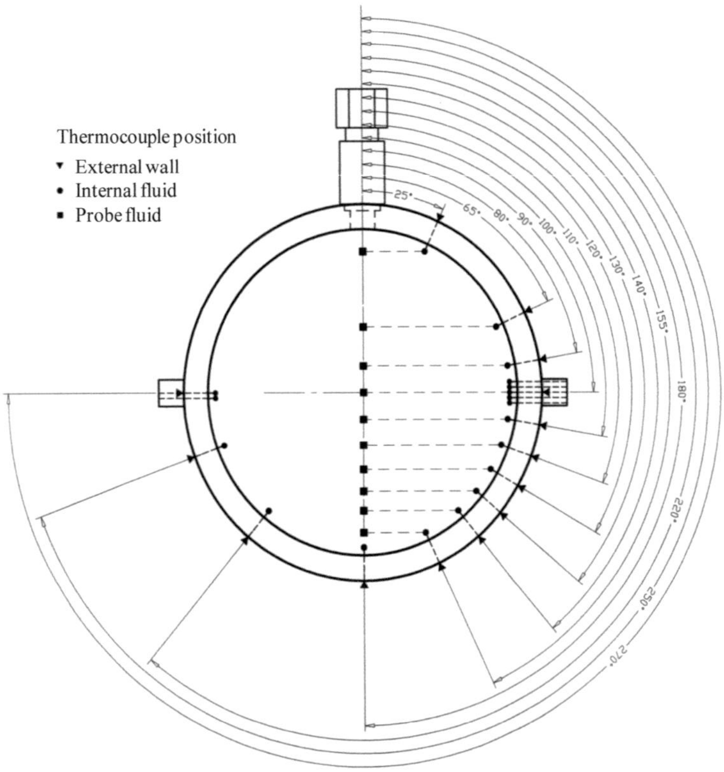

Figure 4. Positions of the thermocouples at Measurement Station II

Figure 5 shows the thermocouple distribution on Measurement Station III. Close to the angular position of 90° a set of 4 internal thermocouples, named from the highest to the lowest vertical position as T3I01, T3I02, T3I03, and T3I04, was positioned 3 mm from the internal wall and displaced 2 mm from each other to measure fluid temperature. A fifth internal thermocouple, named T3I05, was placed at the angular position of 180°, shown in Fig. 5 by circle symbols. Two thermocouples, named T3E01 and T3E02, were brazed on the outside wall of the tube at the angular positions of 90° and 180° respectively, shown by triangle symbol in Fig. 5. Finally, a removable probe was placed along the cross section's vertical diameter containing a set of 6 fluid thermocouples, shown as square symbols in Fig. 5. These probe thermocouples were named T3S01, T3S02, T3S03, T3S04, T3S05 and T3S06 from top to bottom. They were placed respectively at the same vertical positions of thermocouples T2S03, T2S04, T2S05, T2S07, T2S08 and T2S10.

A set of three thermocouples was positioned at Measuring Station A to detect the instant when the injected cold water reaches its position. The thermocouples were placed inside the tube 3 mm from the wall, at the center of the cross section by a probe and at the external wall

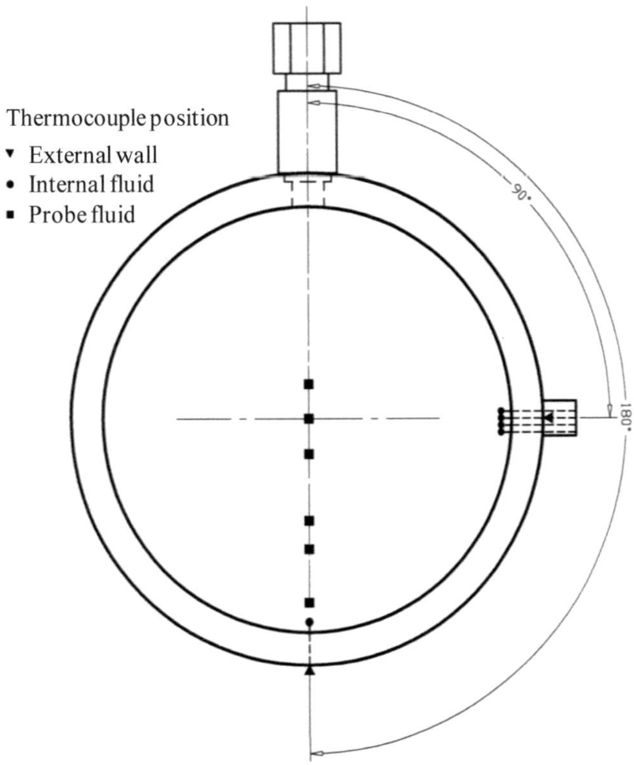

Figure 5. Positions of the thermocouples at Measurement Station III

Figure 6 shows a photograph of the test section pipe after the brazing of the thermocouples. Figure 6 shows in detail the outside of Measuring Station I. The external thermocouples were brazed directly to the pipe and the internal thermocouples were brazed through special stainless steel injection needles. Some aluminum brackets for the thermocouples are seen in the back, which were only used during the assembly of the experimental facility. Figure 7 shows the Measuring Station I internal thermocouples. Figure 7 shows a photograph of the ITET, including the horizontal tube of the injection nozzle, the pressure vessel simulating the steam generator and the cold water tank.

Other measurements performed were:

- injection flow rate of cold water, using a set of orifice plate and differential pressure transmitter;
- water temperature in the cold water tank, using an isolated type K thermocouple of 1 mm in diameter;

- water temperature in the cold water injection pipe, both close to the orifice plate and also close to the point of injection to the nozzle simulation tube, using two isolated type K thermocouples of 1 mm in diameter;
- temperature inside the steam generator simulation vessel, using an isolated type K thermocouple of 1 mm in diameter;
- pressure inside the steam generator simulation vessel, using a gauge pressure transducer;
- pressure in water injection line, using a gauge pressure transducer;
- water level in the cold water tank using a differential pressure transmitter.

Figure 6. The test section's horizontal pipe after the thermocouples brazing, and detail of the measuring station

Figure 7. The internal thermocouples in the Measuring Station I, and the Thermal Stratification Experimental Facility (ITET) during assembly

2.3. The measuring uncertainty

The measuring uncertainties for the main parameters, obtained according to ISO (1993), were:

- 2.4°C for the temperature measurements;
- 2.4 % of the measured value for the flow rate measurements; and,
- 1.5 % for the gauge pressure measurements.

3. Simulation results

The experimental results of thermal stratification developed at the Thermal Hydraulics Laboratory of CDTN/CNEN were used for comparisons with CFD results. In recent theoretical evaluations, CFD (Computational Fluid Dynamic) analysis using three dimensional Reynolds Averaged Navier Stokes (RANS) has been used, which is due to several reasons, from the ease of use of commercial codes and development of low costs computational systems of reasonable processing capacity, to the speed at which results are obtained.

However, before CFD can be considered as a reliable tool for the analysis of thermal stratification there is a need to establish the credibility of the numerical results. Procedures must be defined to evaluate the error and uncertainty due to aspects such as mesh refinement, time step, turbulence model, wall treatment and appropriate definition of boundary conditions. These procedures are referred to as Verification and Validation (V&V) processes (Roache, 2010). In 2009 a standard was published by the American Society of Mechanical Engineers (ASME) establishing detailed procedures for V&V of CFD simulations (ASME, 2009).

According to the Standard for Verification and Validation in Computational Fluid Dynamics and Heat Transfer – V&V 20 (ASME, 2009), the objective of validation is to estimate the modeling error within an uncertainty range. This is accomplished by comparing the result of a simulation (S) and an experiment (D) at a particular validation point. The discrepancy between these two values, called comparison error (E), can be defined by Equation 1 as the combination of the errors of the simulation ($\delta_s = S - \text{True Value}$) and experiment ($\delta_{exp} = D - \text{True Value}$) to an unknown True Value.

$$E = S - D = \delta_s - \delta_{exp} \tag{1}$$

The simulation error can be decomposed in input error (δ_{input}) that is due to geometrical and physical parameters, numerical error (δ_{num}) that is due to the numerical solution of the equations and modeling error (δ_{model}) that is due to assumptions and approximations. Splitting the simulation error in its three components and expanding Equation 1 to isolate the modeling error gives Equation 2.

$$\delta_{model} = E - (\delta_{num} + \delta_{input} - \delta_{exp}) \tag{2}$$

The standard applies then to this analysis the same concepts of error and uncertainty used in experimental data analysis, defining a validation standard uncertainty, u_{val} as an estimate of the standard deviation of the parent population of the combination of the errors in brackets in Equation 2, in such a way that the modeling error falls within the range $[E + u_{val}, E - u_{val}]$, or using a more common notation:

$$\delta_{model} = E \pm u_{val} \tag{3}$$

Supposing that the errors are independent, u_{val} can be defined as Equation 4.

$$u_{val} = \sqrt{u_{num}^2 + u_{input}^2 + u_{exp}^2} \qquad (4)$$

The estimation of these uncertainties is at the core of the process of validation. The experimental uncertainty can be estimated by well established techniques (ISO, 2003). Input uncertainty is usually determined by any propagation techniques or analytically (ASME, 2009). The numerical uncertainty, on the other hand, poses greater difficulties to access.

The estimation of the numerical uncertainty is called verification and is usually split into two categories: code and solution verification. Code verification evaluates the mathematical correctness of the code and is accomplished by simulating a problem that has an exact solution and verifying if that solution is obtained. This activity requires extensive programming access to the core of the code which is not available in commercial codes, due to this it is common practice to take commercial codes as verified by the supplier.

Solution verification is the process of estimating the numerical uncertainty for a particular solution of a problem of interest. The two main sources of errors here are the discretization and iteration processes. The discretization error is the difference between the result of a simulation using a finite grid in time and space and that obtained with an infinitely refined one. The methods developed to evaluate it are based on a systematic grid refinement study where the solution is expected to asymptotically approximate the exact value as the grid is refined, at a rate proportional to the discretization order of the solution. The iteration error is present in codes that use iterative solvers, where the result must converge to the exact value as the iterations develop. It is usually estimated using the residual root mean square (RMS) between subsequent iterations of a variable over all the volumes of the domain.

The numerical simulation of the Experiment 1 shown in Table 2 was performed by Resende et al (2011b and 2011c) using CFX 13.0 (ANSYS, 2010) code in a simplified geometry. The geometry in Fig. 2 was simulated with the omission of the flanges and most of the lower inlet geometry, as shown in Fig. 8. These simplifications have no significant influence on the results. A second flow condition showed in Table 2 (Experiment 2) was also simulated to further evaluate the numerical methodology.

Experiment	Flow rate [kg/s]	P_{gauge} [bar]	T_{hot} [°C]	T_{cold}[°C]
1	0.76	21.1	219.2	31.7
2	1.12	21.4	217.7	28.7
(1)	0.03	0.5	2.4	2.4

(1) Global uncertainty

Table 2. Setup parameters for the experiments and simulations

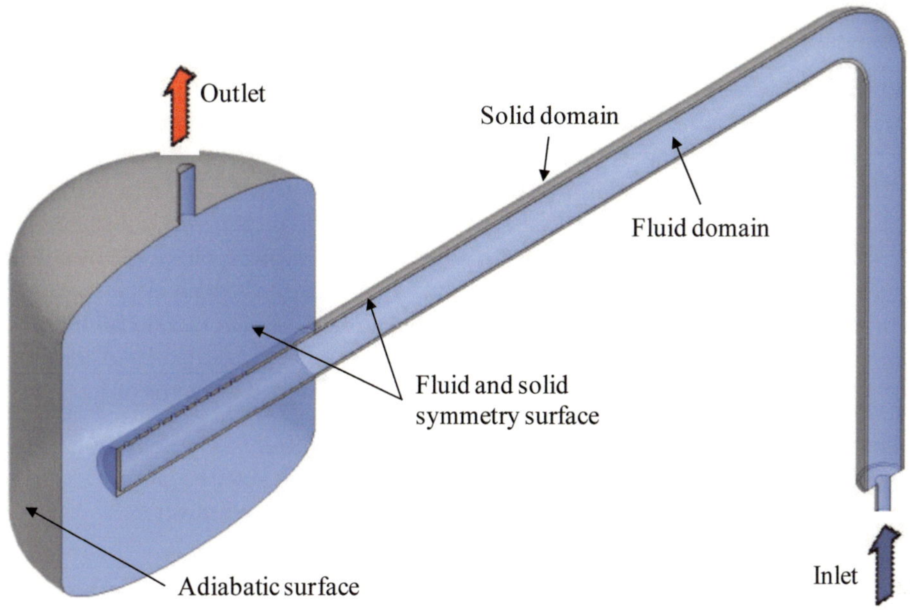

Figure 8. Computational model domains and boundary conditions.

The computational model was generated with two domains: one solid, corresponding to the pipes, and one fluid for the water in its interior. A vertical symmetry plane along the pipe was adopted to reduce the mesh size in one half, minimizing processing time. The walls in the vessel region were considered adiabatic as the external tube walls. Mass flow inlet and outlet conditions were defined at the bottom end of the pipe and high end of the vessel, respectively. Figure 2 shows the computational model's details.

The initial conditions shown in Table 2 were used in the simulations. Water properties like density, viscosity and thermal expansivity were adjusted by regression as function of temperature with data extracted from Table IAPWS-IF97, in the simulation range (25 °C to 221 °C). The RANS - Reynolds Averaging Navier-Stokes equations, the two equations of the RNG *k*- turbulence model, with scalable wall functions, the full buoyancy model and the total energy heat transfer model with the viscous work term were solved. The simulations were performed using parallel processing with up to six workstations with two 4 core processor and 24 GB of RAM. All simulations were performed using the high resolution numerical scheme (formally second order) for the discretization of the conservation and RNG k- turbulence model equations terms and second order backward Euler scheme for the transient terms. A root mean square (RMS) residual target value of 10^{-6} was defined as the convergence criteria for the simulations in double precision. By using this RMS target the interactive error is minimized and can be neglected in the uncertainty evaluation as its contribution are usually many orders lower that of other sources like discretization (Roache, 2010).

A mesh and time step study described in the following section were performed according to ASME V&V 20 standard to assess the numerical uncertainty (ASME, 2009).

A solution verification study was performed according to ASME CFD Verification and Validation standard to evaluate mesh and time step uncertainties (ASME, 2009).

Three gradually refined non-structured tetrahedral meshes with prismatic near wall elements (inflated) were generated for the model presented in Fig. 2 to evaluate mesh related uncertainty. Progressive grid refinements were applied to edge sizing of the piping elements. The ratio between the height of the last prismatic layer and the first tetrahedral was kept equal to 0.5 for all meshes. Three layers of prismatic structured volumes were built close to the surfaces in the solid and fluid domains. The growth factor between prismatic layers was maintained constant with a value of 1.2. A localized mesh edge sizing of 5 mm was applied at the inlet nozzle of the vertical pipe and vessel outlet nozzle for all meshes. At the outlet holes of the horizontal pipe an edge sizing of 2 mm was also used for all meshes. Element sizing in the vessel was set to expand freely with a growth factor of 1.2.

The characteristics of the generate meshes are shown in Table 3. The table includes the resulting grid refinement ratio (r_i) and representative grid edge size (h_i) defined by Equations 5 and 6, respectively. Figure 9 shows some details of the generated meshes.

$$r_i = h_{\text{last coarse mesh } i+1} / h_{\text{present mesh } i} \tag{5}$$

$$h_i = (\text{Model volume} / \text{Number of elements of i mesh})^{1/3} \tag{6}$$

Mesh i	h_i [mm]	No. of elements / nodes	r_i	Element Edge Length [mm]
1	2.84	2,809,114 / 13,533,642	1.83	2.5
2	5.22	583,012 / 2,191,174	1.67	5.0
3	8.70	198,152 / 472,909	-	10.0

Table 3. Meshes characteristics

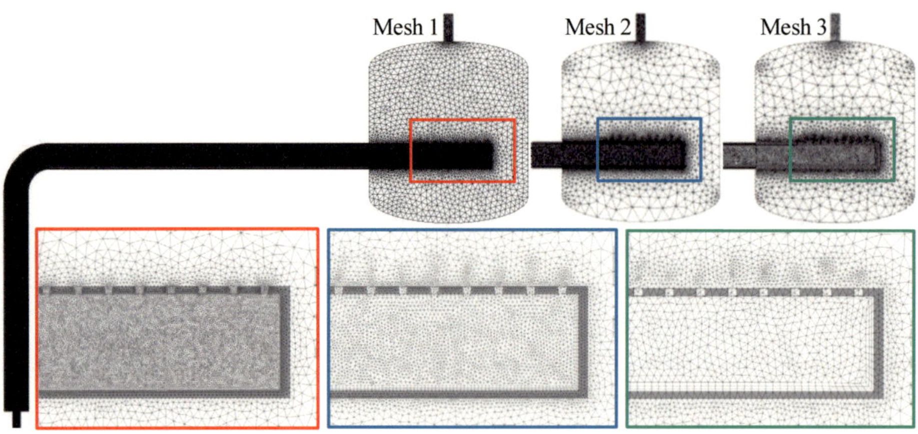

Figure 9. Mesh details.

To evaluate time step related uncertainty, three gradually refined time steps shown in Tab. 4 were used for the simulation of the model with mesh 2 presented in Fig. 3. Table 4 includes the resulting time step refinement ratio (r_j) defined by Equation 7.

$$r_j = t_{last\ coarse\ time\ syep\ j+1} / t_{present\ time\ step} \tag{7}$$

Time j	t_j [s]	r_j
1	0.075	1.51
2	0.113	1.50
3	0.169	-

Table 4. Time steps characteristics

Solution verification was performed using the three generated meshes and three simulated time steps based on the Grid Convergence Index method (GCI) of the ASME V&V 20 standard (ASME, 2009). The theoretical basis of the method is to assume that the results are asymptotically converging towards the exact solution of the equation system as the discretization is refined with an apparent order of convergence (p) that is in theory proportional to the order of the discretization scheme. The objective of the method is to determine p utilizing three systematically refined discretizations and determine relative to the finest discretization result a 95% confidence interval ($\pm U_{num\ 95\%} = \pm GCI$) where the exact solution is. In other word, the objective is to determine the expanded uncertainty interval due to the discretization.

Considering the representative grid edge sizes $h_{i-1}<h_i<h_{i+1}$ and grid refinement ratios $r_i=h_{i+1}/h_i$, the apparent order of convergence p can be determined by Equations 8, 9 and 10. In an analogous manner similar equations can be obtained for time discretization, however these will be omitted for brevity.

$$p_i = \frac{1}{ln(r_i)} \left| ln \left| \varepsilon_{i+1}/\varepsilon_i \right| + q(p_i) \right| \tag{8}$$

$$q(p_i) = ln\left(\frac{r_i^{p_i} - s}{r_{i+1}^{p_i} - s}\right) \tag{9}$$

$$s = 1 \bullet sgn(\varepsilon_{i+1}/\varepsilon_i) \tag{10}$$

where $\varepsilon_{i+1} = \phi_{i+2} - \phi_{i+1}$, $\varepsilon_i = \phi_{i+1} - \phi_i$, ϕ_k denotes the variable solution on the kth grid and sgn is the signal function (sgn(x) = -1 for x < 0; 0 for x = 0 and 1 for x > 0).

It is recommended by the standard ASME (2009) that the obtained value of p be limited to the maximum theoretical value, which for the used high resolution and Euler discretization scheme is 2. Also the value of p can be limited to a minimum of 1 to avoid exaggerations of the predicted uncertainty, however when limited it is recommended that the obtained value is presented for comparison.

With the value of p the expanded uncertainty GCI can be calculated using Equation 11 using an empirical Factor of Safety (Fs), equal to 1.25, that is recommended for studies with more than three meshes (ASME, 2009).

$$GCI_i = \frac{Fs \bullet \varepsilon_i}{r_i^{p_i} - 1} \qquad (11)$$

When the presented procedure is applied to obtain the GCI for local variables, such as a temperature profile, an average value of p should be used as to represent a global order of accuracy.

Mesh and time step uncertainties are considered independent in this study and the total numerical expanded uncertainty is calculated through Equation 12.

$$U_{num} = \sqrt{GCI_{mesh}^2 + GCI_{time\ step}^2} \qquad (12)$$

In this study the temperature profiles along time in several positions of the test section were evaluated. Figure 10 displays the analyzed positions that are equivalent to the thermocouple positions of the experiments.

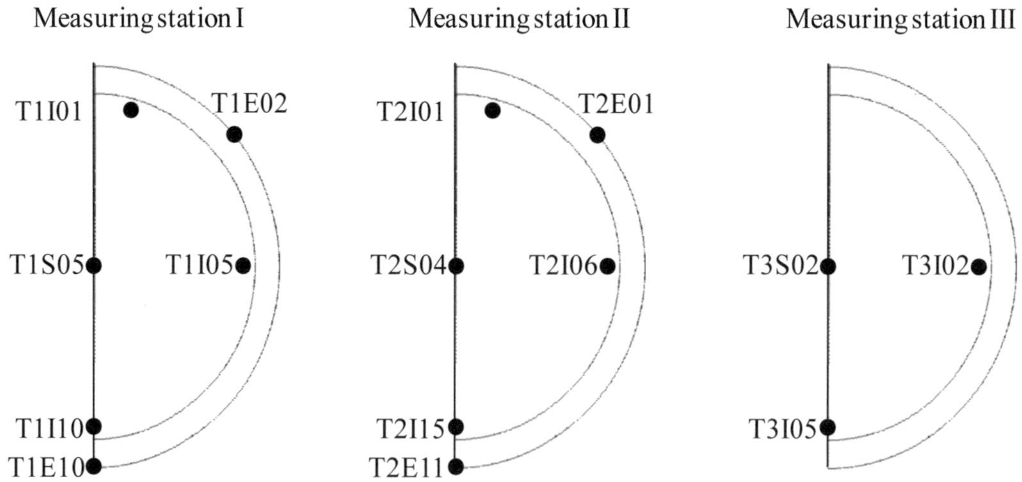

Figure 10. Thermocouples positions

Table 5 shows the some of the obtained results of the performed verification process. Average values for p and GCI are presented as the maximum GCI of the entire profile. These maximums were all located in regions of steep temperature gradients, which explain the very high observed values.

Position in the pipe	Mesh			Time step:		
	p_m*	GCI_m* [°C]	Maximum GCI_m [°C]	p_t*	GCI_t* [°C]	Maximum GCI_t[°C]
Internal						
T1I01	1.58	14.012	41.608	1.00	0.056	0.173
T1I05	1.88	1.174	35.020	1.32	0.415	17.257
T1I10	1.52	0.496	75.830	1.27	0.578	45.339
T2I01	1.32	7.394	22.829	1.31	0.030	0.218
T2I06	1.87	1.377	51.166	1.22	0.594	21.577
T2I15	1.48	1.489	99.554	1.20	0.748	70.006
T3I02	1.80	1.099	64.103	1.21	0.544	13.988
T3I05	1.47	1.220	122.122	1.23	0.584	51.247
Probe						
T1S05	1.64	2.034	60.014	1.23	0.546	14.964
T2S04	1.65	1.766	58.885	1.16	0.902	15.776
T3S02	1.44	2.381	45.358	1.32	0.796	20.076
External						
T1E02	1.61	2.198	6.347	1.05	0.010	0.095
T1E10	1.56	0.199	0.601	1.21	0.087	0.537
T2E01	1.34	0.164	1.115	1.17	0.003	0.015
T2E11	1.16	1.599	3.302	1.47	0.0185	0.621

* Time averaged values.

Table 5. Verification process results for several thermocouple positions.

It can be observed in Table 5 that uncertainties due to the mesh are in average greater than those due to the time step. One reason for these values could be attributed to the course mesh used in the study that could lead to overestimation of the total uncertainty of the refined mesh. In average only thermocouples T1I01 and T2I01 displayed uncertainties above the experimental one of 2.4 °C, both located in the upper region of the vertical tube which indicates that this region is the most affected by the mesh refinement.

An example of the obtained results from the verification process is shown in Fig. 11 that displays the temperature profiles at thermocouple position T2S04 obtained by the simulated meshes and time steps. The figure also shows the uncertainties obtained along the simulated time. It is observed that the mesh contribution to uncertainty is much greater than that of the time step. The grater values of uncertainty were obtained the abrupt temperature drop region and at the subsequent temperature oscillation period.

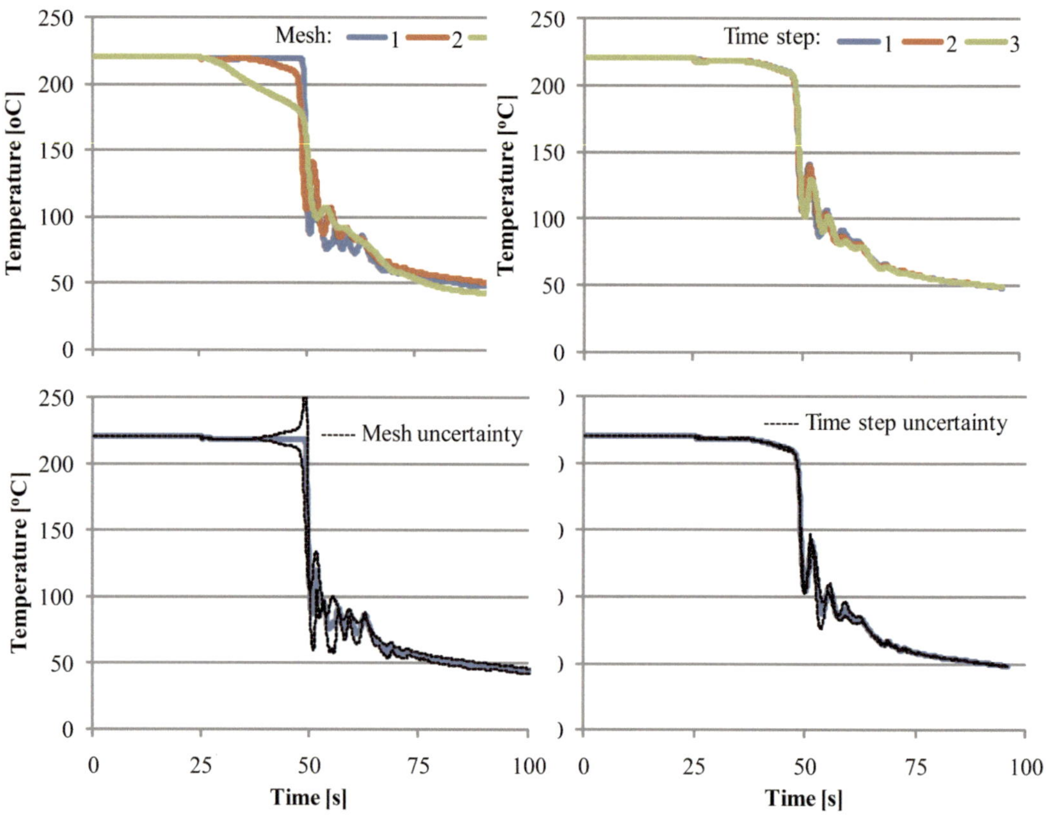

Figure 11. Numerical uncertainty evaluation due to the mesh and time step

The obtained uncertainty prediction through the solution verification process proposed by ASME V&V 20 standard ASME (2009) showed very variable and sometimes incoherent results for the uncertainty prediction. The method takes in account three discretizations for the estimate of GCI and requires that the results between these discretizations be "well behaved" to produce coherent uncertainty estimates. Convergence must be "well behaved" due to the core assumption made by the method that the solution is converging asymptotically as the mesh is refined. This is a very strong assumption as it has been concluded in recent studies that it is safest to assume that the numerical data are not within the asymptotic regime (Eça et al., 2009). It is in fact questionable that even the finest meshes in use today can produce solutions that are in this regime (Lockard, 2010). However, the obtained method gives a good insight as to how results are varying as the mesh is refined and the estimated uncertainty may not be accurate but is a quantification on how "well behaved" is the solution.

Following the solution verification, a validation process was performed comparing the numerical results with experimental data. To determine the validation expanded uncertainty, U_{val} (Eq. 4), only the estimated numerical and experimental uncertainties were considered neglecting the input contribution. Although the input uncertainty is in fact non-neglectable, its evaluation is beyond the purpose of this study as it is extremely complex requiring hundreds of simulations taking in account fabrication tolerances and uncertainties in all measurable variables.

Figures 12 shows a comparison between numerical and experimental results as well as the validation error (E = S – D) and validation uncertainty for the upper thermocouples. Very high validation uncertainty after the beginning of the temperature drop can be observed. This high uncertainty is attributed to the mesh that influences greatly the results in this region. Validation become poor after 150 s of simulation, however before that time numerical and experimental results agree well.

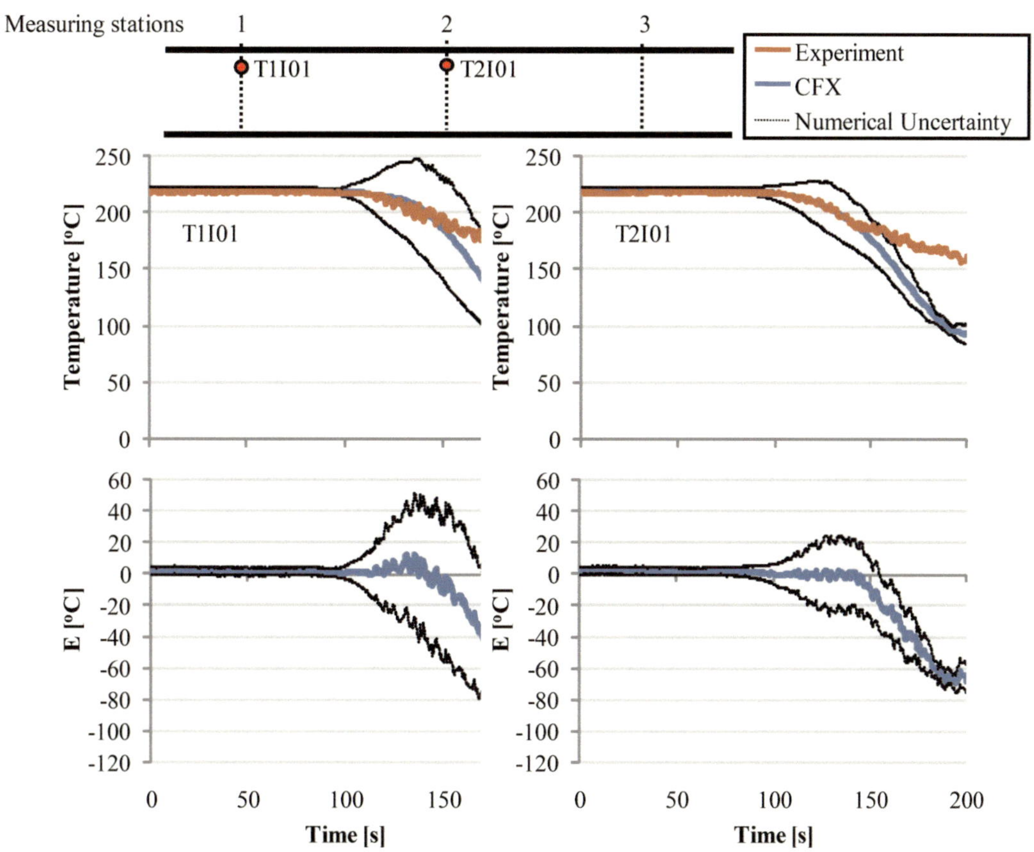

Figure 12. Validation results for the upper thermocouples

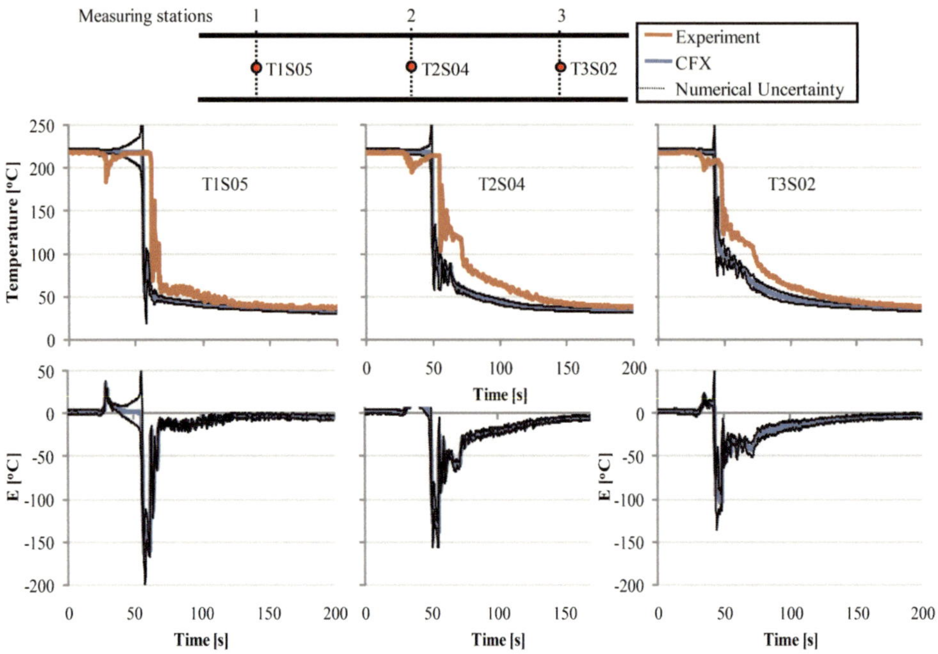

Figure 13. Validation results for the probe thermocouples

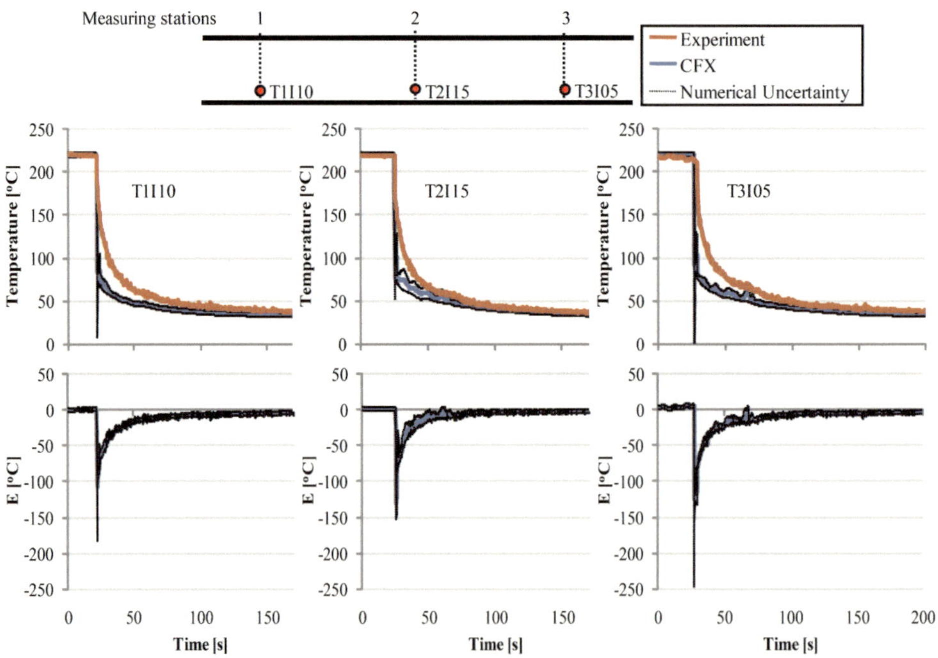

Figure 14. Validation results for the lower thermocouples

Figures 13 and 14 show a comparison between numerical and experimental results with the validation uncertainty and the validation error (E = S – D) for the probe and lower thermocouples, respectively. For both regions results show a high validation error and uncertainty for the beginning of the temperature drop and subsequent oscillations. It is observed that at the cold water front reaches the center of the pipe (probe thermocouples) before the experiment and that the temperature drop in the lower region of the pipe is quicker in the simulation. Although considerable validation error is observed the qualitative agreement between experiment and simulation can be considered good as most of the behavior observed was reproduced.

Figure 15 shows the evolution of the temperature differences between the average temperatures on the highest and lowest positions of the horizontal tube calculated through the Equation 13 for the internal and external thermocouples.

$$DT = \left[\left(T_1^u + T_2^u\right) - \left(T_1^l + T_2^l\right)\right] / 2 \tag{13}$$

where the superscripts u and l are relative to the upper and the lower positions and subscripts 1 and 2 to the first and the second measuring station of the horizontal tube.

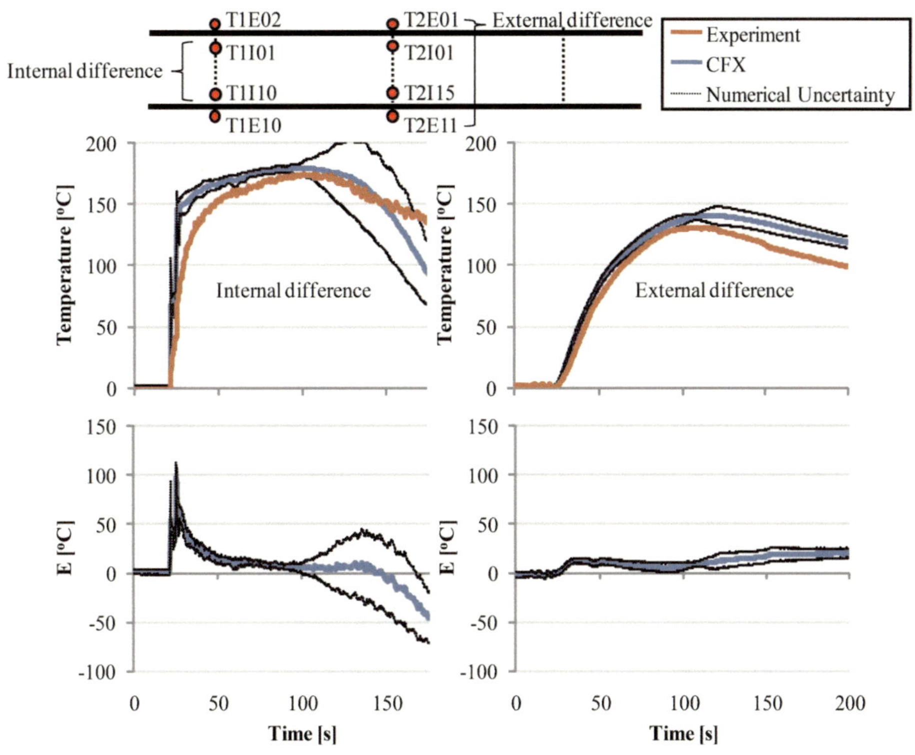

Figure 15. Validation results for the temperature difference between upper and lower thermocouples positioned internally and externally.

Figure 15 shows that for the region of highest temperature difference, and therefore, most critical for the piping integrity, the validation error is relatively low and well predicted. It is also observed that the external temperature difference agreement between experimental and numerical results is very good during the evaluated time.

The performed validation process showed the importance of proper quantitative evaluation of numerical results. In past studies a qualitative evaluation of the results would be considered sufficient and the present model would be (as it has been) considered very good for the prediction and study of thermal stratification. However, with the present V&V study it was possible to identify objectively the strengths and weaknesses of the model.

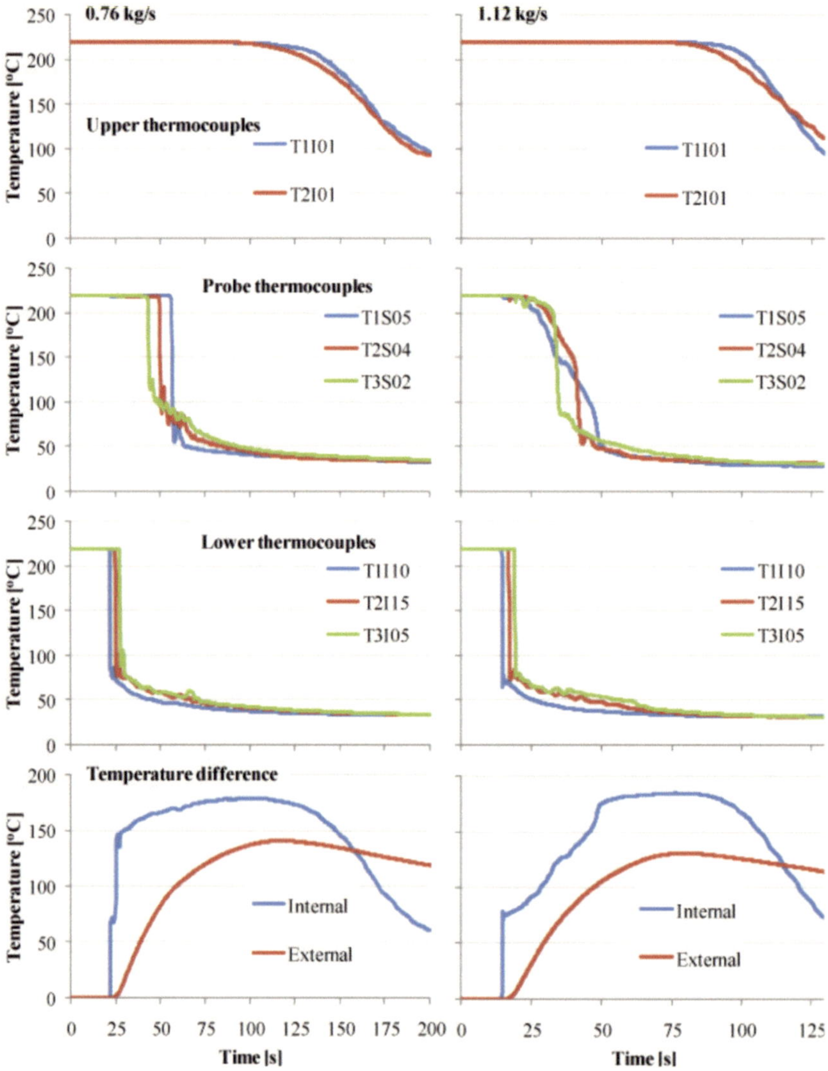

Figure 16. Comparison of numerical results obtained for two flow conditions

Figure 16 shows a comparison between numerical results obtained by the presented model for two flow conditions. By the results it is possible to conclude that thermal stratification occurs for both flow rates with similar intensity and temperature differences levels.

Figure 17 shows the cold water front evolution obtained numerically for the flow rate of 1.12kg/s. It can be observed that a cold water "head" is formed as the cold water front advances in the horizontal pipe. It can also be observed a change in the direction of the cold water front after reaching the end of the tube.

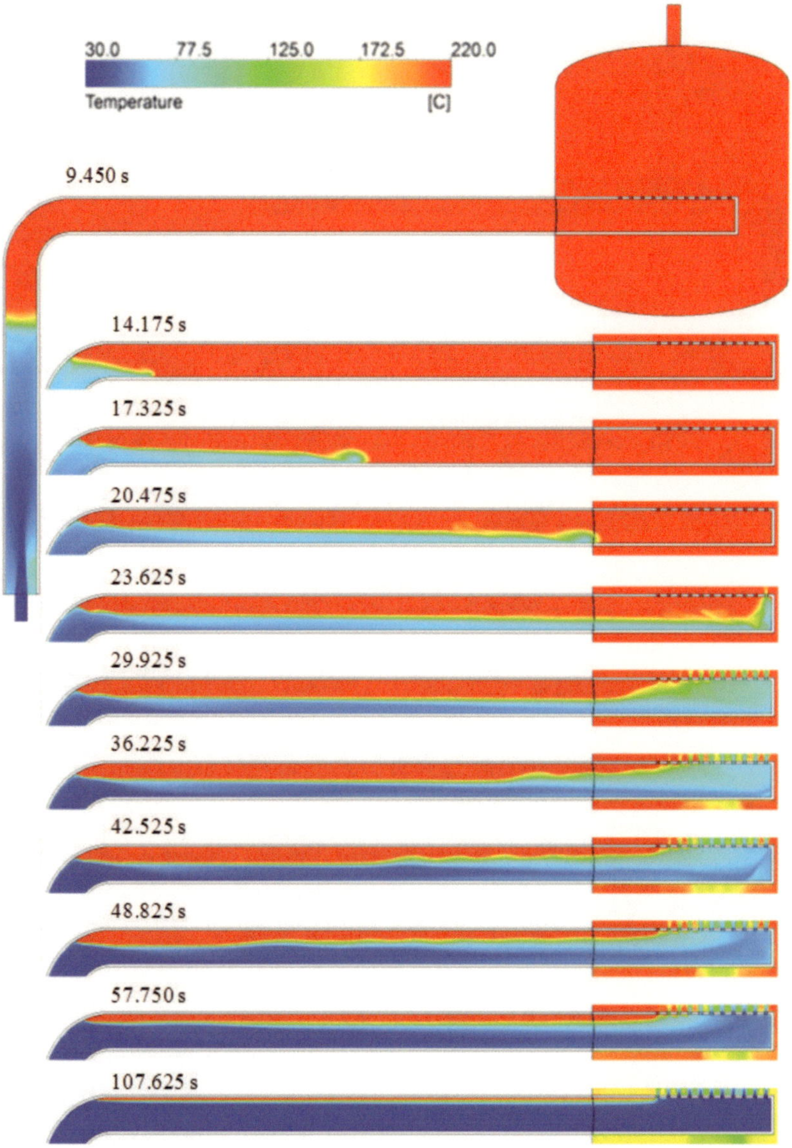

Figure 17. Temperature contours along time for flow rate 1.12 kg/s.

Figure 18 shows details of the flow behavior and flow velocity evolution in the simulation of flow rate 1.12 kg/s. It can be observed that as the cold water front enters the horizontal pipe it accelerates due to stratification and that the front induces a recirculation flow of the hot water at the top of the pipe, as mass must be conserved.

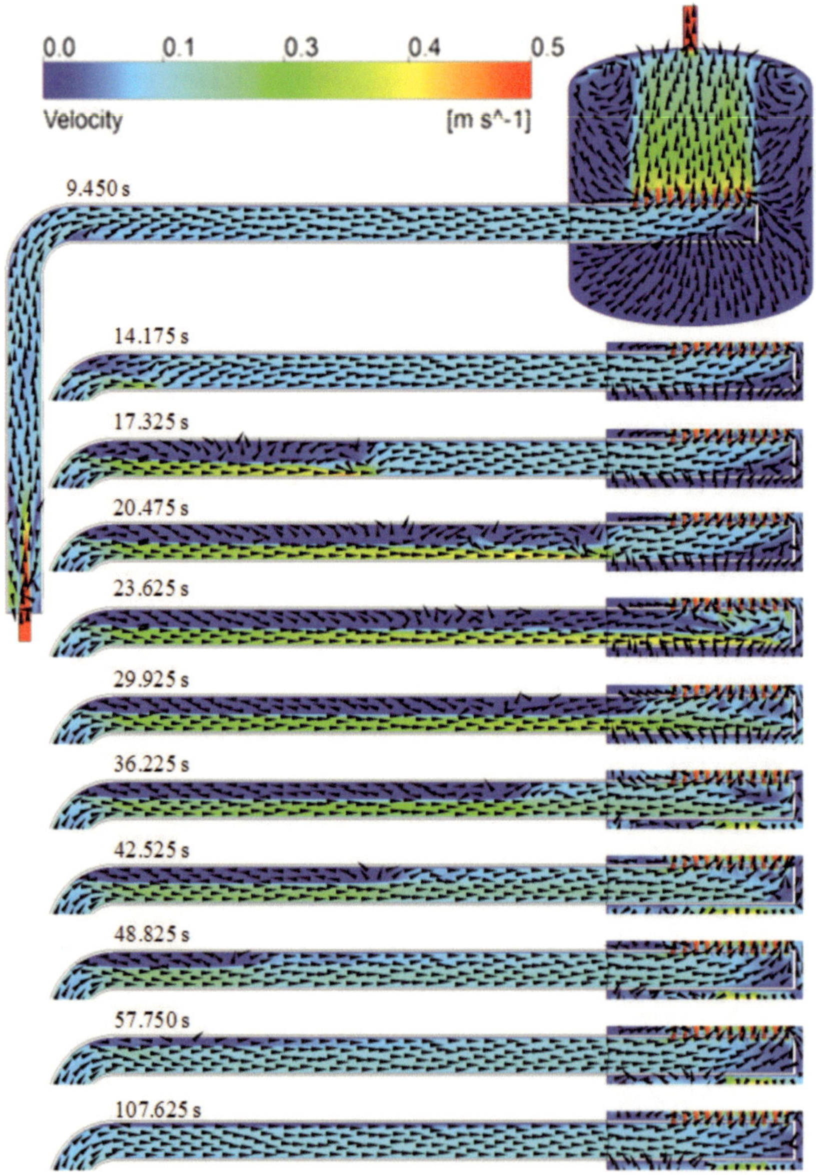

Figure 18. Flow velocity and behavior along time for flow rate 1.12 kg/s.

http://dx.doi.org/10.5772/52614

Figure 18 highlights the previously observed behavior, i.e., as the cold front reaches the end of the pipe it starts filing the pipe in the inversed direction eliminating almost all of the recirculating hot water. However, some hot water remains imprisoned at the top of the pipe as the injected cold water takes control of all water exits. This phenomenon is observed experimentally and causes the thermal stratification at the top of the pipe to persist for many minutes depending on the flow rate

4. Conclusion

The numerical simulation of one phase thermally stratified flow experiments in a pipe, similar to the steam generator injection nozzle at the secondary loop of a Pressurized Water Reactor (PWR), was proposed. The simulations were done using CFD codes (Rezende et al. 2011b).

A V&V evaluation of the numerical CFD methodology based on ASME (2009) standard was performed. Solution verification was also performed using three progressively refined meshes and time steps. Temperature profiles in several positions inside and outside the piping system were evaluated. In average the uncertainties due to the mesh were greater than those due to the time step. One reason for these values could be attributed to the course mesh used in the study that could lead to overestimation of the total uncertainty of the refined mesh. In average only thermocouples located in the upper region of the vertical tube displayed uncertainties above the experimental one (2.4 °C), which indicates that this region is the most affected by the mesh refinement.

The performed validation process showed the importance of proper quantitative evaluation of numerical results. In past studies a qualitative evaluation of the results would be considered sufficient and the present model would be (as it has been) considered satisfactory for thermal stratification prediction and study. However, with the present V&V study it was possible to identify objectively the strengths and weaknesses of the model.

Although considerable validation error was observed the qualitative agreement between experiment and simulation can be considered good as most of the behavior observed was reproduced. The performed validation process showed the importance of proper quantitative evaluation of numerical results.

Acknowledgment

This research project is supported by the following Brazilian institutions: Nuclear Technology Development Centre (CDTN), Brazilian Nuclear Energy Commission (CNEN), Research Support Foundation of the State of Minas Gerais (FAPEMIG), and Brazilian Council for Scientific and Technological Development (CNPq).

Author details

Hugo Cesar Rezende[1], André Augusto Campagnole dos Santos[1], Moysés Alberto Navarro[1], Amir Zacarias Mesquita[1] and Elizabete Jordão[2]

1 Nuclear Technology Development Center / Brazilian Nuclear Energy Commission (CDTN/CNEN), Brazil

2 Faculty of Chemical Engineering / University of Campinas (FEQ/UNICAMP), Brazil

References

[1] ANSYS(2010). CFX-13.0 User Manuals, ANSYS Inc., Canonsburg, Pennsylvania, USA.

[2] ASME(2009). Standard for Verification and Validation in Computational Fluid Dynamics and Heat Transfer- V&V 20.

[3] Eça, L., Hoekstra, M., Roache, P., & Coleman, H. (2009). Code verification, solution verification and validation: an overview of. the 3rd Lisbon workshop, AIAA.

[4] Häfner, W. (2004). Thermische Schichit-Versuche im Horizontalen Rohr, Kernforschungszentrum Karlsruhe GmbH, Karlsruhe, Germany, 238 p.

[5] ISO,(2003). Guide to the Expression of Uncertainty in Measurement. ISO, Geneva, Switzerland.

[6] Launder, B. E., & Spalding, D. B. (1974). The Numerical Computation of Turbulent Flows,. Computer Methods in Applied Mechanics and Engineering, , 3, 269-289.

[7] Lockard, D. P. (2010). In Search of Grid Converged Solutions. *Procedia Engineering*, 6, 224-233.

[8] Maliska, C. R. (2004). Transferência de Calor e Mecânica dos Fluidos Computacional, LTC- Livros Técnicos e Científicos Editora S. A., Rio de Janeiro, RJ- Brasil.

[9] Rezende, H. C. (2012). Theoretical and Experimental Study of Thermal Stratification in Single Phase Horizontal Pipe., ScD. Thesis, Universidade Estadual de Campinas, São Paulo. (in Portuguese).

[10] Rezende, H. C., Santos, A. A. C., Navarro, M. A., & Jordão, E. (2012). Verification and Validation of a thermal stratification experiment CFD simulation. Nuclear Engineering and DesignPrint), http://dx.doi.org/10.1016/j.nucengdes.2012.03.044., 1, 1-10.

[11] Rezende, H. C., Santos, A. A. C., & Navarro, A. M. (2011a). Thermal Hydraulics special theme for CFD codes- Thermal stratification experiments, Special Theme- INAC 2011.

[12] Rezende, H. C., Santos, A. A. C., & Navarro, M. A. (2011b). THSPSimulation of a Thermal Stratification Experiment Using CFD Codes- CDTN. Proceedings of International Nuclear Atlantic Conference (INAC 2011). Belo Horizonte., 1.

[13] Rezende, H. C., Navarro, M. A., Mesquita, A. Z., Santos, A. A. C., & Jordão, E. (2011c). Experiments On One-Phase Thermally Stratified Flows In Nuclear Reactor Pipe Lines. RevistaCientífica ESIME Redalyc, 1665-0654, 15, 17-24.

[14] Roache, P. J. (2010). Fundamentals of Verification and Validation. Hermosa Publishers.

[15] Schuler, X., & Herter, K. H. (2004). Thermal Fatigue due to Stratification and Thermal Schock Loading of Piping, 30th MPA- Seminar in conjunction with the 9th German-Japanese Seminar, Stuttgart, Oct. 6- 7, , 6.

Chapter 2

New Methods in Doppler Broadening Function Calculation

Daniel Artur P. Palma, Alessandro da C. Gonçalves, Aquilino Senra Martinez and Amir Zacarias Mesquita

Additional information is available at the end of the chapter

http://dx.doi.org/10.5772/52464

1. Introduction

In all nuclear reactors some neutrons can be absorbed in the resonance region and, in the design of these reactors, an accurate treatment of the resonant absorptions is essential. Apart from that, the resonant absorption varies with fuel temperature, due to the Doppler broadening of the resonances (Stacey, 2001). The thermal agitation movement of the reactor core is adequately represented in microscopic cross-section of the neutron-core interaction through the Doppler Broadening function. This function is calculated numerically in modern systems for the calculation of macro-group constants, necessary to determine the power distribution in a nuclear reactor. This function has also been used for the approximate calculations of the resonance integrals in heterogeneous fuel cells (Campos and Martinez, 1989). It can also be applied to the calculation of self-shielding factors to correct the measurements of the microscopic cross-sections through the activation technique (Shcherbakov and Harada, 2002). In these types of application we can point out the need to develop precise analytical approximations for the Doppler broadening function to be used in the codes that calculates the values of this function. Tables generated from such codes are not convenient for some applications and experimental data processing.

This chapter will present a brief retrospective look at the calculation methodologies for the Doppler broadening function as well as the recent advances in the development of simple and precise analytical expressions based on the approximations of Beth-Plackzec according to the formalism of Briet-Wigner.

2. The Doppler broadening function

Let us consider a medium with a temperature where the target nuclei are in thermal movement. In a state of thermal equilibrium for a temperature T, the velocities are distributed according to Maxwell-Boltzmann distribution (Duderstadt and Hamilton, 1976),

$$f(V) = N\left(\frac{M}{2\pi kT}\right)^{\frac{3}{2}} e^{-MV^2/2KT}, \tag{1}$$

where N is the total number of nucleus, M is the mass of the nucleus and k is Boltzmann's constant.

Considering the neutrons as an ideal gas in thermal equilibrium, it is possible to write the average cross-section for neutron-nucleus interaction taking into consideration the movement of the neutrons and of the nucleus as:

$$\bar{\sigma}(v,T) = \frac{1}{vN}\int d^3V(|\vec{v}-\vec{V}|)\sigma(|\vec{v}-\vec{V}|)f(\vec{V}), \tag{2}$$

where $f(\vec{V})$ is the distribution function of Maxwell-Boltzmann as given by equation (1) and $\vec{V} = V\hat{\Omega}$ is the velocity of the target nuclei. Denoting $\vec{v}_r = \vec{v} - \vec{V}$ the relative velocity between the movement of the neutron and the movement of the target nucleus and considering the isotropic case, that is, with no privileged direction, it is possible to separate the integration contained in equation (2) in the double integral:

$$\bar{\sigma}(v,T) = \frac{1}{vN}\int_0^\infty dVV^2 f\left(\vec{V}\right)\int_{4\pi} v_r\sigma(v_r)d\hat{\Omega}. \tag{3}$$

It is possible to see clearly in equation (3) that the cross-section depends of the relative velocity between the neutrons and the target nuclei. As the nuclei are in thermal movement, the relative velocity can increase or decrease. This difference between relative velocities rises to the Doppler deviation effect in cross-section behaviour. After integrating equation (3) in relation to the azimuthal angle (ϕ) the average cross-section for neutron-nucleus interaction can be written thus:

$$\bar{\sigma}(v,T) = \frac{2\pi}{vN}\int_0^\infty dVV^2 f(\vec{V})\int_0^\pi v_r\sigma(v_r)\sin\theta d\theta. \tag{4}$$

Denoting $\mu = \cos\theta$ so that $d\mu = -\sin\theta d\theta$, equation (4) takes the form of:

$$\bar{\sigma}(v,T) = \frac{2\pi}{vN}\int_0^\infty dVV^2 f(\vec{V})\int_{-1}^1 v_r\sigma(v_r)d\mu. \tag{5}$$

From the definition of the relative velocity one has the relation,

$$v_r^2 = v^2 + V^2 - 2vV\mu, \tag{6}$$

and, as a result,

$$d\mu = -\frac{v_r dv_r}{vV}. \tag{7}$$

With the aid of a simple substitution, using relations (6) and (27), equation (5) is thus written as:

$$\bar{\sigma}(v,T) = \frac{2\pi}{v^2 N}\int_0^{\infty} dVVf(\vec{V})\int_{|v-V|}^{|v+V|} v_r^2 \sigma(v_r)dv_r. \tag{8}$$

In equation (8), the limits of integration are always positive due to the presence of the module. As a result, one should separate the integral found in equation (8) into two separate integrals, as follows,

$$\bar{\sigma}(v,T) = \frac{2\pi}{v^2 N}\left[\int_0^{v} dVVf(\vec{V})\int_{v-V}^{v+V} v_r^2 \sigma(v_r)dv_r + \int_v^{\infty} dVVf(\vec{V})\int_{V-v}^{v+V} v_r^2 \sigma(v_r)dv_r\right]. \tag{9}$$

It is possible to modify the limits of integration for equation (9) taking into account that the mass of the target nucleus is much larger than the mass of the incident neutron. In terms of relative velocity, equation (9) can be written as:

$$\bar{\sigma}(v,T) = \frac{2\pi}{v^2 N}\left[\int_{v-v_r}^{v+v_r} v_r^2 \sigma(v_r)dv_r \int_0^{v} dVVf(\vec{V}) + \int_{v_r-v}^{v+v_r} v_r^2 \sigma(v_r)dv_r \int_v^{\infty} dVVf(\vec{V})\right]. \tag{10}$$

In replacing the expression of the Boltzmann distribution function, equation (1), in equation (10) one has:

$$\bar{\sigma}(v,T) = \frac{2\beta^3}{v^2\sqrt{\pi}}\left[\int_{v-v_r}^{v+v_r} v_r^2 \sigma(v_r)dv_r \int_0^{v} dVVe^{-\beta^2 V^2} + \int_{v_r-v}^{v+v_r} v_r^2 \sigma(v_r)dv_r \int_v^{\infty} dVVe^{-\beta^2 V^2}\right], \tag{11}$$

where it was defined $\beta^2 \equiv \frac{M}{2kT}$. Introducing the variables for reduced velocities $\varpi_r = \beta v_r$ and $\varpi = \beta v$, equation (11) is written by:

$$\bar{\sigma}(v,T) = \frac{2\beta^2}{\varpi^2\sqrt{\pi}} \times \left[\int_0^{\varpi/\beta} \varpi_r^2 \sigma(\varpi_r)d\varpi_r \int_{(\varpi-\varpi_r)/\beta}^{(\varpi+\varpi_r)/\beta} dVVe^{-\beta^2 V^2} + \int_{\varpi/\beta}^{\infty} \varpi_r^2 \sigma(\varpi_r)d\varpi_r \int_{(\varpi_r-\varpi)/\beta}^{(\varpi+\varpi_r)/\beta} dVVe^{-\beta^2 V^2}\right]. \tag{12}$$

Integrating equation (12) in relation to V one gets to the expression:

$$\bar{\sigma}(v,T)=\frac{1}{\varpi^{2}\sqrt{\pi}}\int_{0}^{\infty}\varpi_{r}^{2}\sigma(\varpi_{r})\left[e^{-(\varpi-\varpi_{r})^{2}}-e^{-(\varpi+\varpi_{r})^{2}}\right]d\varpi_{r}. \tag{13}$$

For resonances (that is, for the energy levels of the composed nucleus) it is possible to describe the energy dependence of the absorption cross-section by a simple formula, valid for $T=0K$, known as Breit-Wigner formula for resonant capture, expressed in function of the energy of the centre-of-mass by,

$$\sigma_{\gamma}(E_{CM})=\sigma_{0}\frac{\Gamma\gamma}{\Gamma}\left(\frac{E_{0}}{E_{CM}}\right)^{1/2}\frac{1}{1+\frac{4}{\Gamma^{2}}(E_{CM}-E_{0})^{2}}, \tag{14}$$

where E_0 is the energy where the resonance occurs and E_{CM} is the energy of the centre-of-mass of the neutron–nucleus system. Apart from that, we find in equation (14) the term σ_0, that is the value of the total cross-section $\sigma_{total}(E)$ in resonance energy E_0 that can be written in terms of the reduced wavelength λ_0 by:

$$\sigma_{0}=4\pi\lambda_{0}^{2}\frac{\Gamma_{n}}{\Gamma}g=2.608\times10^{6}\frac{(A+1)^{2}}{A^{2}E(eV)}\frac{\Gamma_{n}}{\Gamma}g, \tag{15}$$

where the statistical spin factor g is given by the expression:

$$g=\frac{2J+1}{2(2I+1)}, \tag{16}$$

where I is the nuclear spin and J is the total spin (Bell and Glasstone, 1970).

In replacing the expression (14) in equation (13) one finds an exact expression for the average cross-section, valid for any temperature:

$$\bar{\sigma}_{\gamma}(v,T)=\sigma_{0}\frac{\Gamma_{\gamma}}{\Gamma}\frac{\beta^{2}}{\sqrt{\pi}v^{2}}\times$$
$$\int_{0}^{\infty}dv_{r}\left(\frac{E_{0}}{E_{CM}}\right)^{1/2}\frac{v_{r}^{2}}{1+\frac{4}{\Gamma^{2}}(E_{CM}-E_{0})^{2}}\left[e^{-\beta^{2}(v-v_{r})^{2}}-e^{-\beta^{2}(v+v_{r})^{2}}\right], \tag{17}$$

In a system with two bodies it is possible to write the kinetic energy in the centre-of-mass system, by

$$E_{CM} = \frac{M_R v_r^2}{2}, \tag{18}$$

where $M_R = \frac{mM}{m+M}$ is the reduced mass of the system.

For the problem at hand, of a neutron that is incident in a thermal equilibrium system with a temperature T, it is a good approximation to assume that $v \approx v_r$. Thus, the ratio between the kinetic energy of the incident neutron and the kinetic energy of the centre-of-mass system is thus written

$$\frac{E_{CM}}{E_0} = \frac{A+1}{A}, \tag{19}$$

where A is the atomic mass of the target core. Resulting:

$$\begin{aligned} y &= \frac{2}{\Gamma}\left(E_{CM} - E_0\right) \text{ (a)} \\ x &= \frac{2}{\Gamma}\left(E - E_0\right), \text{ (b)} \end{aligned} \tag{20}$$

and denoting $\beta^2 = \frac{1}{2v_{th}^2}$ one finally obtains the expression for the cross-section of radioactive capture near any isolated resonance with an energy peak E_0, as written by:

$$\bar{\sigma}_\gamma(E,T) = \sigma_0 \frac{\Gamma_\gamma}{\Gamma}\left(\frac{E_0}{E}\right)^{1/2} \Psi(x,\xi), \tag{21}$$

where

$$\Psi(x,\xi) = \frac{\xi}{2}\int_{-2E/\Gamma}^{+\infty} \frac{dy}{1+y^2}\left[\exp\left(-\frac{(v-v_r)^2}{2v^2_{th}}\right) - \exp\left(-\frac{(v+v_r)^2}{2v^2_{th}}\right)\right], \tag{22}$$

where v_r is the module for relative neutron-nucleus velocity, v is the module for neutron velocity, and

$$\xi \equiv \frac{\Gamma}{\Gamma_D}. \tag{23}$$

The Doppler width for resonance Γ_D is expressed by:

$$\Gamma_D = (4E_0kT / A)^{1/2}. \tag{24}$$

All the other nuclear parameters listed below are well established in the literature,

- A= mass number;
- T= absolute temperature;
- E= energy of incident neutron;
- E_{CM}= energy of centre-of-mass;
- E_0= energy where the resonance occurs;
- Γ= total width of the resonance as measured in the lab coordinates;
- $\Gamma_D=(4E_0kT / A)^{1/2}$= Doppler width of resonance;
- v= neutron velocity module;
- $v_r = |v - V|$ = module of the relative velocity between neutron movement and nucleus movement;
- $v_{th} = \sqrt{\frac{2kT}{M}}$ = module of the velocity for each target nucleus.

3. The Bethe and Placzek approximations

The expression proposed by Bethe and Placzek for the Doppler broadening function $\psi(x, \xi)$ is obtained from some approximations, as follows:

1. one neglects the second exponential in equation (22), given that it decreases exponentially and is negligible in relation to first integral in equation (22) given that $(v+v_r)^2 >> (v-v_r)^2$.
2. it is a good approximation to extend the lower limit for integration down to $-\infty$ in equation (22), given that the ratio between the energy of neutron incidence and the practical width is big.
3. being E_{CM} the energy of the system in the centre-of-mass system and E the energy of the incident neutron, the following relation is always met:

$$E_{CM}^{1/2} = E^{1/2}\left(1+\frac{E_{CM}-E}{E}\right)^{1/2} = E^{1/2}(1+\eta)^{1/2}, \tag{25}$$

where it was denoted that $\eta = \frac{E_{CM} - E}{E}$. Equation (25) can be expanded in a Taylor series and, to the first order, is written by

$$E_{CM}^{1/2} = E^{1/2}\left(1 + \frac{\eta}{2} - \frac{\eta^2}{4} + \ldots\right) \approx E^{1/2}\left(1 + \frac{E_{CM} - E}{2E}\right) \tag{26}$$

In terms of the masses and velocities, equation (26) is written as follows:

$$\left(\frac{M_R v_r^2}{2}\right)^{1/2} = \left(\frac{mv^2}{2}\right)^{1/2}\left(1 + \frac{\frac{M_R v_r^2}{2} - \frac{mv^2}{2}}{mv^2}\right), \tag{27}$$

where M_R is the reduced mass of the system. For heavy nucleus $M_R \approx m$ and equation (27) can be written as:

$$v_r = \frac{v_r^2 + v^2}{2v}, \tag{28}$$

so that,

$$v - v_r = v - \frac{v_r^2 + v^2}{2v} = \frac{v^2 - v_r^2}{2v}. \tag{29}$$

In replacing approximation equation (29) in the remaining exponential of equation (22) one finally obtains the Doppler broadening function that will be approached in this chapter,

$$\psi(x,\xi) \approx \frac{\xi}{2\sqrt{\pi}} \int_{-\infty}^{+\infty} \frac{dy}{1+y^2} \exp\left[-\frac{\xi^2}{4}(x-y)^2\right]. \tag{30}$$

The approximations made in this section apply in almost all the practical cases, and are not applicable only in situations of low resonance energies ($E < 1eV$) and very high temperatures.

4. Properties of the Doppler broadening function $\psi(x, \xi)$

The function $\psi(x, \xi)$ as proposed by the approximation of Bethe and Placzek has an even parity, is strictly positive and undergoes a broadening as that variable ξ diminishes, that is, varies inversely with the absolute temperature of the medium. For low temperatures, that is, when temperature in the medium tend to zero, the Doppler broadening function can be represented as shown below:

$$\lim_{T\to 0}\psi(x,\xi)=\frac{\xi}{2\sqrt{\pi}}\lim_{T\to 0}\int_{-\infty}^{+\infty}\frac{dy}{1+y^2}\exp\left[-\frac{\xi^2}{4}(x-y)^2\right]=\frac{1}{1+x^2} \tag{31}$$

Equation (31) is known as an asymptotic approximation of the Doppler broadening function. For high temperatures, that is, when the temperature of the medium tends to infinite, the Doppler broadening function can be represented through the Gaussian Function, given that:

$$\lim_{T\to\infty}\psi(x,\xi)=\frac{\xi}{2\sqrt{\pi}}\lim_{T\to\infty}\int_{-\infty}^{+\infty}\frac{dy}{1+y^2}\exp\left[-\frac{\xi^2}{4}(x-y)^2\right]=\frac{\xi}{2\sqrt{\pi}}\exp\left(-\frac{\xi^2}{4}x^2\right). \tag{32}$$

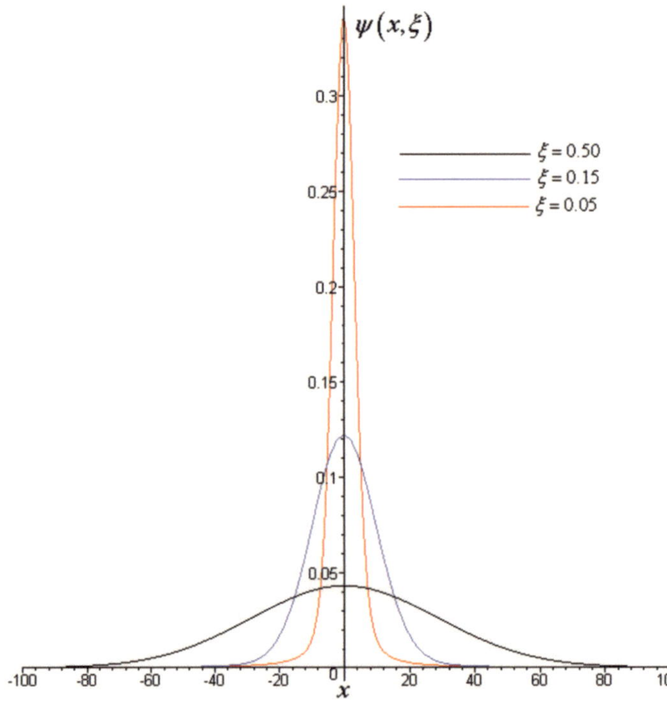

Figure 1. The Doppler broadening function for $\xi=0.05$, 0.15 and 0.5.

The area over the curve of the Doppler Broadening function is written as below and, as it consists of separable and known integers it is possible to write:

$$\int_{-\infty}^{+\infty}\psi(x,\xi)dx=\frac{\xi}{2\sqrt{\pi}}\int_{-\infty}^{+\infty}\frac{dy}{1+y^2}\int_{-\infty}^{+\infty}e^{-\frac{\xi^2}{4}(x-y)^2}dx=\frac{\xi}{2\sqrt{\pi}}(\pi)\left(\sqrt{\pi}\frac{2}{\xi}\right)=\pi. \tag{33}$$

From equation (33) once concludes that the area over the curve of the Doppler Broadening function is constant for the intervals of temperature and energy of interest in thermal reactors. This property is valid even for broadened resonances as shown in Figure 1, considering the different values of variable ξ.

5. Analytical approximations for the Doppler broadening function

This section describes the main approximation methods for the Doppler broadening function, according to the approximation of Bethe and Placzek, equation (30).

5.1. Asymptotic expansion

A practical choice to calculate the Doppler broadening function is its asymptotic expression resulting from the expansion of the term $\frac{1}{1+y^2}$ in equation (30) in a Taylor series around $y = x$.

$$\frac{1}{1+y^2} = \frac{1}{1+x^2} - \frac{2x}{\left(1+x^2\right)^2}(y-x) + \frac{-1+3x^2}{\left(1+x^2\right)^2}(y-x)^2 - \frac{4x\left(-1+x^2\right)}{\left(1+x^2\right)^4}(y-x)^3 + \ldots \tag{34}$$

In replacing equation (34) in equation (30) and integrating term by term, one obtains the following the asymptotic expansion:

$$\psi(x,\xi) = \frac{1}{1+x^2}\left\{\begin{array}{l} 1 + \frac{2}{\xi^2}\frac{\left(3x^2-1\right)}{\left(1+x^2\right)^2} + \\ \frac{12}{\xi^4}\frac{\left(5x^4-10x^2-1\right)}{\left(1+x^2\right)^4} + \ldots \end{array}\right\}. \tag{35}$$

Despite equation (35) being valid only for $|x.\xi| > 6$, it is quite useful to determine the behaviour of the Doppler Broadening function in specific conditions. For high values of x, it is possible to observe that function $\psi(x, \xi)$ presents the following asymptotic form:

$$\psi(x,\xi) \approx \frac{1}{1+x^2} \tag{36}$$

5.2. Method of Beynon and Grant

Beynon and Grant (Beynon and Grant, 1963) proposed a calculation method for the Doppler broadening function that consists of expanding the exponential part of the integrand of the Doppler broadening function $\psi(x, \xi)$ in the Chebyshev polynomials and integrate, term by term, the resulting expression, which allows writing:

$$\psi(a,\,b)=\frac{1}{a}\left\{\sqrt{\pi}\cos(ab)\left[1-E_2(a)\right]e^{a^2}+J(a,b)\right\}e^{-\frac{1}{4}b^2}, \tag{37}$$

where $a=\dfrac{1}{2}\xi$ and $b=\xi\cdot x$ and still,

$$\begin{aligned}J(a,b)=&\frac{1}{a}\left\{\frac{1}{2!}(ab)^2-\frac{1}{4!}(ab)^4+\frac{1}{6!}(ab)^6\cdots\right\}\\&+\frac{1}{2a^3}\left\{\frac{1}{4!}(ab)^4-\frac{1}{6!}(ab)^6\cdots\right\}+\\&\cdots+\frac{1}{\sqrt{\pi}a^{2n+1}}\Gamma\left(\frac{2n+1}{2}\right)\cdot\left\{\frac{1}{(2n+2)!}(ab)^{2(n+1)}\cdots\right\}+\end{aligned} \tag{38}$$

and,

$$E_2(a)=\frac{2}{\sqrt{\pi}}\int_0^a e^{-y^2}\,dy. \tag{39}$$

For values where the condition $|x.\xi|>6$ was met (Beynon and Grant, 1963) is recommended the use of the asymptotic expression of the function $\psi(x,\,\xi)$, equation (36). It should be pointed that the results obtained by this method have become a reference in several works on the Doppler broadening function.

5.3. Method of Campos and Martinez

The core idea of the method proposed by (Campos and Martinez, 1987) is to transform the Doppler broadening function from its integral form into a differential partial equation subjected to the initial conditions. Differentiating equation (30) in relation to x one obtains:

$$\begin{aligned}\frac{\partial\psi(x,\xi)}{\partial x}=&\frac{\xi^2}{2}\times\\&\left\{-\frac{\xi x}{2\sqrt{\pi}}\int_{-\infty}^{\infty}\frac{dy}{1+y^2}\exp\left[-\frac{\xi^2}{4}(x-y)^2\right]+\frac{\xi}{2\sqrt{\pi}}\int_{-\infty}^{\infty}\frac{ydy}{1+y^2}\exp\left[-\frac{\xi^2}{4}(x-y)^2\right]\right\}.\end{aligned} \tag{40}$$

Acknowledging in equation (40) the very Doppler broadening function and the term of interference as defined by the integral:

$$\chi(x,\xi)=\frac{\xi}{\sqrt{\pi}}\int_{-\infty}^{+\infty}\frac{ydy}{1+y^2}\exp\left[-\frac{\xi^2}{4}(x-y)^2\right], \tag{41}$$

it is possible to write:

$$\frac{\partial\psi(x,\xi)}{\partial x}=\frac{\xi^2}{2}\left[-x\psi(x,\xi)+\frac{\chi(x,\xi)}{2}\right]. \tag{42}$$

Deriving equation (42) again in relation to x, after expliciting function $\chi(x,\ \xi)$ in the same equation, one has:

$$\begin{aligned}&\frac{\partial^2\psi(x,\xi)}{\partial x^2}+x\xi^2\frac{\partial\psi(x,\xi)}{\partial x}+\frac{\xi^2}{4}\left(x^2\xi^2+2\right)\psi(x,\xi)\\&=\frac{\xi^5}{8\sqrt{\pi}}\int_{-\infty}^{\infty}\frac{y^2dy}{1+y^2}\exp\left[-\frac{\xi^2}{4}(x-y)^2\right].\end{aligned} \tag{43}$$

The right side of equation (43) can be written in another way, given that $\dfrac{y^2}{1+y^2}=1-\dfrac{1}{1+y^2}$,

$$\int_{-\infty}^{\infty}\frac{y^2dy}{1+y^2}\exp\left[-\frac{\xi^2}{4}(x-y)^2\right]=\frac{2\sqrt{\pi}}{\xi}\left[1-\psi(x,\xi)\right]. \tag{44}$$

In replacing the result obtained in (44) in equation (43) one obtains the differential equation where Campos and Martinez based themselves to obtain an analytical approximation for the broadening function $\psi(x,\ \xi)$:

$$\frac{4}{\xi^2}\frac{\partial^2\psi(x,\xi)}{\partial x^2}+4x\frac{\partial\psi(x,\xi)}{\partial x}+\left(\xi^2x^2+\xi^2+2\right)\psi(x,\xi)=\xi^2, \tag{45}$$

subjected to the initial conditions:

$$\begin{aligned}&\psi(x,\xi)|_{x=0}\equiv\psi_0=\frac{\xi\sqrt{\pi}}{2}\exp\left(\frac{\xi^2}{4}\right)\left[1-erf\left(\frac{\xi}{2}\right)\right] &\text{(a)}\\&\frac{\partial\psi(x,\xi)}{\partial x}|_{x=0}=0. &\text{(b)}\end{aligned} \tag{46}$$

Admitting that function $\psi(x,\ \xi)$ may be expanded in series,

$$\psi(x,\xi)=\sum_{n=0}^{\infty}c_n(\xi)\,x^n \tag{47}$$

and in replacing-se equation (47) in the differential equation as given by equation (45), one obtains after some algebraic manipulation the following polynomial equation:

$$\begin{aligned}&\left[\frac{8}{\xi^2}c_2+\left(\xi^2+2\right)c_0\right]+\left[\frac{24}{\xi^2}c_3+\left(\xi^2+6\right)c_1\right]x+\\&+\sum_{n=2}^{\infty}\left[\frac{4}{\xi^2}(n+2)(n+1)c_{n+2}+\left(4n+\xi^2+2\right)c_n+\xi^2c_{n-2}\right]x^n=\xi^2,\end{aligned} \tag{48}$$

where:

$c_0 = \psi_0,$

$c_1 = \frac{\xi^2}{8}\left[\xi^2 - (\xi^2 + 2)\psi_0\right],$

and all the other terms are calculated from the following relation of recurrence:

$$c_{n+1} = -\frac{\xi^2}{4}\frac{(4n + \xi^2 + 2)c_n + \xi^2 c_{n-1}}{(n+2)(n+1)}$$

The representation in series for the Doppler broadening function, as given by equation (47), is valid only for $|x.\xi| < 6$. For the cases where $|x.\xi| > 6$, (Campos and Martinez, 1987) used the asymptotic form as given by equation (35), as well as proposed by Beynon and Grant.

5.4. Four order method of Padé

The Padé approximation is one of the most frequently used approximations for the calculation of the Doppler broadening function and its applications and can efficiently represent functions, through a rational approximation, that is, a ratio between polynomials. For the four-order Padé approximation (Keshavamurthy& Harish, 1993) they proposed the following polynomial ratio:

$$\psi(x,\xi) = h\frac{a_0 + a_2(hx)^2 + a_4(hx)^4 + a_6(hx)^6}{b_0 + b_2(hx)^2 + b_4(hx)^4 + b_6(hx)^6 + b_8(hx)^8}, \quad (49)$$

whose coefficients are given in Tables 1 and 2.

$p_0 = \sqrt{\pi}$	$q_1 = \frac{\sqrt{\pi}(-9\pi + 28)}{2(6\pi^2 - 29\pi + 32)}$
$p_1 = \frac{-15\pi^2 + 88\pi - 128}{2(6\pi^2 - 29\pi + 32)}$	$q_2 = \frac{36\pi^2 - 195\pi + 256}{6(6\pi^2 - 29\pi + 32)}$
$p_2 = \frac{\sqrt{\pi}(33\pi - 104)}{6(6\pi^2 - 29\pi + 32)}$	$q_3 = \frac{\sqrt{\pi}(-33\pi + 104)}{6(6\pi^2 - 29\pi + 32)}$
$p_3 = \frac{-9\pi^2 + 69\pi - 128}{3(6\pi^2 - 29\pi + 32)}$	$q_4 = \frac{9\pi^2 - 69\pi + 128}{3(6\pi^2 - 29\pi + 32)}$

Table 1. Coefficients p and q of the four-order Padé Approximation

http://dx.doi.org/10.5772/52464

$h = \xi/2$
$a_0 = (p_0 + p_1 h - p_2 h^2 - p_3 h^3)(1 - q_1 h - q_2 h^2 + q_3 h^3 + q_4 h^4)$
$a_2 = (p_2 + 3p_3 h)(1 - q_1 h - q_2 h^2 + q_3 h^3 + q_4 h^4) + (p_0 + p_1 h - p_2 h^2 - p_3 h^3)(q_2 - 3q_3 h - 6q_4 h^2) + (-p_1 + 2p_2 h + 3p_3 h^2)(q_1 + 2q_2 h - 3q_3 h^2 - 4q_4 h^3)$
$a_4 = q_4(p_0 + p_1 h - p_2 h^2 - p_3 h^3) + (p_2 + 3p_3 h)(q_2 - 3q_3 h - 6q_4 h^2) - p_3(q_1 + 2q_2 h - 3q_3 h^2 - 4q_4 h^3) + (-p_1 + 2p_2 h + 3p_3 h^2)(q_3 + 4q_4 h)$
$a_6 = q_4(p_2 + 3p_3 h) - p_3(q_3 + 4q_4 h)$
$b_0 = (1 - q_1 h - q_2 h^2 + q_3 h^3 + q_4 h^4)^2$
$b_2 = 2(1 - q_1 h - q_2 h^2 + q_3 h^3 + q_4 h^4)(q_2 - 3q_3 h - 6q_4 h^2) + (q_1 + 2q_2 h - 3q_3 h^2 - 4q_4 h^3)^2$
$b_4 = (q_2 - 3q_3 h - 6q_4 h^2)^2 + 2q_4(1 - q_1 h - q_2 h^2 + q_3 h^3 + q_4 h^4) + 2(q_1 + 2q_2 h - 3q_3 h^2 - 4q_4 h^3)(q_3 + 4q_4 h)$
$b_6 = 2q_4(q_2 - 3q_3 h - 6q_4 h^2) + (q_3 + 4q_4 h)^2$
$b_8 = q_4^2$

Table 2. Coefficients h, a and b of the four-order Padé Approximation

From the coefficients of Tables 1 and 2, and of equation (49), one obtains in the end the following analytical approximation for function $\psi(x, \xi)$, according to the four-order Padé approximation:

$$\psi(x,\xi) = \frac{\eta(x,\xi)}{\omega(x,\xi)}, \tag{50}$$

where $\eta(x, \xi)$ and $\omega(x, \xi)$ are the following polynomials:

$$\begin{aligned}\eta(x,\xi) = 2\xi \cdot \big(&7,089815404 \cdot 10^{22} + 1,146750844 \cdot 10^{23}\xi + 8,399725059 \cdot 10^{22}\xi^2 \\ &+3,622207053 \cdot 10^{22}\xi^3 + 9,957751740 \cdot 10^{21}\xi^4 + 1,749067258 \cdot 10^{21}\xi^5 \\ &+1,835165213 \cdot 10^{20}\xi^6 + 8,940072699 \cdot 10^{18}\xi^7 - 2,539736657 \cdot 10^{21}\xi^2 x^2 \\ &+2,069483991 \cdot 10^{21}\xi^3 x^2 + 3,972393548 \cdot 10^{21}\xi^4 x^2 + 1,919319560 \cdot 10^{21}\xi^5 x^2 \\ &+3,670330426 \cdot 10^{20}\xi^6 x^2 + 2,682021808 \cdot 10^{19}\xi^7 x^2 + 1,048748026 \cdot 10^{19}\xi^4 x^4 \\ &+1,702523008 \cdot 10^{20}\xi^5 x^4 + 1,835165209 \cdot 10^{20}\xi^6 x^4 + 2,682021806 \cdot 10^{19}\xi^7 x^4 \\ &+8,940072688 \cdot 10^{18}\xi^7 x^6\big),\end{aligned} \tag{51}$$

and

$$\begin{aligned}\eta(x,\xi) = \big(&3,490642925\cdot 10^{23}\xi + 3,464999381\cdot 10^{23}\xi^2 + 2,050150991\cdot 10^{23}\xi^3\\
&+7,933771118\cdot 10^{22}\xi^4 + 3,670330427\cdot 10^{20}\xi^7 x^6 + 1,788014539\cdot 10^{19}\xi^8 x^8\\
&+3,670330426\cdot 10^{20}\xi^7 + 3,533894806\cdot 10^{21}\xi^6 + 1,788014541\cdot 10^{19}\xi^8\\
&+2,062859460\cdot 10^{22}\xi^5 + 3,426843796\cdot 10^{22}\xi^2 x^2 + 5,586613630\cdot 10^{22}\xi^4 x^2\\
&+2,649703323\cdot 10^{22}\xi^5 x^2 + 6,613512625\cdot 10^{22}\xi^3 x^2 + 1,101099129\cdot 10^{21}\xi^7 x^2\\
&+7,301013353\cdot 10^{21}\xi^6 x^2 + 3,590774413\cdot 10^{21}\xi^4 x^4 + 1,101099125\cdot 10^{21}\xi^7 x^4\\
&+5,868438581\cdot 10^{21}\xi^5 x^4 + 4,000342261\cdot 10^{21}\xi^6 x^4 + 7,152058156\cdot 10^{19}\xi^8 x^2\\
&+2,332237305\cdot 10^{20}\xi^6 x^6 + 1,072808721\cdot 10^{20}\xi^8 x^4 + 7,152058152\cdot 10^{19}\xi^8 x^6.\end{aligned} \tag{52}$$

5.5. Frobenius method

In this method the homogeneous part of the differential equation that rules the Doppler broadening function, equation (45), is solved using the Frobenius Method (Palma et. al., 2005) that consists fundamentally of seeking a solution of the differential equation in the form of series around the point $x = x_0$, with a free parameter, that is, as follows:

$$\psi(x,\xi) = x^s \sum_{n=0}^{\infty} c_n(\xi)\, x^n = \sum_{n=0}^{\infty} c_n(\xi)\, x^{n+s}, \tag{53}$$

with $c_0 \neq 0$ and where s is the parameter that grants the method flexibility.

Deriving equation (53) and replacing it in the homogeneous equation associated to equation (45) one obtains, after grouping the similar terms:

$$\begin{aligned}&\sum_{n=0}^{\infty} c_n (n+s)(n+s-1)\, x^{n+s-2} + \sum_{n=0}^{\infty} c_n \xi^2 \left[(n+s) + \frac{\xi^2+2}{4}\right] x^{n+s}\\
&+\frac{\xi^2}{4}\sum_{n=0}^{\infty} c_n\, x^{n+s-2} = 0.\end{aligned} \tag{54}$$

The initial equation of the problem, obtained when $n = 0$, remembering that $c_0 \neq 0$ is

$$c_0 s(s-1) = 0. \tag{55}$$

From equation (55), as $c_0 \neq 0$, one obtains that $s = 0$ or $s = 1$. Using first $s = 0$ and $c_0 \neq 0$ one obtains the following relations of recurrence:

$$c_n = -\frac{\xi^2\left(4n + \xi^2 - 6\right)}{4n(n+1)} c_{n-2}, \text{ valid for } n = 2 \text{ or } n = 3 \tag{56}$$

$$c_n = -\frac{\xi^2\left[c_{n-2}\left(4n + \xi^2 - 6\right) + c_{n-4}\xi^2\right]}{4n(n+1)}, \text{valid for } n \geq 4. \tag{57}$$

Considering the case where $s=1$, one obtains the other series linearly independent with the first term, not null, denoted by $\tilde{c}_0$:

$$c_n = -\frac{\xi^2\left(4n+\xi^2-2\right)}{4n\left(n+1\right)}c_{n-2},\ \text{valid for } n \geq 4 \tag{58}$$

$$c_n = -\frac{\xi^2\left[c_{n-2}\left(4n+\xi^2-2\right)+c_{n-4}\xi^2\right]}{4n\left(n+1\right)},\ \text{valid for } n \geq 4 \tag{59}$$

With this the homogeneous solution assumes the following form:

$$\psi_h\left(x,\xi\right) = \left(c_0 + c_2x^2 + c_4x^4 + \ldots\right) + \left(\tilde{c}_0x + \tilde{c}_2x^3 + \tilde{c}_4x^5 + \ldots\right), \tag{60}$$

where the coefficients are all known from the relations of recurrence, equations (56) to (59). In writing function $\psi_h(x, \xi)$ as:

$$\begin{aligned}\psi_h\left(x,\xi\right) &= \exp\left(-\frac{\xi^2x^2}{4}\right)\sum_{n=0}^{\infty}A_nx^n = \left(1-\frac{\xi^2x^2}{4}+\frac{\xi^4x^4}{32}+\ldots\right)\sum_{n=0}^{\infty}A_nx^n \\ &= A_0 + A_1x + \left(A_2 - \frac{\xi^2}{4}A_0\right)x^2 + \left(A_3 - \frac{\xi^2}{4}A_1\right)x^3 + \left(A_4 - \frac{\xi^2}{4}A_2 + \frac{\xi^4}{32}A_0\right)x^4 + \\ &\left(A_5 - \frac{\xi^2}{4}A_3 + \frac{\xi^4}{32}A_1\right)x^5 + \left(A_6 - \frac{\xi^2}{4}A_4 + \frac{\xi^4}{32}A_2 - \frac{\xi^6}{384}A_0\right) + \left(A_7 - \frac{\xi^2}{4}A_5 + \frac{\xi^4}{32}A_3 - \frac{\xi^6}{384}A_1\right) + \ldots.\end{aligned} \tag{61}$$

it is possible to determine all the coefficients A_n equalling, term by term, equations (60) and (61) so to write:

$$\begin{aligned}\psi_h\left(x,\xi\right) &= k_1\exp\left(-\frac{\xi^2x^2}{4}\right)\left[1-\frac{1}{2}\left(\frac{\xi^2x}{2}\right)^2+\frac{1}{24}\left(\frac{\xi^2x}{2}\right)^4+\frac{1}{720}\left(\frac{\xi^2x}{2}\right)^6+\ldots\right] \\ &+k_2\exp\left(-\frac{\xi^2x^2}{4}\right)\left[\frac{\xi^2x}{2}-\frac{1}{6}\left(\frac{\xi^2x}{2}\right)^3+\frac{1}{120}\left(\frac{\xi^2x}{2}\right)^5+\frac{1}{5040}\left(\frac{\xi^2x}{2}\right)^7+\ldots\right],\end{aligned} \tag{62}$$

Acknowledging the expansion of the cosine and sine functions, one obtains an analytical form to solve the homogeneous part of the differential equations that rule the Doppler broadening function:

$$\psi_h\left(x,\xi\right) = \exp\left(-\frac{\xi^2x^2}{4}\right)\left[k_1\cos\left(\frac{\xi^2x}{2}\right)+k_2\sin\left(\frac{\xi^2x}{2}\right)\right]. \tag{63}$$

In order to obtain the particular solutions of equation (45), and consequently its general solution, it is possible to apply the method of parameter variation from the linearly independent solutions:

$$\psi_1(x,\xi) = \exp\left(-\frac{\xi^2 x^2}{4}\right)\cos\left(\frac{\xi^2 x}{2}\right) \tag{64}$$

$$\psi_2(x,\xi) = \exp\left(-\frac{\xi^2 x^2}{4}\right)\sin\left(\frac{\xi^2 x}{2}\right), \tag{65}$$

Supposing a solution thus,

$$\psi_p(x,\xi) = u_1\psi_1(x,\xi) + u_2\psi_2(x,\xi) \tag{66}$$

where functions $u_1(x)$ and $u_2(x)$ are determined after the imposition of the initial conditions expressed by equations (46a) and (46b) and of the imposition of the nullity of the expression:

$$u_1^{'}(x)\psi_1(x,\xi) + u_2^{'}(x)\psi_2(x,\xi) = 0, \tag{67}$$

That, along with the condition,

$$u_1^{'}(x)\psi_1^{'}(x,\xi) + u_2^{'}(x)\psi_2^{'}(x,\xi) = \frac{\xi^4}{4}, \tag{68}$$

Which results from the very equation (45), form a linear system whose solution is given by the equations:

$$u_1^{'}(x) = -\frac{\xi^2}{2}\exp\left(\frac{\xi^2 x^2}{4}\right)\sin\left(\frac{\xi^2 x}{2}\right) \Rightarrow u_1(x) = -\frac{\xi^2}{2}\int_0^x dx^{'}\exp\left(\frac{\xi^2 x^{'2}}{4}\right)\sin\left(\frac{\xi^2 x^{'}}{2}\right) \tag{69}$$

$$u_2^{'}(x) = \frac{\xi^2}{2}\exp\left(\frac{\xi^2 x^2}{4}\right)\cos\left(\frac{\xi^2 x}{2}\right) \Rightarrow u_2(x) = \frac{\xi^2}{2}\int_0^x dx^{'}\exp\left(\frac{\xi^2 x^{'2}}{4}\right)\cos\left(\frac{\xi^2 x^{'}}{2}\right). \tag{70}$$

Integrating equations (69) and (70),

$$u_1(x) = \frac{\xi\sqrt{\pi}}{4}\exp\left(\frac{\xi^2}{4}\right)\left[erf\left(\frac{i\xi x - \xi}{2}\right) - erf\left(\frac{i\xi x + \xi}{2}\right) + 2erf\left(\frac{\xi}{2}\right)\right] \tag{71}$$

$$u_2(x) = -i\frac{\xi\sqrt{\pi}}{4}\exp\left(\frac{\xi^2}{4}\right)\left[erf\left(\frac{i\xi x-\xi}{2}\right)+erf\left(\frac{i\xi x+\xi}{2}\right)\right], \tag{72}$$

it is possible to write the solution particular of equation (45) as follows:

$$\begin{aligned}\psi_p(x,\xi) = -i\frac{\xi\sqrt{\pi}}{4}\sin\left(\frac{\xi^2 x}{2}\right)\exp\left[-\frac{\xi^2}{4}(x^2-1)\right]\left[erf\left(\frac{i\xi x-\xi}{2}\right)+erf\left(\frac{i\xi x+\xi}{2}\right)\right]+\\ \frac{\xi\sqrt{\pi}}{4}\cos\left(\frac{\xi^2 x}{2}\right)\exp\left[-\frac{\xi^2}{4}(x^2-1)\right]\left[erf\left(\frac{i\xi x-\xi}{2}\right)-erf\left(\frac{i\xi x+\xi}{2}\right)+2erf\left(\frac{\xi}{2}\right)\right].\end{aligned} \tag{73}$$

As the general solution of differential equation (45) is the sum of the solution of the homogeneous and particular equations, the initial conditions are imposed, as expressed by equations (46a) and (46b), to determine the constants:

$$k_1 = \frac{\xi\sqrt{\pi}}{2}\exp\left(\frac{\xi^2}{4}\right)\left[1-erf\left(\frac{\xi}{2}\right)\right] \tag{74}$$

$$k_2 = 0. \tag{75}$$

Finally, according to the Frobenius Method, the Doppler Broadening function can be written thus:

$$\begin{aligned}\psi(x,\xi) = \frac{\xi\sqrt{\pi}}{2}\exp\left[-\frac{1}{4}\xi^2\left(x^2-1\right)\right]\cos\left(\frac{\xi^2 x}{2}\right)\times\\ \left\{1+\mathrm{Re}\,\phi(x,\xi)+\tan\left(\frac{\xi^2 x}{2}\right)\mathrm{Im}\,\phi(x,\xi)\right\},\end{aligned} \tag{76}$$

where $\phi(x,\xi)=erf\left(\frac{i\xi x-\xi}{2}\right)$.

5.6. Fourier transform method

In doing the transformation of variables $u=\frac{\xi}{2}(x-y)$ in the full representation of the Doppler broadening function, equation (30), one obtains the expression

$$\psi(\xi,x) = \frac{1}{\sqrt{\pi}}\int_{-\infty}^{+\infty}\frac{e^{-u^2}du}{1+\left(x-2\frac{u}{\xi}\right)^2}, \tag{77}$$

that can be mathematically interpreted as the convolution of the Lorentzian function with a gaussian function, as exemplified by the equation below:

$$\psi(\xi, x) = f * g \equiv \int_{-\infty}^{+\infty} g(u)\, f(x-u)\, du \,, \tag{78}$$

where $f(x-u) = \dfrac{1}{1+\left(x-2\dfrac{u}{\xi}\right)^2}$ is the lorentzian function and $g(u) = \dfrac{1}{\sqrt{\pi}} e^{-u^2}$ the gaussian function. Function $f(x-u)$ admits a full representation through the Fourier cosine transform (Polyanin and Manzhirov, 1998), as being

$$f(x-u) = \int_0^{\infty} e^{-w} \cos\left[\left(x-2\frac{u}{\xi}\right)w\right] dw. \tag{79}$$

In replacing-se equation (79) in the integer of convolution, as given by equation (78), applying the properties of the integrals of convolution one gets to the following expression:

$$\psi(\xi, x) = f * g \equiv \int_0^{\infty} e^{-w} \int_{-\infty}^{+\infty} g(u) \cos\left[\left(x-2\frac{u}{\xi}\right)w\right] du\, dw = \int_0^{\infty} e^{-w}\, I(w)\, dw \,, \tag{80}$$

where,

$$\begin{aligned} I(w) &\equiv \frac{1}{\sqrt{\pi}} \int_{-\infty}^{+\infty} e^{-u^2} \cos\left[\left(x-2\frac{u}{\xi}\right) w\right] du \\ &= \frac{1}{\sqrt{\pi}} \cos(xw) \int_{-\infty}^{+\infty} e^{-u^2} \cos\left[\left(2\frac{u}{\xi}\right) w\right] du = e^{-\frac{w^2}{\xi^2}} \cos(xw). \end{aligned} \tag{81}$$

In replacing equation (81) in the equation (80), one obtains a new full representation of the Doppler broadening function, interpreted as a Fourier cosine transform (Gonçalves et. al., 2008):

$$\psi(\xi,x) = \int_0^{\infty} e^{-\frac{w^2}{\xi^2} - w} \cos(wx)\, dw = \frac{1}{2}\left[\int_0^{\infty} e^{-\frac{w^2}{\xi^2} - 2wa} dw + \int_0^{\infty} e^{-\frac{w^2}{\xi^2} - 2wb} dw\right], \tag{82}$$

where $a \equiv \dfrac{(1-ix)}{2}$ and $b \equiv \dfrac{(1+ix)}{2}$.

The integrals on the right side of equations (3.25) and (3.26) are known as complementary error functions, in which case one can conclude that:

$$\int_0^{\infty} e^{-\frac{w^2}{\xi^2}-2wa} dw = \frac{\xi\sqrt{\pi}}{2} e^{\frac{(xi-1)^2}{4}\xi^2} \text{erfc}\left(\frac{\xi - i\xi x}{2}\right) \tag{83}$$

$$\int_0^{\infty} e^{-\frac{w^2}{\xi^2}-2wb} dw = \frac{\xi\sqrt{\pi}}{2} e^{\frac{(xi+1)^2}{4}\xi^2} \text{erfc}\left(\frac{\xi + i\xi x}{2}\right). \tag{84}$$

In replacing equations (83) and (84) in equation (82) it is possible to write the following expression for the Doppler broadening function:

$$\begin{aligned} \psi(x,\xi) &= \frac{\xi\sqrt{\pi}}{4} e^{\frac{(xi-1)^2\xi^2}{4}} \left[1 + erf\left(\frac{i\xi x - \xi}{2}\right)\right] \\ &+ \frac{\xi\sqrt{\pi}}{4} e^{\frac{(xi+1)^2\xi^2}{4}} \left[1 - erf\left(\frac{i\xi x + \xi}{2}\right)\right]. \end{aligned} \tag{85}$$

With some algebraic manipulation it is easy to prove that the Fourier transform method and the Frobenius method, equations (85) and (76) respectively, provide identical results.

5.7. Fourier series method

From the representation of the Doppler broadening function in a Fourier cosine transform, equation (82), it is possible to write

$$\psi(\xi,x) = \int_0^{\infty} e^{-\frac{w^2}{\xi^2}-w} \cos(wx)\, dw = \int_0^{\infty} G(w) e^{-w} \cos(wx)\, dw, \tag{86}$$

where function $G(w)=e^{-\frac{w^2}{\xi^2}}$ is even and can be expanded into a Fourier series in cosines:

$$G(w) = \frac{a_0}{2} + \sum_{n=1}^{\infty} a_n \cos\left(\frac{n\pi w}{L}\right), \tag{87}$$

where

$$a_0 = \frac{\xi\sqrt{\pi}}{L} erf\left(\frac{L}{\xi}\right) \tag{88}$$

$$a_n = \frac{\xi\sqrt{\pi}}{2L} e^{-\left(\frac{n\pi\xi}{2L}\right)^2} \left[erf\left(\frac{2L + n\pi\xi^2 i}{2\xi L}\right) + erf\left(\frac{2L - n\pi\xi^2 i}{2\xi L}\right)\right]. \tag{89}$$

In replacing equation (87) in equation (86) and integrand, it is possible to write the following expression for the Doppler broadening function in the form of Fourier series:

$$\psi(x,\xi)=\frac{\xi\sqrt{\pi}}{2L\left(1+x^2\right)}erf\left(\frac{L}{\xi}\right)+\frac{\xi\sqrt{\pi}}{L}\sum_{n=1}^{\infty}F_n(x,\xi,L)\,\mathrm{Re}\left[Z(\xi,L)\right], \tag{90}$$

where

$$F_n(x,\xi,L)\equiv\frac{\left[(n\pi)^2+L^2\left(1+x^2\right)\right]e^{-\left(\frac{n\pi\xi}{2L}\right)^2}}{L^2\left(1+x^2\right)^2+(n\pi)^2\left(2-2x^2+(n\pi/L)^2\right)}, \tag{91}$$

$$Z(n,\xi,L)=erf\left(\frac{n\pi\xi^2 i+2L^2}{2\xi L}\right). \tag{92}$$

5.8. Representation of function $\psi(x,\xi)$ using Salzer expansions

Although the formulations obtained for function $\psi(x,\xi)$ from the Frobenius method, Fourier transform and Fourier series methods only contain functions that are well-known in literature, it can be inconvenient to work with error functions that contain an imaginary argument. One of the ways to overcome this situation is to calculate the real and imaginary parts of function $\phi(x,\xi)$ using the expansions proposed by Salzer (Palma and Martinez, 2009)

$$\phi(x,\xi)=erf\left(\frac{i\xi x-\xi}{2}\right)=\mathrm{Re}\,\phi(x,\xi)+\mathrm{Im}\,\phi(x,\xi)i, \tag{93}$$

where:

$$\mathrm{Re}\,\phi(x,\xi)\cong-erf\left(\frac{\xi}{2}\right)+\exp\left(-\frac{\xi^2}{4}\right)\times \left\{\frac{1}{\pi\xi}\left[\cos\left(\frac{\xi^2 x}{2}\right)-1\right]+\frac{2}{\pi}\sum_{n=1}^{n\max}\frac{\exp\left(-n^2/4\right)}{n^2+\xi^2}f_n(x,\xi)\right\} \tag{94}$$

$$\mathrm{Im}\,\phi(x,\xi)\cong\exp\left(-\frac{\xi^2}{4}\right)\left\{\frac{1}{\pi\xi}\sin\left(\frac{\xi^2 x}{2}\right)+\frac{2}{\pi}\sum_{n=1}^{n\max}\frac{\exp\left(-n^2/4\right)}{n^2+\xi^2}g_n(x,\xi)\right\}, \tag{95}$$

where auxiliary functions $f_n(x,\xi)$ and $g_n(x,\xi)$ are written by:

$$f_n(x,\xi)=-\xi+\xi\cosh\left(\frac{n\xi x}{2}\right)\cos\left(\frac{\xi^2 x}{2}\right)-n\sinh\left(\frac{n\xi x}{2}\right)\sin\left(\frac{\xi^2 x}{2}\right), \tag{96}$$

$$g_n(x,\xi) = \xi \cosh\left(\frac{n\xi x}{2}\right)\sin\left(\frac{\xi^2 x}{2}\right) + n\sinh\left(\frac{n\xi x}{2}\right)\cos\left(\frac{\xi^2 x}{2}\right). \tag{97}$$

5.9. The Mamedov method

Mamedov (Mamedov, 2009) put forward an analytical formulation to calculate function $\psi(x, \xi)$, based on its representation in the form of a Fourier transform, equation (82). Using the expansions in series of the exponential and cosine functions,

$$e^{-x} = \lim_{N\to\infty}\sum_{i=1}^{N}\frac{(-1)^i x^i}{i!}, \tag{98}$$

$$\cos x = \lim_{N\to\infty}\sum_{i=0}^{N}\frac{(-1)^i x^{2i}}{(2i)!}, \tag{99}$$

and the well-known binomial expansion

$$(x \pm y)^n = \lim_{N\to\infty}\sum_{m=0}^{N}(\pm 1)^m F_m(n) x^{n-m} y^m, \tag{100}$$

Mamedov proposed the following expressions for the Doppler broadening function:

for $\xi > 1$ and $x > 1$

$$\begin{aligned}\psi(x,\xi) = {} & \frac{\xi}{2\sqrt{\pi}}\lim_{L\to\infty}\sum_{i=0}^{L}F_i(-1)\Bigg\{\frac{1}{\left(x^2+1\right)^{i+1}}\sum_{j=0}^{i}F_j(i)\left[1+(-1)^j\right] \\ & \times\frac{2^{2i}x^j}{\xi^{2i-j+1}}\gamma\left(\frac{2i-j+1}{2},\frac{\xi^2\left(x^2+1\right)^2}{4}\right)+\left(x^2+1\right)^i\lim_{M\to\infty}\sum_{k=0}^{M}F_k(-1-i) \\ & \times\left[1-(-1)^k\right]\frac{x^k\xi^{2i+k+1}}{2^{2i+2}}\Gamma\left(-\frac{2i+k+1}{2},\frac{\xi^2\left(x^2+1\right)^2}{4}\right)\Bigg\},\end{aligned} \tag{101}$$

for $\xi \le 1$ and $x \le 15$

$$\psi(x,\xi) = \frac{\xi}{2\sqrt{\pi}}\lim_{L\to\infty}\sum_{i=0}^{L}\sum_{j=0}^{M}\frac{(-1)^{i+j}x^{2j}\xi^{2j+i+1}}{2i!(2j!)}\Gamma\left(\frac{i+2j+1}{2}\right), \tag{102}$$

for $\xi \le 1$ and $x = 0$

$$\psi(x,\xi) = \frac{\xi}{2\sqrt{\pi}} \lim_{L\to\infty} \sum_{i=0}^{L} \frac{(-1)^i \xi^{i+1}}{2i!} \Gamma\left(\frac{i+1}{2}\right), \tag{103}$$

where $\Gamma(x, \xi)$, $\gamma(x, \xi)$ and $\Gamma(x)$ are the well-know incomplete Gamma functions and $F_m(n)$ are binomials coefficients defined by:

$$F_m(n) = \begin{cases} \dfrac{n(n-1)...(n-m+1)}{m!}, & \text{for integer n} \\ \dfrac{(-1)^m \Gamma(m-n)}{m!\Gamma(-n)}, & \text{for non-integer n} \end{cases} \tag{104}$$

6. Numerical calculation of function $\psi(x, \xi)$

The numerical calculation of the Doppler broadening function consists of calculating a defined integral. There are many methods in the literature for this calculation, but in this chapter we will describe a numerical reference method based on the Gauss-Legendre quadrature. In basic terms, the Gauss-Legendre quadrature method consists of approximating a defined integer through the following expression:

$$\int_{-1}^{1} f\left(\frac{b-a}{2}\eta + \frac{b+a}{2}\right) dx \approx \frac{b-a}{2} \sum_{i=1}^{N} w_i f\left(\frac{b-a}{2}\eta_i + \frac{b+a}{2}\right), \tag{105}$$

where N is the order of the quadrature, $\eta_i$is the point of the quadrature and w_i the weight corresponding to the point of quadrature. The points of the Gauss-Legendre quadrature are the roots of the polynomials of Legendre (Arfken, 1985) in the interval $[-1, 1]$, as generated from the Rodrigues' formula,

$$P_n(x) = \frac{1}{2^n n!} \frac{d^n}{dx^n} \left\{ (x^2 - 1)^n \right\}. \tag{106}$$

for an isotope at a given temperature, that is, for a fixed value for variable ξ, the function $\psi(x, \xi)$ decreases rapidly and a very high value is not necessary for what we will consider our numerical infinite. This fact can be evidenced at Figure 1.

For that an adequate numerical infinite (x=5000) was considered, as well as a high-order quadrature (N =15), whose points of Legendre and respective weights are found in Table 3. The results obtained with this method, whose handicap is the high computing cost, can be seen in Table 4.

http://dx.doi.org/10.5772/52464

i	x_i	$\omega(x_i)$
1	0.9879925	0.0307532
2	0.9372734	0.0703660
3	0.8482066	0.1071592
4	0.7244177	0.1395707
5	0.5709722	0.1662692
6	0.3941513	0.1861610
7	0.2011941	0.1984315
8	0.0000000	0.2025782
9	-0.2011941	0.1984315
10	-0.3941513	0.1861610
11	-0.5709722	0.1662692
12	-0.7244177	0.1395707
13	-0.8482066	0.1071592
14	-0.9372734	0.0703660
15	-0.9879925	0.0307532

Table 3. Points of Legendre η_i and respective w_i weights.

ξ / x	0	0.5	1	2	4	6	8	10	20	40
0.01	0.00881	0.00881	0.00881	0.00881	0.00881	0.00880	0.00880	0.00879	0.00873	0.00847
0.02	0.01753	0.01753	0.01752	0.01752	0.01750	0.01746	0.01742	0.01735	0.01685	0.01496
0.03	0.02614	0.02614	0.02614	0.02612	0.02605	0.02594	0.02578	0.02557	0.02393	0.01836
0.04	0.03466	0.03466	0.03465	0.03461	0.03445	0.03418	0.03381	0.03333	0.02965	0.01857
0.05	0.04309	0.04308	0.04306	0.04298	0.04267	0.04216	0.04145	0.04055	0.03380	0.01639
0.10	0.08384	0.08379	0.08364	0.08305	0.08073	0.07700	0.07208	0.06623	0.03291	0.00262
0.15	0.12239	0.12223	0.12176	0.11989	0.11268	0.10165	0.08805	0.07328	0.01695	0.00080
0.20	0.15889	0.15854	0.15748	0.15331	0.13777	0.11540	0.09027	0.06614	0.00713	0.00070
0.25	0.19347	0.19281	0.19086	0.18325	0.15584	0.11934	0.08277	0.05253	0.00394	0.00067
0.30	0.22624	0.22516	0.22197	0.20968	0.16729	0.11571	0.07043	0.03881	0.00314	0.00065
0.35	0.25731	0.25569	0.25091	0.23271	0.17288	0.10713	0.05726	0.02816	0.00289	0.00064
0.40	0.28679	0.28450	0.27776	0.25245	0.17360	0.09604	0.04569	0.02110	0.00277	0.00064
0.45	0.31477	0.31168	0.30261	0.26909	0.17052	0.08439	0.03670	0.01687	0.00270	0.00064
0.50	0.34135	0.33733	0.32557	0.28286	0.16469	0.07346	0.03025	0.01446	0.00266	0.00063

Table 4. Reference values for Doppler Broadening Function $\psi(x, \xi)$.

7. Conclusion

A brief retrospective look at the calculation methodologies for the Doppler broadening function considering the approximations of Beth-Plackzec according to the formalism ofBriet-Wigner was presented in this chapter.

Acknowledgements

This research project is supported by Instituto Alberto Luiz Coimbra de Pós-graduação e PesquisaemEngenharia/Universidade Federal do Rio de Janeiro (COPPE/UFRJ), Comissão-Nacional de Energia Nuclear (CNEN), Centro de Desenvolvimento de Tecnologia Nuclear (CDTN). It was financially supported by through National Institute of Science and Technology of Innovative Nuclear Reactors,Brazilian Council for Scientific and Technological Development (CNPq,) and Research Support *Foundation of the State of Minas Gerais* (FAPEMIG).

Author details

Daniel Artur P. Palma[1], Alessandro da C. Gonçalves[2], Aquilino Senra Martinez[2] and Amir Zacarias Mesquita[3]

1 Comissão Nacional de Energia Nuclear (CNEN), , Brasil

2 Instituto Alberto Luiz Coimbra de Pós-graduação e Pesquisa em Engenharia – Universidade Federal do Rio de Janeiro (COPPE/UFRJ), Brasil

3 Centro de Desenvolvimento de Tecnologia Nuclear/ Comissão Nacional de Energia Nuclear (CDTN/CNEN), Brasil

References

[1] Arfken, G. (1985). Mathematical Method for Physicists. Academic Press Inc, London.

[2] Bell, G.I., Glasstone, S. (1970).Nuclear Reactor Theory. Van Nostrand Reinhold Co., New York.

[3] Campos, T.P.R. & Martinez, A.S. (1989). Approximate Calculation of the Resonance Integral for Isolated Resonances. *Journal of Nuclear Science and Technology,* Vol. 102, No. 3, (July 1989), pp. 211-218, ISSN 0029-5639.

[4] Campos, T.P.R., Martinez, A.S. (1987). The dependence of pratical width on temperature. *Annals of Nuclear Energy,* Vol. 34, No. 1-2, (May 1987), pp. 68-82, ISSN 0306-4549.

[5] Duderstadt, J. J, Hamilton, L.J. (1976). Nuclear Reactor Analysis. John Wiley and Sons, New York.

[6] Gonçalves, A.C., Martinez, A.S., Silva, F.C. (2008). Solution of the Doppler broadening function based on the Fourier cosine transform. *Annals of Nuclear Energy*, Vol. 35, No. 10, (June 2008), pp. 1871-1881, ISSN 0306-4549.

[7] Keshavamurthy, R.S., Harish, R. (1993). Use of Padé Approximations in the Analytical Evaluation of the J(θ,β) Function and its Temperature Derivative. *Nuclear Science and Engineering*, Vol. 115, No. 1, (September 1993), pp. 81-88, ISSN 0029-5639.

[8] Mamedov, B.A. (2009). Analytical evaluation of Doppler functions arising from resonance effects in nuclear processes. *Nuclear Instruments & Methods in Physics Research. Section A, Accelerators, Spectrometers, Detectors and Associated*, Vol. 608, (July 2009), pp. 336-338, ISSN 0306-4549.

[9] Palma, D.A.P., Martinez, A. S (2009). A faster procedure for the calculation of the *J(ζ,β)). Annals of Nuclear Energy*, Vol. 36, N0.10, (October 2009), pp. 1516-1520, ISSN 0168-9002.

[10] Palma, D.A.P., Martinez, A.S., Silva, F.C. (2005). The Derivation of the Doppler Broadening Function using Frobenius Method. *Journal of Nuclear Science and Technology*, Vol. 43, No. 6, (December 2005), pp. 617-622, ISSN 0022-3131.

[11] Polyanin, A.D &Manzhirov, A.V. (1998). Handbook of Integral Equation. CRC Press, New York.

[12] Shcherbakov, O., Harada, H. (2002). Resonance Self-Shielding Corrections for Activations Cross Section Measurements. *Journal of Nuclear Science and Technology*, Vol. 39, No. 5, (May 2002), pp. 548-553, ISSN 0022-3131.

[13] Stacey, W.M. (2001). Nuclear Reactor Physics. John Wiley and Sons, New York.

Chapter 3

Isothermal Phase Transformation of U-Zr-Nb Alloys for Advanced Nuclear Fuels

Rafael Witter Dias Pais,
Ana Maria Matildes dos Santos,
Fernando Soares Lameiras and
Wilmar Barbosa Ferraz

Additional information is available at the end of the chapter

http://dx.doi.org/10.5772/53821

1. Introduction

Significance in the study of ternary alloys in the U-Zr-Nb system is due to the scenario for their use as nuclear fuel not only for research and test reactors, but also for advanced high flux, power and fast reactors. The importance of this family of alloys is mainly due to the high thermal conductivity, ease of fabrication and good compatibility with the fuel cladding, besides the fact it has high density. Evidently, knowing the basic thermodynamic and diffusion characteristics of this system makes it considerably easier to solve many important problems including the optimization of the fuel composition and their stabilized phase under in-pile conditions. An especially important role in solving such problems is based in the isothermal phase transformation characteristics, which clearly establish the behavior of ternary systems.

The uranium, with a melting point of approximately 1132 °C has three allotropic phases: (i) γ, high temperature phase with body centered cubic structure, stable between 1132° and 775 °C (ii) β phase stable between 775 °C and 668 °C, complex tetragonal structure with 30 atoms per unit cell and (iii) α phase, present at temperatures below 668 °C with orthorhombic structure [1].

For the particular case of U-Zr-Nb system, the introduction of elements zirconium and niobium can result in several structures such as (i) cubic γ^s phase, obtained by quenching in water from the high temperature γ phase, (ii) tetragonal γ^0 produced by the aging of γ^s, (iii) tetragonal α', phase transition of α and (iv) the monoclinic α'', transition of α precipitated by aging γ^0 [2-5].

Since both physical and chemical properties of the nuclear fuel are intimately associated with the current phase, it is impossible to ensure the stability of metallic nuclear fuel without the knowledge of the kinetics of phase transformation in a broad temperature range. For the phase transformations in isothermal conditions, the kinetic behavior of the phase transformation curves is obtained by Time–Temperature-Transformation diagram, known as TTT diagram.

For system U-Zr-Nb, the alloy of composition U-2.5Zr-7.5Nb is the one that has received the largest volume of study in recent decades and different TTT diagrams have been determined as can be seen in Figure 1.

A first investigation in order to determine the TTT diagram for the alloy U-2.5Zr-7.5Nb was performed by Peterson [3] as shown in Figure 1a. This author determined the TTT curves using two main techniques: (i) Rockwell hardness C and (ii) electrical resistivity.

Later, in 1969, Dean [2] made some progress in determining the TTT diagram (Figure 1b) using metallographic techniques and the Vickers hardness tests for phases quantification. Some years later, in an independent and simultaneous way Giraud-Heraud [4] (Figure 1c) and Karnowsky [5] (Figure 1d) determined other two diagrams. The Giraud-Heraud curves, unlike Karnowsky curves, appear to be an improvement of the diagram obtained by Dean as observed on the curves of each diagram.

Besides the determination of the TTT curves, Karnowsky also studied the kinetics of phase transformations based on the equation of Johnson–Mehl–Avrami–Kolmogorov (JMAK) [6] and determined the equation coefficients. An investigation of the kinetics of phase transformation based on equation JMAK was also performed in 1974 by Vandermeer [7] whose TTT curves is shown in Figure 1e. This author introduced an innovation adding a fourth stage in the time-temperature-transformation diagram.

The state of the art on the development of TTT diagrams of U-2.5Zr-7.5Nb alloy still has many gaps and therefore it is necessary to continue parametric investigations and quantifications more precisely of the points near of the TTT diagrams curves. There is already a reasonable consensus on the identification of the phases present in the upper region of the TTT curves for this alloy in the region of temperature between about 400° and 650 °C. But in the range of temperature from about 100° to 400 °C, there are disagreements on the identification and characterization by metallographic technique with regard to the presence of phase α''(monoclinic). As reported in the literature, these disagreements may be due to experimental difficulties to reveal their bainitic microstructure.

The alloys of U-Zr-Nb with monoclinic structure α'' are important to be a stable phase, with excellent mechanical properties and corrosion resistance. Thus, it becomes essential to use during operation of the reactor power in the region of 300 °C. It is observed that for other U-Zr-Nb alloy compositions, unlike U-2.5Zr-7.5Nb, there is an absence of information for construction of the TTT diagrams.

This chapter covers a description of the methodology employed in the study of phase transformation in uranium based alloys with special reference to transformation kinetics under isothermal conditions of system U-Zr-Nb. The determination of the TTT curves includes in general, the following stages: (i) uranium alloys obtention, (ii) isothermal treatments, (iii) microstructural and phase characterization and (iv) construction of the time-temperature-transformation curve and study of the kinetics of isothermal phase transformation.

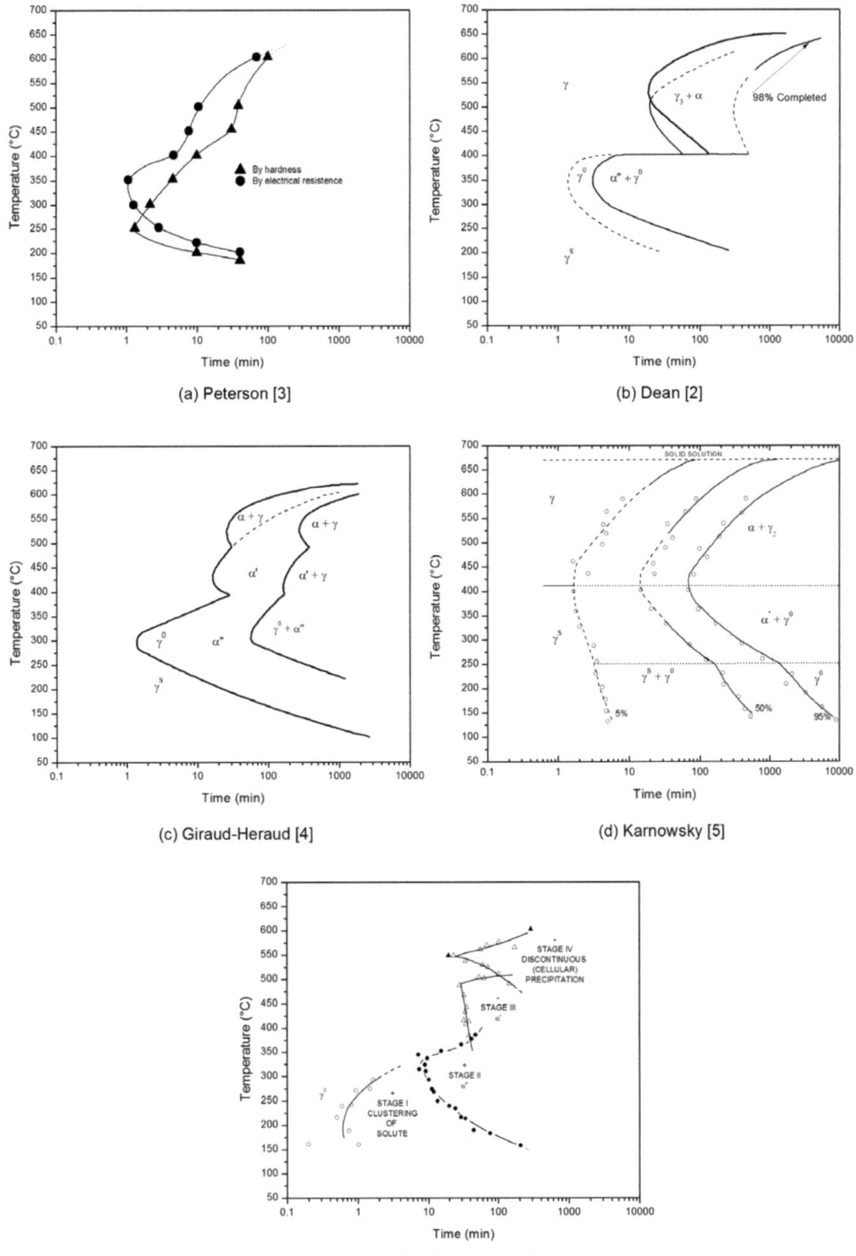

Figure 1. TTT diagram of the system U-2.5Zr-7.5Nb obtained by different authors.

2. Uranium alloys obtention

The determination of time-temperature-transformation curves for the U-Zr-Nb system requires alloys with chemical and microstructural homogeneity. Several equipments and methodologies can be employed to achieve this goal. In this item we covered a standard strategy employing equipments that are common to many laboratories. Bellow it is presented a simplified methodology and the main challenges to obtain the uranium alloys samples.

2.1. Melting of uranium alloys

For uranium alloys obtention it was employed a vacuum induction melting (VIM) furnace. The constituents uranium and alloy elements are melting in a graphite crucible under vacuum up to $1.33x10^{-3}$ Pa at temperatures of about 1500 °C. The time of charge effervescent must be controlled to minimize the carbon contamination and to homogenize the melting charge. The formation of carbides in the alloy structure occurs even in minimal levels of carbon contamination which does not interfere in the TTT diagrams.

After the homogenization, the charge is poured into mold and the casting ingot is cooled at room temperature. The mold can be made of copper or graphite in cylindrical or rectangular shapes.

2.2. Alloy homogenization

The alloy homogenization is realized through thermomechanical followed by stress-relieving heat treatments whose goal is to break the alloy cast structure obtaining a homogeneous structure with uniform distribution of chemical elements and equiaxed grains. The stress-relieving heat treatment is conducted at high vacuum ($1.33x10^{-3}$ Pa) and high temperature about 1000 °C and a plateau of approximately 25h.

Among the several hot forming processes employed for breaking the cast structure, the rolling process is undoubtedly the most widely used not only for the uranium-based alloys but for most metals. In this process, breaking of the cast structure is achieved by thickness reduction (about 40%) of the ingot in steps of hot rolling. For these thermomechanical treatments, there are important recommendations in each step as: (i) to avoid excessive oxidation of the ingot using an inert atmosphere and (ii) to prevent warping of the ingot.

3. Isothermal treatments

Isothermal treatments in order to determine the transformation-temperature-time curves are schematically shown in Figure 2. These treatments contain several steps in the following sequence named segment A-B, segment B-C, segment C-D and segment E-F.

In the segment A-B, the sample must be heated to the high temperature phase that is the stable phase for the composition of interest and the heating rate in this case is not relevant.

For the U-Zr-Nb system, the stable phase at high temperatures is the γ phase that. For the composition U-2.5Zr-7.5Nb in weight percent, for example, the temperature of 800 °C corresponds to the stable γ phase. Segment B-C is the region of constant temperature which extension is the time required to complete the stabilization of the high temperature phase. For U-2.5Zr-7.5Nb alloy, the high temperature is the stable γ phase and segment B-C corresponds to about 1 hour.

After segment B-C, the samples follow different conditions according to the isothermal temperature of interest. This temperature is always lower to the segment B-C. Thus, the decreasing in the temperature is performed quickly (down-quenching) until that the chosen temperature T1 is reached, avoiding phase transformations. The quickly down-quenching is especially critical for the isotherms in the knee region of low temperatures (about 300 °C) where the onset of phase transformation occurs in tens of seconds. Usually, this down-quenching is achieved by dipping the sample into metal bath as for example a tin bath or a mixture of lead-tin-bismuth.

Upon reaching the temperature T1, the sample remains in the isotherm for a time t1 and is rapidly cooled in order to retain the formed structure (segment E-F). The same procedure is adopted for the chosen times as t2, t3, t4, t5 etc. (Figure 2). Due to the nature of the process of isothermal transformation, the times are given on a logarithmic scale.

The retention structure is realized rapidly by transferring the sample to the room temperature in water or liquid nitrogen. After this quenching, the structure of the sample is characterized to determine the current transformed phase fractions. Thus it is possible to investigate the phase transformations in function of times of isothermal treatments. For practical purposes, the phases are determined in the range from 5 to 95% of transformation. The TTT diagrams exhibit C-curves or unusual double-C curves.

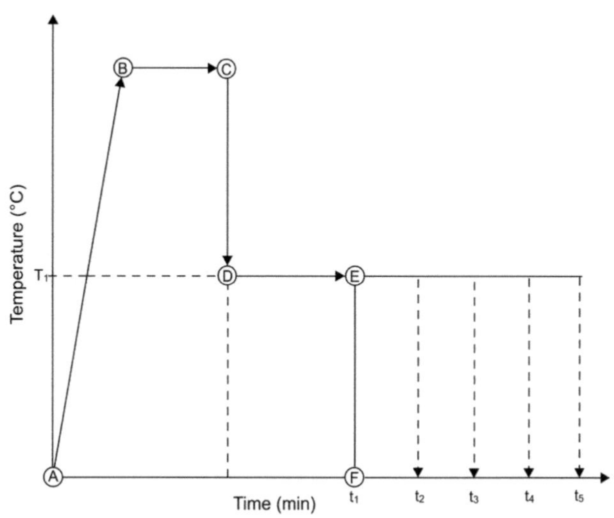

Figure 2. Schematic configuration of the isothermal treatments.

3.1. Installation for isothermal treatments developed at CDTN

We developed in the Researh Institute - CDTN (Centro de Desenvolvimento da Tecnologia Nuclear) an innovative isothermal treatments installation with precise monitoring of all steps involved in the procedures adopted to realize the isothermal treatments as described above.

This isothermal treatments installation comprises of three main parts: (i) high temperature zone, (ii) isothermal treatment zone and (iii) cooling zone as shown in Figure 3. The different zones are interconnected by a steel tube that is cooled between the zones by a water cooling system.

The tube ends are coupled to the left a pressure gauge, a vacuum pump and a gas outlet valve and to the right a gas control valve and a system that allows movement of the samples whose function is to transfer the sample between the different zones. This system has a haste connected to the sample crucible whose displacement is controlled by software what allows a rapid and accurate haste shift, so that the sample can be moved from the high temperature zone to cooling zone in in a few seconds. Also to the right there is a tempering container for quenching in water or oil, or liquid nitrogen.

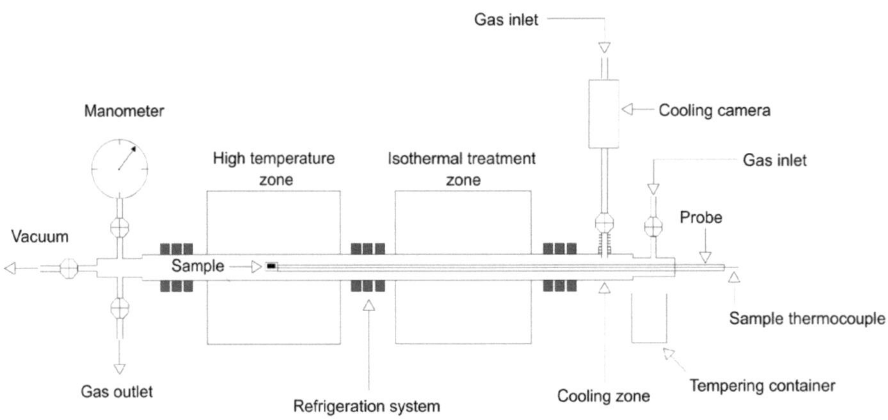

Figure 3. Isothermal treatments installation for phase transformations.

In order to perform the experiments, the tube is evacuated to about 1.33x10^4 Pa and in the following inert gas is injected. This procedure is carried out five times to ensure the removal of oxygen inside of the tube. Next, the high temperature zone is activated reaching the temperature of 800 °C, at a rate of 10 °C/min. At this time, the isothermal treatment zone is heated up to the desired isothermal treatment temperature. After reaching the temperature of 800 °C, the sample is shifted to the high temperature zone. The sample remains in this region for an hour in order to stabilize the high temperature γ phase. In the following, the sample is cooled until isothermal temperature and time programed. Finally, once reached the time of isothermal treatment the sample is quenched in water.

This innovative isothermal treatments installation eliminates the quenching in tin bath or mixture of lead-tin-bismuth between the zones of high temperature and isothermal temperature. Furthermore, this automated installation ensures a great reproducibility of the experiments.

4. Microstructural and phase characterizations

A variety of different techniques can be used for microstructural and phase characterizations as metallography, Vickers microhardness, X-ray diffraction, dilatometry, electrical resistivity, thermal measurement techniques, magnetic permeability among others. Generally, more than one technique can be used in a complementary way.

Below are described some techniques used to obtain the TTT curves for the system U-Zr-Nb showing some limitations and difficulties as also the samples preparation applied on each technique.

The most common techniques used to determine the TTT curves that are available in most laboratories are: (i) optical microscopy, (ii) scanning electron microscopy (SEM) coupled with energy dispersive spectroscopy X-ray (EDS), (iii) Vickers microhardness (HV), (iv) X-ray diffraction (DRX) associated with the Rietveld method.

4.1. Sample preparation

In the first time, the samples should be taken randomly and prepared by conventional metallographic techniques as cutting, mounting, grinding and mechanical and electrolytic polishing until to obtain a specular surface.

Optical microscopy is employed to observe mainly precipitates, phases and grains. To reveal the grains it is utilized electrolytic etching with a common solution 10wt% of oxalic acid in water. The revelation of the grains is a difficult task for the U-Zr-Nb alloys. The major difficulty is the revelation of phases formed on the isotherms for low-temperature region where microstructural characteristics of phase transformation are not clear. In order to overcome this problem it is necessary an extensive study aimed at determining optimized solutions of chemical attack for this alloys. For the 600 °C isotherm, for example, the microstructure is easily revealed, and there is no problem in identifying the progress of the transformation. The electrolytic attack and their solutions are well known in these conditions.

For X-ray diffraction analysis, the sample preparation is identical to metallographic procedure presented above but without electrolytic etching. Due to the high density of the uranium in these alloys, analysis by X-ray diffraction is restricted only to some atomic layers. Thus, care must be taken in metallographic preparation to avoid mechanical deformations induced on the sample surface. These deformations are manifested as broadening of the Bragg reflections making it difficult to analysis phases mainly because the uranium metastable phases have peaks very close together.

The sample preparation is identical to metallographic procedure above for the techniques as Vickers microhardness, scanning electron microscopy and energy dispersive spectroscopy X-ray.

4.2. Optical microscopy

The optical microscopy is one of the techniques used in the construction of TTT diagrams. As mentioned above, good quality of the micrographs is related to an adequate sample surface which is not always easy to obtain for the U-Zr-Nb system, especially for low-temperature isotherms.

Some optical micrographies are presented in Figure 4 as examples of the evolution of the TTT diagrams for U-3Zr-9Nb alloy treated isothermally at 600 °C for times of 100, 1000 and 10,000 min, utilizing the isothermal treatment installation (Figure 3). Similar microstructural characteristics are also observed for the U-2.5Zr-7.5Nb system submitted to the same thermal treatment conditions.

As can be seen in the Figure 4, the phase transformation $\gamma \rightarrow \alpha + \gamma$ occurs during up to 10,000 min. The γ phase appears as a single phase at 100 minutes (Figure 4a) and in Figure 4b there is clearly distinct regions, i.e., untransformed γ phase and a perlitic phase ($\alpha + \gamma$). Already at 10,000 minutes (Figure 4c) a perlitic structure is present in the entire sample. Thus, the perlitic structure grows gradually over the whole sample.

(a)

(b)

(c)

Figure 4. Optical micrographies of U-3Zr-9Nb alloy isothermally transformed at 600 °C for (a) 100 min, (b) 1000 min and (c) 10,000 min.

The fraction of transformed phases for construction of the TTT diagram can be performed through software image analyzer. The Quantikov® software [8], which was developed at CDTN is utilized for quantification of the phases of U-Zr-Nb alloys and includes modules for digital image processing, geometric measurements, graphics and hypertexts.

4.3. Scanning electron microscopy and X-ray energy dispersive spectrometer

Scanning electron microscopy coupled with energy dispersive spectroscopy of X-rays is another technique employed to construct the TTT diagrams for the U-Zr-Nb system. A wide range of high resolution and depth of field of the SEM analysis allows to obtain images with three-dimensional appearance showing microstructural details not revealed in the optical microscope. Figure 5 shows some SEM micrographies for U-3Zr-9Nb alloy obtained from studies in U-Zr-Nb alloys in our laboratory. From this figure, the lamellae can be observed in more detail compared with the optical micrographies. Figure 5b emphasizes the lamellar structure during the isothermal treatment at 10,000 minutes.

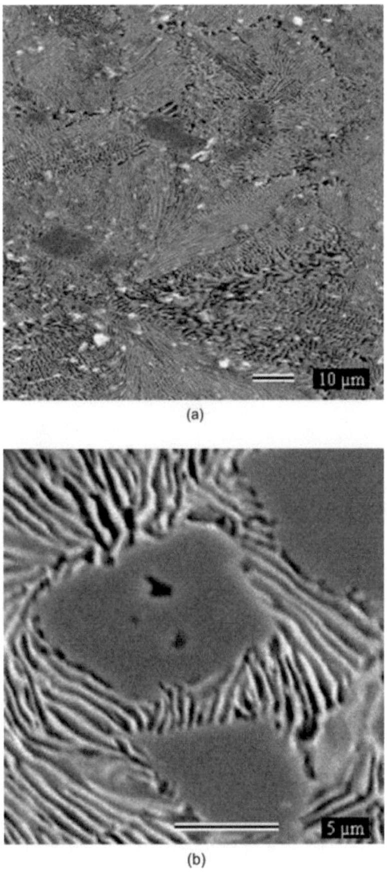

Figure 5. SEM micrographies of U-3Zr-9Nb alloy isothermally transformed at 600 °C for 10,000 min with different magnifications.

The chemical microanalysis of each element in the alloys are simultaneously detected by the EDS technique. This technique has advantage to allow chemical analysis on a very small region on the samples. This chemical analysis can be performed by two ways: (i) stationary, when the electron probe remains in a given position until complete X-ray events are recorded by the detector and (ii) scanning, when the electron probe moves over a line on the samples covering a determined length [9]. For the system U-Zr-Nb both modes of analysis are important. The stationary mode may be employed, for example, for microanalysis of precipitates and different chemical structures observed in the sample. The scanning mode is particularly useful to verify the concentration of a particular element across different regions.

An example of EDS analysis is shown in Figure 6 where the concentration of the elements uranium, niobium and zirconium are monitored along a given line length. On the left upper side of the figure, the micrograph of the sample shows, the presence of a precipitate (dark region) embedded in the matrix (light region). The drawn arrow passes through the matrix and precipitate indicating the direction path of electron probe. It is observed in this graph a sharp change in elements composition in the precipitate region showing the predominance of the elements niobium and zirconium.

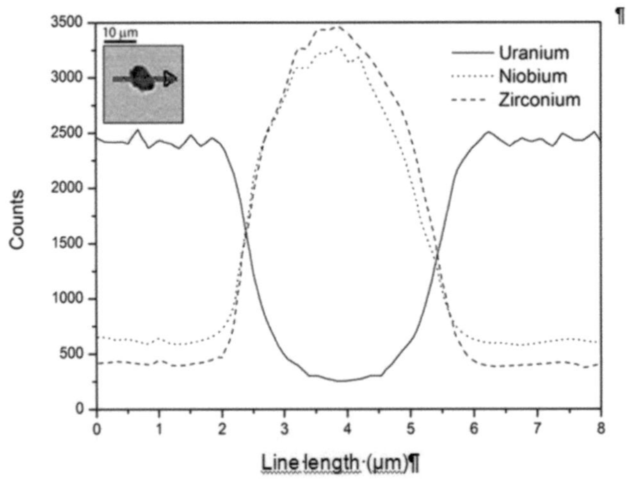

Figure 6. EDS chemical microanalysis on the U-2.5Zr-7.5Nb alloy.

4.4. Vickers microhardness

The Vickers microhardness is a technique widely used to study phase transformations in alloys under isothermal conditions especially due to their sensitivity to distinguish different phases. For the U-Zr-Nb system , Vickers microhardness has a particular importance for assessing the evolution of the phase transformations especially for a low temperature isotherm region. As mentioned earlier, this region is difficult to be analyzed by microscopy due to the difficulty in revealing the structure.

http://dx.doi.org/10.5772/53821

Vickers microhardness results, as shown in Figure 7, present an increase in the microhardness from the high-temperature isothermal phase transformations to regions of low temperature (about 300 °C). In this region of low temperature, the microhardness continuously increases with aging time. These results have similar behavior with Dean results [2]. But, in the high temperature region, approximately 600 °C, Dean microhardness results show an abrupt increasing conflicting very much from our results. Faced the differences mentioned it is important to mention that the microhardness results are very scarce in the literature for the U-Zr-Nb alloys what it makes necessary to investigate the microhardness in these alloy in a systematic way covering this shortcoming.

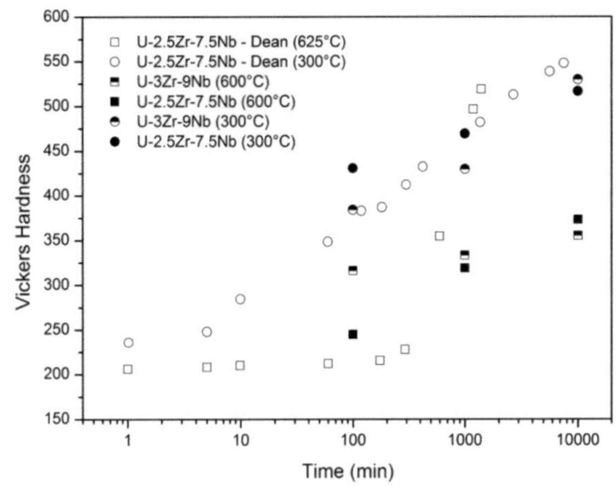

Figure 7. Vickers microhardness results for U-Zr-Nb alloys.

4.5. X-ray Diffraction

The X-ray diffraction is the main technique used to study the phase transformations in materials. A comprehensive study of X-ray diffraction for the U-2.5Zr-7.5Nb alloy was conducted by Yakel in 1976 [10], where the main metastable phases were identified. Uranium based alloys have a high absorption coefficient for X-ray which limits diffraction analysis in the surface layers of samples making it dependent on the surface preparation as discussed above. Additionally, patterns of X-ray diffraction peaks exhibit excessive enlargement which also contributes to increase the difficulties of the analysis particularly where the proximity of Bragg reflections are very close.

In order to overcome these problems, the Rietveld method emerges as a powerful technique for analyzing the X-ray diffraction data. The Rietveld method works to reduce the difference between the observed experimental diffraction pattern and the calculated diffraction pattern from a known structural model. In other words, the calculated diffraction pattern is modified by adjusting specific parameters to be as close as possible to the experimental pattern. This allows extract information of material structure, e.g., adjusted lattice parameter and quantitative analysis of phases, making the method a relevant tool in the studies of the phase transformation in the system U-Zr-Nb. A complete description of the mathematical involved in the Rietveld analysis can be obtained in the reference [11].

Figure 8 shows a typical Rietveld analysis for the U-2.5Zr-7.5Nb and U-3Zr-9Nb alloys. Both alloys were subjected to isothermal treatment at 600 °C for 10,000 minutes. As can be seen in the patterns of Figure 8, only perlitic structure are present indicating the occurrence of γ phase transformation to α + γ. In this analysis, the red points corresponds the experimental diffractograms, the green line represents the calculated diffractogram and the magenta line is the difference between the observed and calculated patterns. Black and red marks correspond to Bragg reflections of γ and α phases, respectively. Through the intensity of the main peaks, it can be inferred that the α phase transformation is highest in the alloy with a lower content of alloying elements, i.e., U-2.5Zr-7.5Nb alloy. The accurate quantification of the percentage of transformed phase and the evaluation of the lattice parameter of the structures formed are the major contribution of the Rietveld method to study the isothermal transformation system in U-Zr-Nb.

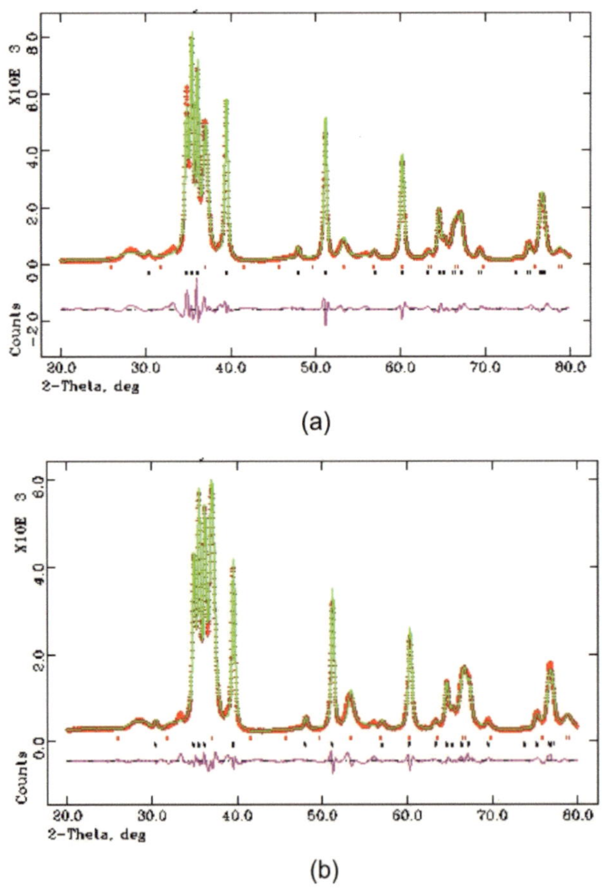

Figure 8. Rietveld refinements of the alloys aged 24 hours at 600 °C: (a) U-2.5Zr-7.5Nb, (b) U-3Zr-9Nb.

http://dx.doi.org/10.5772/53821

5. Construction of the time-temperature-transformation curve and study the kinetics of phase transformation

We reported that isothermal treatments in function of time are needed to determine the transformation-temperature-time curves being the time given on a logarithmic scale due to the nature of the process of isothermal transformation. It is also reported the techniques utilized for determining the percentage of phase transformed for construction of TTT diagrams. Thus, the TTT diagrams are constructed through isothermal treatments accomplished by phase characterization.

For kinetics studies of phase transformations, it is necessary to construct curves, usually referred to S-curves. In these curves, the transformed phase fractions are given as a function of time of transformation and important informations about the isothermal as nucleation time, start--transformation time, half-transformation time half and end-transformation time.

The S-curves for the U-2.5Zr-7.5Nb alloy adapted from Vandermeer [7] and Karnowsky [5] at different temperatures are shown in Figure 9 and it can be noted that the temperature change the shape of the transformation S-curves as a result of its influence on the kinetics of phase transformation.

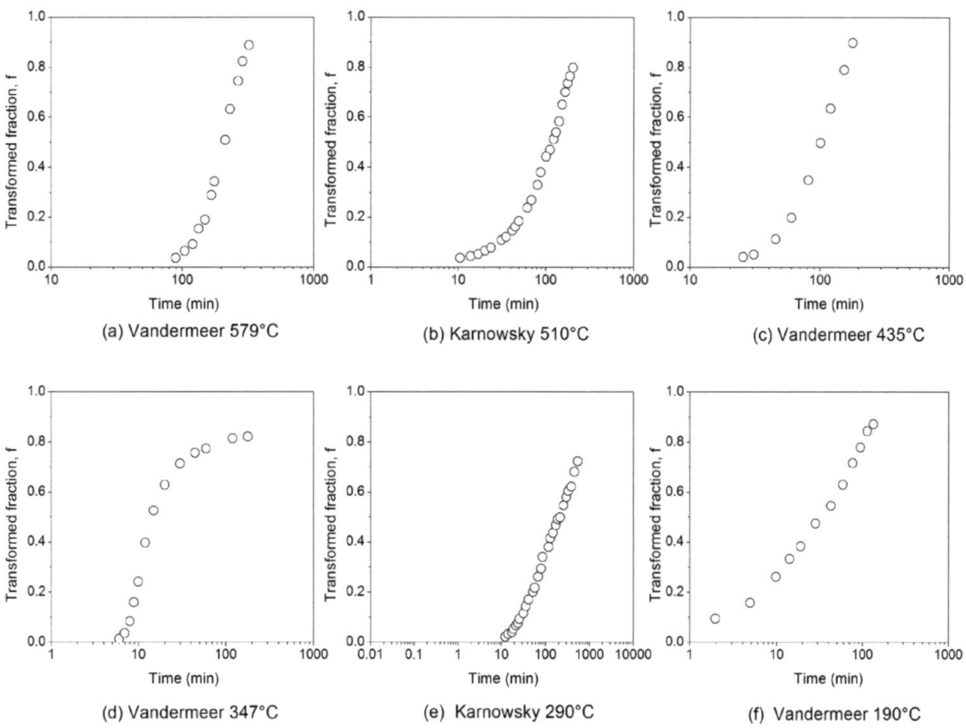

Figure 9. S-curves for the U-2.5Zr-7.5Nb alloy. (a), (c) (d) and (f) adapted from [7]; (b) and (e) from [5].

The determination of S-curves is generally accompanied by a study of the kinetics of phase transformation under isothermal conditions through the phenomenological model described by Johnson-Mehl-Avrami-Kolmogorov and represented by the following equation:

$$f = 1 - exp\left(-kt^{n}\right) \tag{1}$$

where f is the fraction transformed, t is the aging time and k and n are constants. The value of n constant allows to perform a series of inferences on the mechanism of phase transformation under isothermal conditions. The constants n and k can be obtained by linearization of this equation where n is the slope of the curve and k is the interception of y axis. Figure 10 shows the linearized curves of the isothermal temperatures of 579° C and 510 ° C which S-curves are shown in Figures 9a and 9b. Through the determined n values, the mechanisms of phase transformations can be inferred in Christian work [6].

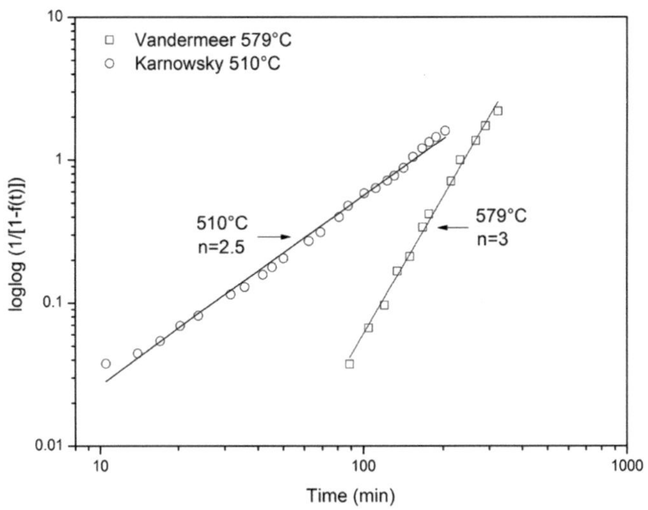

Figure 10. Linearized curves of the isothermal temperatures of 579 °C and 510 °C (adapted from [5] and [7]).

If the kinetics of phase transformation is governed by only one mechanism, the curve loglog (1/(1-f)) vs. log t shows a linear behavior as that shown in Figure 10 for the U-2.5Zr-7.5Nb alloy for the isotherms of high temperature. However, in the case of low isothermal temperature the U-2.5Zr-7.5Nb alloy does not exhibit a curve of linear behavior. Instead of the curves present two regions with different slopes indicating the occurrence of two mechanisms governing the kinetics of isothermal phase transformations. The complete elucidation of the mechanisms present in the phase transformations for U-2.5Zr-7.5Nb alloy is still at a preliminary stage and needs to be extensively investigated.

The construction of the TTT diagrams can be accomplished by the transfer of the S- curves as shown schematically in Figure 11.

http://dx.doi.org/10.5772/53821

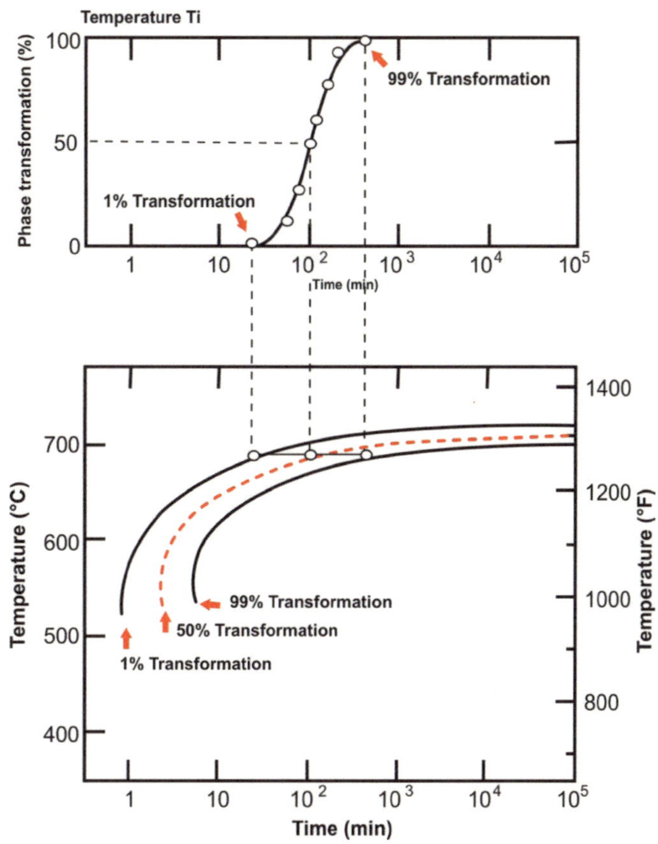

Figure 11. Scheme of construction of the TTT diagrams from S- curves (adapted from [12]).

6. Conclusions

This Chapter presents the state of the art of TTT diagrams of the U-2.5Zr-7.5Nb alloy as well as the development about this subject in the CDTN. It was observed that the study of the phase transformations and TTT diagrams is more focused on U2.5Zr7.5Nb alloy and even thus there are many disagreements in the phase characterization mainly in the phases present in the isothermal regions of lower temperatures. Moreover, a huge shortcoming exists in the literature about the development of TTT diagrams in other compositions of the U-Zr-Nb system.

In this context, orientations to study the isothermal phase transformations and to develop the TTT diagrams for the U-Zr-Nb system were presented. Furthermore, the methods and techniques employed as well as the care involved in each experimental step are also shown.

Summarizing the main informations in this Chapter are listed below:

- It was shown the main techniques utilized in the development of TTT diagrams for the U-Zr-Nb system;
- Experimental results carried out at CDTN for isotherms at 300 ° C and 600 °C were presented;
- Finally, it was described an innovative installation for isothermal treatments developed at CDTN. This allows that the study of isothermal phase transformations in the U-Zr-Nb system will be performed in a more accurate way by monitoring of the involved stages in the procedures adopted. Thus, further diagrams will be constructed with isotherms in the temperature range between about 100 ° to 650 ° C for alloys of U-2.5Zr-7.5Nb and U-3Zr-9Nb.

Acknowledgements

The financial support of the Conselho Nacional de Desenvolvimento Científico e Tecnológico (CNPq), Coordenação de Aperfeiçoamento de Pessoal de Nível Superior (CAPES) and Fundação de Amparo à Pesquisa do Estado de Minas Gerais (FAPEMIG) are gratefully acknowledged.

Author details

Rafael Witter Dias Pais, Ana Maria Matildes dos Santos, Fernando Soares Lameiras and Wilmar Barbosa Ferraz

Nuclear Technology Development Centre (CDTN/CNEN), Belo Horizonte, MG, Brazil

References

[1] Burke, J.J. Colling, D.A., Gorum, A.E., Greenspan, J. Physical Metallurgy of Uranium Alloys. Massachusetts in Cooperation with the Metalls and Ceramics Information Center, Columbus, Ohio: Brook Hill Co., Chestnut Hill; 1976.

[2] Dean, C.W. A Study of the Time-Temperature-Transformation Behaviour of a Uranium-7.5 weight per cent Niobium 2.5 weight per cent Zirconium Alloy. Tennesse: Union Carbide Corporation - Nuclear Division OAK Ridge Y-12 Plant, 1969.

[3] Peterson, C.A.W. Vandervoort, R.R. The Properties of a Metastable Gamma-Phase Uranium-Base Alloys: U2.5Zr7.5Nb. Lawrence Radiation Laboratory, University of California. 1964. UCRL-7869.

[4] Giraud-Heraud, F. et Guillaumin, J. Formation de Phases de Transition Dans L'Alliage U-7,5% Nb-2,5% Zr. Acta Metallurgica. 1973, Vol. 21, pp. 1243-1252.

[5] Karnowsky, M.M., Rohde, R.W. The Transformation Behaviour of a U-16.4At %Nb-5.6At%Zr Alloys. Journal of Nuclear Materials. 1973/74; v.49, p. 81-90.

[6] Christian, J.W. The Theory of Transformations in Metals and Alloys, Pergamon, 2002.

[7] Vandermeer, R.A. Recent Observations of Phase Transformation in U-Nb-Zr Alloy In J.J., Colling, D.C., Gorum, A.E. Greenspan, J. Burke. Physical Metallurgy of Uranium Alloys. Colorado: s.n., 1974, p. 219-257.

[8] Pinto, L.C.M. Quantikov: um analisador microstructural para o ambiente WindowsTM. Tese de Doutorado, Instituto de Pesquisas Energéticas e Nucleares. São Paulo: s.n., 1996.

[9] Leng, Yang. Materials characterization: introduction to microscopic and spectroscopic method. Singapore: John Wiley & Sons, 2008.

[10] Yakel, H.L. A Review of X-Ray Diffraction Studies in Uranium Alloys. The Physical Metallurgy of Uranium Alloys Conference Sponsored by the AEC Army Materials and Mechanical Research Center, Vail, Colorado. 1974.

[11] Young, R.A. The Rietveld Method. New York : Oxford University Press, 1995.

[12] Boyer, H, Editor. Atlas of Isothermal Transformation and Cooling Transformation Diagrams. American Society for Metals, 1977.

Chapter 4

Enriched Gadolinium Burnable Poison for PWR Fuel – Monte Carlo Burnup Simulations of Reactivity

Hugo M. Dalle, João Roberto L. de Mattos and
Marcio S. Dias

Additional information is available at the end of the chapter

http://dx.doi.org/10.5772/53381

1. Introduction

Nuclear power plant utilities have to make business in a very competitive market. Therefore, some of them may have interest to increase the competitiveness through the use of nuclear fuels which could achieve very high burnups, beyond 70 GWd/Ton. In order to reach and surpass this value may be necessary to work with uranium enrichments higher than the traditional 5% limit. Make real such nuclear fuels demands heavy investments in R&D.

Advanced PWR nuclear fuels that will be able to burn well beyond 70 GWd/Ton necessarily need to have, among other things, uranium enrichments higher than 5%, thus leading to the necessity of using extremely efficient burnable poisons. Gadolinium oxide, also known as Gadolinia (Gd_2O_3) is currently used as burnable poison for PWR fuel. The fabrication processes, properties and behavior of Gadolinia mixed in UO_2 fuel are well known and well established. Seven Gadolinium isotopes naturally exist on earth. From these, only two isotopes, the Gd-155 and Gd-157 have extremely high thermal neutron absorption cross sections.

Burnable poisons containing gadolinium have the undesirable effect of reducing the thermal conductivity of UO_2-Gd_2O_3 fuel and thus leading to higher temperature profiles in the fuel. In order to avoid such hot spots the currently available PWR rods use lower U-235 enrichment in all fuel pellets containing gadolinium. The use of gadolinium enriched in the most important isotopes, Gd-155 and Gd-157, to absorb neutrons, may permit to reduce the content of Gadolinia (Gd_2O_3) in the pellets and thus to improve the thermal conductivity of the fuel rods. This could permit to use the same U-235 enrichment in both types of fuel pellets, the ones with UO_2-Gd_2O_3 and those with only UO_2.

This Chapter has the objective to theoretically investigate the neutronic behavior of PWR fuels that would use UO_2-Gd_2O_3 compositions in which the gadolinium content is enriched from the natural abundance to a 100% Gd-155 and Gd-157 isotopes. The Monte Carlo burnup codes system *Monteburns* [1] is used to perform the simulations. They will evaluate only the effects that modifications in the fuel composition produce over the curves of reactivity x burnup. Indeed, the infinite neutron multiplication factor (k_{inf}) is used to plot the curves instead reactivity. The first set of these simulations aims to estimate the effects of variations in the contents of natural gadolinium and in the uranium enrichment. In the next set, the gadolinium content and the uranium enrichment are fixed, and the gadolinium isotopic compositions are modified by changing the enrichment of Gd-155 and Gd-157 until 100% to each isotope. This procedure is repeated to uranium enrichments going from 2.85% to 15%. After all, one last set of simulations evaluates the influence of the 100% Gd-155 and 100% Gd-157 enriched Gadolinia content. All the results are presented graphically and discussed. Finally, the optimized fuel composition is presented.

2. Physical model

2.1. Geometry and materials data

Figure 1 shows the 3x3 lattice of infinite fuel pin cells used in the simulations. One fuel rod containing the burnable poison (UO_2-Gd_2O_3) is located at the centre of the lattice and it is surrounded by seven rods containing standard fuel (UO_2) and one empty position (control rod guide tube). This geometry is kept to all the simulations. Only the isotopic compositions of the UO_2 and UO_2-Gd_2O_3 pellets are changed to the several cases simulated. As the 3x3 lattice is infinite the reflexive boundary condition was assumed.

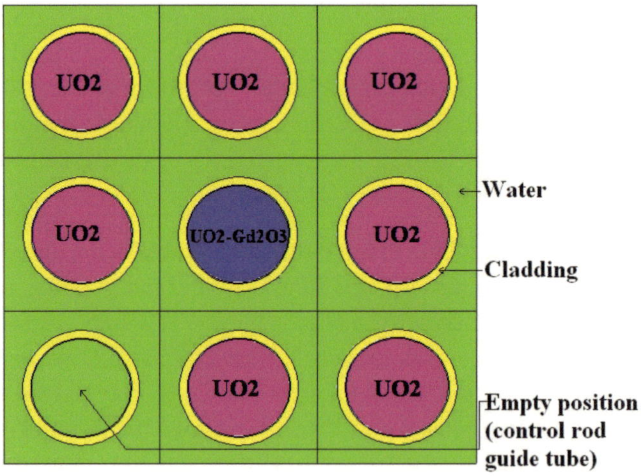

Figure 1. Infinite fuel pin cells 3x3 lattice.

http://dx.doi.org/10.5772/53381

The simulations were done considering that the fuel rods have the same geometric characteristics of the rods used in the AREVA FOCUS 16x16 fuel elements. Design data for this fuel is presented in [2]. Table 1 show some of the characteristics of the fuel rods adopted in the simulation model. Table 2 shows the thermal power and the uranium mass loaded in a typical 1300 MWe PWR reactor. These parameters are necessary to calculate the average specific power used in the burnup calculations. Table 3 presents the atomic number densities used in the model for water and cladding material (Zircalloy-4 was used instead of PCA-2B due the lack of information for this alloy).

Parameter	Unit	Value
Fuel Rods		
Outer diameter of cladding	cm	1.0750
Inner diameter of cladding	cm	0.9300
Active length	cm	390.00
Pitch	cm	1.430
Cladding Material		PCA-2B
Fuel Pellets		
Outer diameter	cm	0.9110
Guide tubes		
Outer diameter	cm	1.38
Thickness	cm	0.0700
Cladding Material		PCA-2B

Table 1. Geometry data of the fuel rods and guide tubes.

Parameter	Value
Thermal power (MW)	4000
Uranium mass (Ton)	~ 111

Table 2. Thermal power and uranium mass of a typical 1300 MWe PWR reactor

MATERIAL Nuclide	**Atoms/barn.cm**
ZIRCALOY-4	
Natural Cromium	7.5891e-05
Natural Iron	1.4838e-04
Natural Zirconium	4.2982e-02
WATER	
H-1	6.6242e-02
O-16	3.3121e-02

Table 3. Atomic number densities of cladding and cooling materials.

Finally, Table 4 presents the theoretical compositions of the several simulated fuels. These fuels are the alloys in which the uranium and gadolinium enrichments and the contents of Gadolinia and uranium dioxide are modified. In the simulations these compositions refer to the UO_2-Gd_2O_3 rod in the center of the 3x3 lattice and the UO_2 rods surrounding it. The calculation of the atomic number densities of the nuclides of Uranium, Gadolinium and Oxygen in fresh fuels to the several inputs was automatized by a MS Excel electronic spreadsheet.

Important to mention that the inputs 0050 (0.0% Gadolinia and 5.0% U235 enrichment for UO_2-Gd_2O_3 rod) and 7028 (7.0% gadolínia and 2.8% U235 enrichment for UO_2-Gd_2O_3 rod) refer to real (commercial) FOCUS [2] fresh fuels and are always used to compare the behavior of the others theoretical fuels simulated.

Some studies regarding sensitivities of the neutron cross sections to temperatures were performed and presented in [3, 4]. These studies led to the conclusion that the MCNP [5] pointwise neutron cross sections data library.66c can be used satisfactorily by Monteburns to simulate the fuel burnup of the lattice in spite of this set of cross sections data be processed at room temperature (293 K). As shown in [4] the results for overall production and destruction of the main nuclides during the burnup are similar if using library.66c or other libraries processed at higher temperatures.

2.2. Burnup intervals, important nuclides and neutron cross sections

In the simulations the rods were irradiated until around 90 GWd/ton. This total burnup was divided into several intervals with the burnup step values increasing since 0.5 until 25 days. Nuclides which the atom fraction, weight fraction, fraction of absorption, and fraction of fission greater than 10^{-4} were considered important nuclides and therefore its one group cross section were calculated for each burnup step. Sensitivities studies were performed to determine this *importance fraction*. However, there are a few nuclides with the importance fraction greater than 10^{-4} to which the one group cross sections were not calculated due to the lack of data in the MCNP standard libraries. The MCNP.66c pointwise neutron cross sections data were used preferably.

	UO_2-Gd_2O_3				UO_2
Input Name	Gd_2O_3 Content(%)	Composition	(%U-235)	UO_2 Content (%)	(%U-235)
0050	0	Natural	5.0	100	5.0
1050	1.0	Natural	5.0	99.0	5.0
2045	2.0	Natural	4.5	98.0	5.0
3038	3.0	Natural	3.8	97.0	5.0
4035	4.0	Natural	3.5	96.0	5.0
5033	5.0	Natural	3.3	95.0	5.0
6031	6.0	Natural	3.1	94.0	5.0
7028	7.0	Natural	2.8	93.0	5.0
hgd4015	2.0	40%Gd-155 and 15% Gd-157	2.8	98.0	5.0
hgd6010	2.0	60%Gd-155 and 10% Gd-157	2.8	98.0	5.0
hgd8005	2.0	80%Gd-155 and 05% Gd-157	2.8	98.0	5.0
hgd9900	2.0	100%Gd-155 and 0% Gd-157	2.8	98.0	5.0
hgd1540	2.0	15%Gd-155 and 40% Gd-157	2.8	98.0	5.0
hgd1060	2.0	10%Gd-155 and 60% Gd-157	2.8	98.0	5.0
hgd0580	2.0	05%Gd-155 and 80% Gd-157	2.8	98.0	5.0
hgd0099	2.0	0%Gd-155 and 100% Gd-157	2.8	98.0	5.0
gd4015	2.0	40%Gd-155 and 15% Gd-157	5.0	98.0	5.0
gd6010	2.0	60%Gd-155 and 10% Gd-157	5.0	98.0	5.0
gd8005	2.0	80%Gd-155 and 05% Gd-157	5.0	98.0	5.0
gd9900	2.0	100%Gd-155 and 0% Gd-157	5.0	98.0	5.0
gd1540	2.0	15%Gd-155 and 40% Gd-157	5.0	98.0	5.0
gd1060	2.0	10%Gd-155 and 60% Gd-157	5.0	98.0	5.0
gd0580	2.0	05%Gd-155 and 80% Gd-157	5.0	98.0	5.0
gd0099	2.0	0%Gd-155 and 100% Gd-157	5.0	98.0	5.0
gdx4015	2.0	40%Gd-155 and 15% Gd-157	6.0	98.0	6.0
gdx6010	2.0	60%Gd-155 and 10% Gd-157	6.0	98.0	6.0
gdx8005	2.0	80%Gd-155 and 05% Gd-157	6.0	98.0	6.0
gdx9900	2.0	100%Gd-155 and 0% Gd-157	6.0	98.0	6.0
gdx1540	2.0	15%Gd-155 and 40% Gd-157	6.0	98.0	6.0
gdx1060	2.0	10%Gd-155 and 60% Gd-157	6.0	98.0	6.0
gdx0580	2.0	05%Gd-155 and 80% Gd-157	6.0	98.0	6.0
gdx0099	2.0	0%Gd-155 and 100% Gd-157	6.0	98.0	6.0
gdd4015	2.0	40%Gd-155 and 15% Gd-157	10.0	98.0	10.0
gdd6010	2.0	60%Gd-155 and 10% Gd-157	10.0	98.0	10.0
gdd8005	2.0	80%Gd-155 and 05% Gd-157	10.0	98.0	10.0
gdd9900	2.0	100%Gd-155 and 0% Gd-157	10.0	98.0	10.0
gdd1540	2.0	15%Gd-155 and 40% Gd-157	10.0	98.0	10.0
gdd1060	2.0	10%Gd-155 and 60% Gd-157	10.0	98.0	10.0

	UO_2-Gd_2O_3				UO_2
Input Name	**Gd_2O_3 Content(%)**	**Composition**	**(%U-235)**	**UO_2 Content (%)**	**(%U-235)**
gdd0580	2.0	05%Gd-155 and 80% Gd-157	10.0	98.0	10.0
gdd0099	2.0	0%Gd-155 and 100% Gd-157	10.0	98.0	10.0
gdq4015	2.0	40%Gd-155 and 15% Gd-157	15.0	98.0	15.0
gdq6010	2.0	60%Gd-155 and 10% Gd-157	15.0	98.0	15.0
gdq8005	2.0	80%Gd-155 and 05% Gd-157	15.0	98.0	15.0
gdq9900	2.0	100%Gd-155 and 0% Gd-157	15.0	98.0	15.0
gdq1540	2.0	15%Gd-155 and 40% Gd-157	15.0	98.0	15.0
gdq1060	2.0	10%Gd-155 and 60% Gd-157	15.0	98.0	15.0
gdq0580	2.0	05%Gd-155 and 80% Gd-157	15.0	98.0	15.0
gdq0099	2.0	0%Gd-155 and 100% Gd-157	15.0	98.0	15.0
1551510	1.0	100%Gd-155	15.0	99.0	15.0
1551515	1.5	100%Gd-155	15.0	98.5	15.0
1551520	2.0	100%Gd-155	15.0	98.0	15.0
1551525	2.5	100%Gd-155	15.0	97.5	15.0
1551530	3.0	100%Gd-155	15.0	97.0	15.0
1551540	4.0	100%Gd-155	15.0	96.0	15.0
1571510	1.0	100% Gd-157	15.0	99.0	15.0
1571515	1.5	100% Gd-157	15.0	98.5	15.0
1571520	2.0	100% Gd-157	15.0	98.0	15.0
1571525	2.5	100% Gd-157	15.0	97.5	15.0
1571530	3.0	100% Gd-157	15.0	97.0	15.0
1571540	4.0	100% Gd-157	15.0	96.0	15.0

Table 4. Enrichments and Gadolinia contents of the fuels.

3. Results and analysis

The results of the various simulations will be graphically presented in the next pages. The subtitles, on right side of the graphs, indicate the input simulated as showed in Table 4. The focus of the analyses is to estimate the behavior of the fuel reactivity as a function of the burnup for the 3x3 lattice of fuel rods. Figure 2 shows the evolution of the infinite neutron multiplication factor (k_{inf}) for fuels with natural Gadolinia content from 0 to 9.9% and different uranium enrichment. As expected, the higher reactivity peak found is to the fuel without any Gadolinia (input 0050) and the peak occurs at zero burnup. The greater is the content of Gadolinia in the fuel the smaller is the reactivity of the fresh fuel. In addition, for all fuels containing Gadolinia the reactivity peak is observed after some days of irradiation and not at zero burnup condition. These peaks moves to the right (to higher burnups) as the Gadolinia content increases.

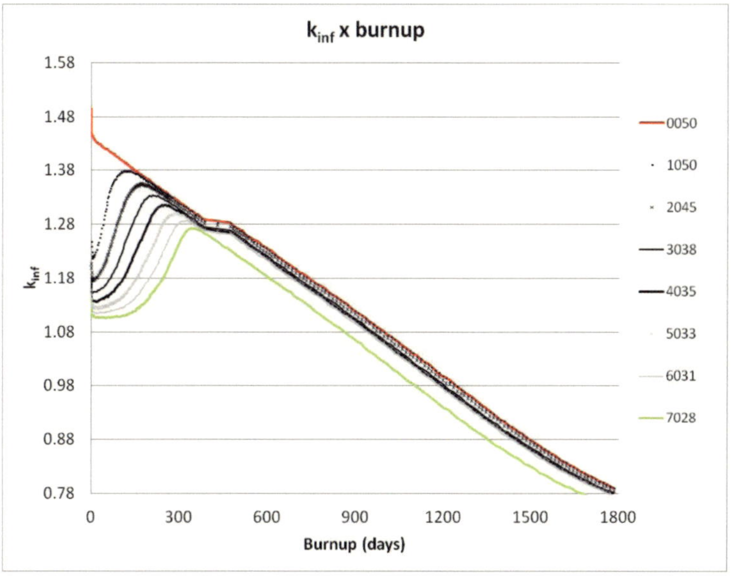

Figure 2. Infinite neutron multiplication factor as a function of burnup – changing Gadolinia content and %U-235.

3.1. Evolution of the infinite neutron multiplication factor as a function of the burnup – Gadolinia content fixed in 2% and changing the enrichments of Gd155, Gd157 and U235

The results presented in this section refer to Gadolinia contents in fuels lower than 7%. Usually this content is the limit to commercial fuels in order to avoid stronger effects affecting the thermal conductivity. As a starting point is used 2.0% Gadolinia content like recommended in [6]. In addition, the natural composition of the gadolinium in the simulations is modified to represent enrichments of Gd-155 and Gd-157 in the range from 5.0% to 100%.

Figure 3 (zoomed to the first 450 days of burnup in Figure 4) shows the evolution of k_{inf} as a function of burnup to the 3x3 lattice to: UO_2-Gd_2O_3 fuel rod, enriched 2.8% in U-235, and the seven UO_2 fuel rods surrounding it enriched 5.0% in U-235. The zoomed Figure 4 shows clearly that all simulated fuels have the reactivity peak lower than the peak observed to the reference fuel 0050 that is free of Gadolinia. The lowest reactivity peak was found to the other reference fuel (7028 - natural Gadolinia) and its curve is very close to the curve of the fuel with 100% Gd-155. Such results are very exciting to proceed simulating fuels having higher U-235 enrichments until to find a composition to which the reactivity peak is close to the 0050 fuel. Figures 5 to 15 show the results of these simulations to U-235 enrichments in the range of 5.0% to 15.0% (subtitles are always as Table 4).

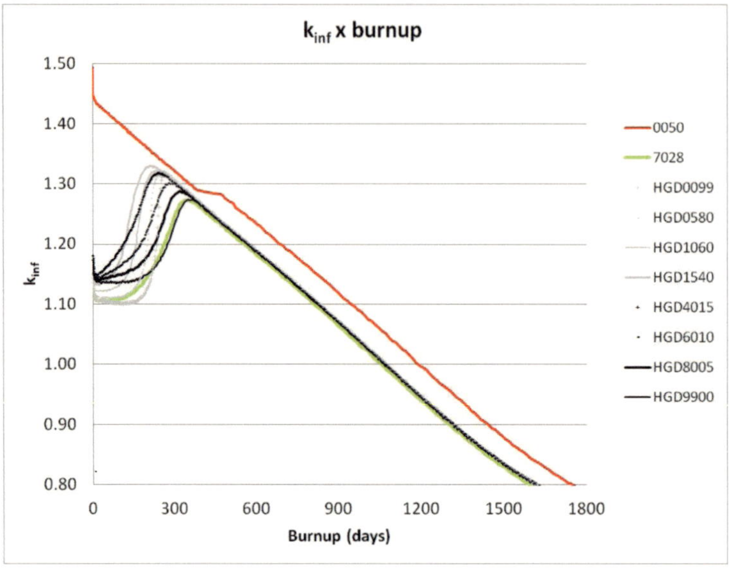

Figure 3. Infinite neutron multiplication factor as a function of burnup – 2.0% Gd_2O_3, 2.8% U-235, changing Gd-155 and Gd-157 enrichments.

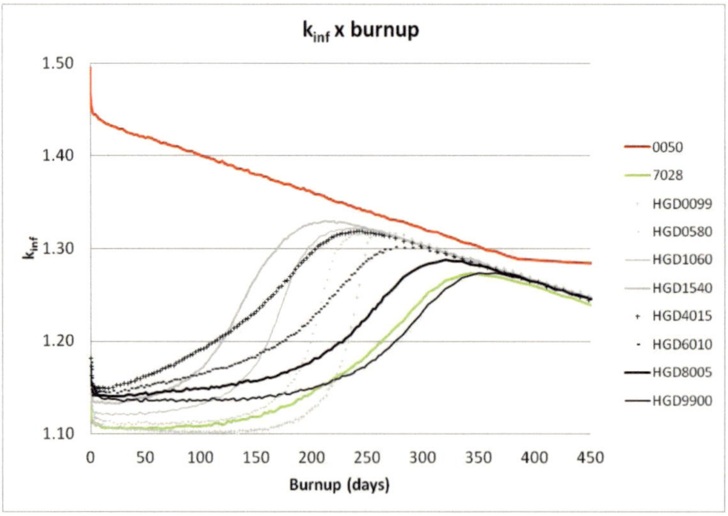

Figure 4. Zoom of Figure 3 (beginning of burnup interval).

Increasing the U-235 enrichment of the UO_2-Gd_2O_3 fuel rod to 5.0%, Figures 5 and 6 are obtained. One can notice a slight growth in reactivity of the compositions simulated and the curves for the eight inputs, after the peaks region, started to decouple from 7028 curve towards the 0050 curve.

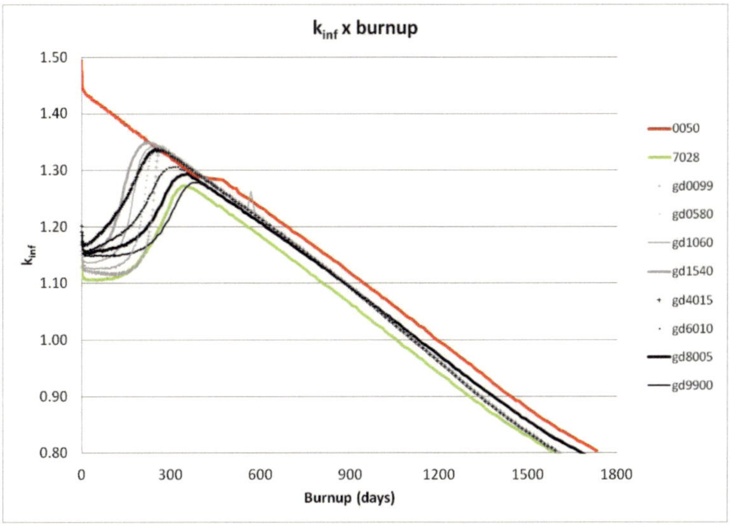

Figure 5. Infinite neutron multiplication factor as a function of burnup – 2.0% Gd_2O_3, 5.0% U-235, changing Gd-155 and Gd-157 enrichments.

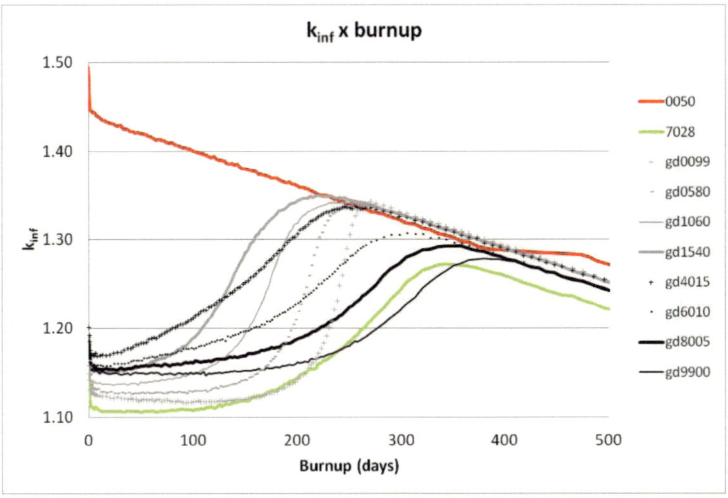

Figure 6. Zoom of Figure 5 (beginning of burnup interval).

Next step of the simulations considered 6.0% of U-235 enrichment in the fuel. Figure 7 plots the results (Figure 8 zoomed it until 600 days). As expected, the reactivities of the fuels containing enriched gadolinium continue growing and surpass the 0050 curve after a few hundred days of burnup. However, the reactivity peak for the 0050 is, so far, the greatest.

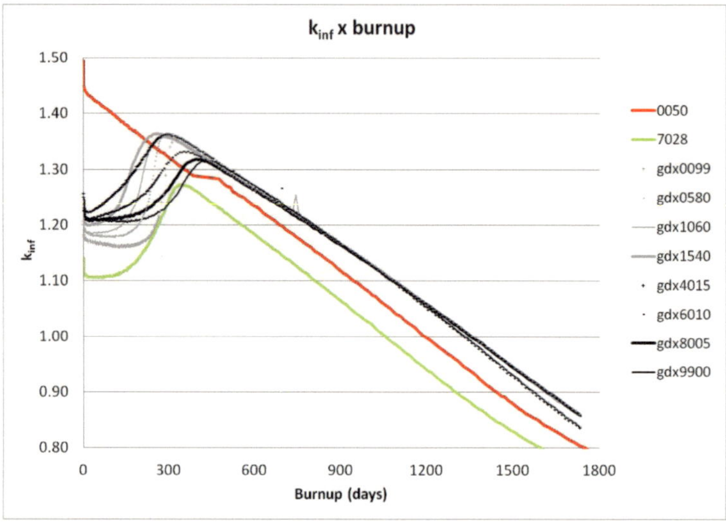

Figure 7. Infinite neutron multiplication factor as a function of burnup – 2.0% Gd_2O_3, 6.0% U-235, changing Gd-155 and Gd-157 enrichments.

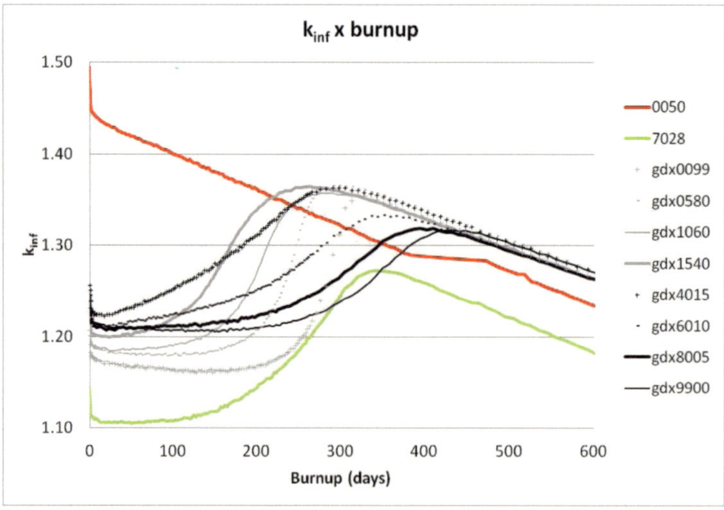

Figure 8. Zoom of Figure 7 (beginning of burnup interval).

The simulations are going to continue while the reactivity peak of the 0050 reference fuel is not exceeded. Figure 9 (and Fig. 10) shows the results for fuel compositions with 10% U-235. The reference fuel 0050 still has the greatest reactivity peak but the others gadolinium enriched compositions are getting closer. Furthermore, it is clearly evident the reactivity gain of the gadolinium enriched fuels for higher burnups as well as the flattening of its reactivity peaks.

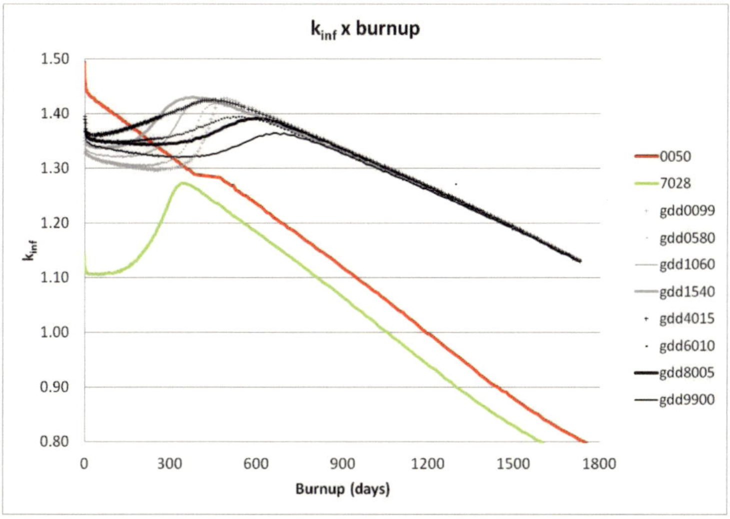

Figure 9. Infinite neutron multiplication factor as a function of burnup – 2.0% Gd_2O_3, 10.0% U-235, changing Gd-155 and Gd-157 enrichments.

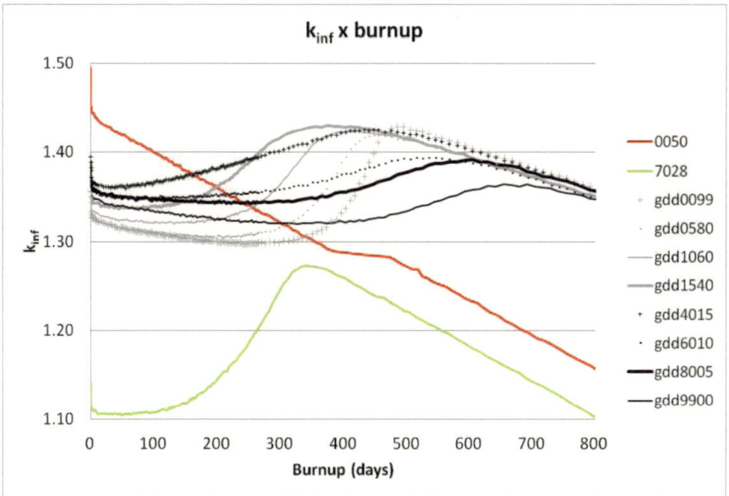

Figure 10. Zoom of Figure 9 (beginning of burnup interval).

The next three figures are plots of the gadolinium enriched fuels having 15.0% of U-235. Figure 11 shows the whole burnup interval, and the Figures 12 and Fig. 13 are the zoomed, respectively, in the beginning and at the end of the burnup interval. They show that the reactivity peaks of the gadolinium enriched fuels finally come close to the 0050 peak. In addition, Figure 11 shows very clearly the reactivity gain of these fuels, compared to the reference fuels, in the whole burnup interval. Reference fuel 0050, for instance, after around 500 days of burnup, has its reactivity reduced to a value that the gadolinium enriched fuels will achieve only after around 1700 days. Finally, it is extremely important to notice that the fuel compositions with higher enrichments in Gd-155 have the reactivity peak more flattened than the others and the reactivity peaks occurs at zero burnup, just like the fuels without burnable poison. Therefore, the next analyses are going to focus in compositions with 100% Gd-155 (compositions with 100% Gd-157 will be studied too, just for comparison).

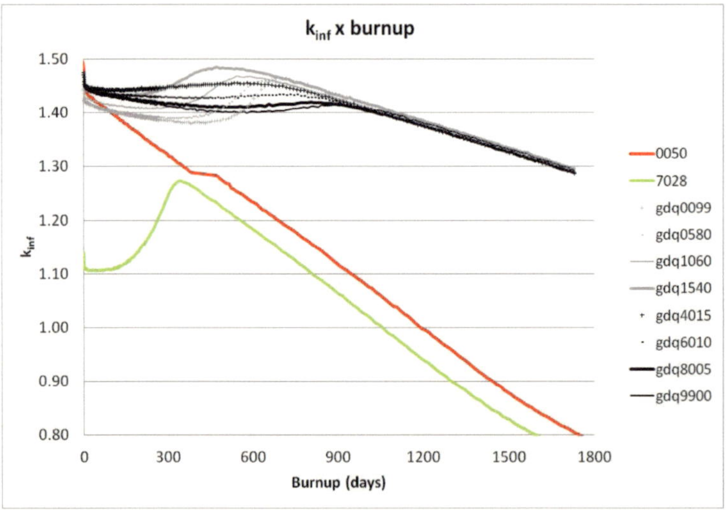

Figure 11. Infinite neutron multiplication factor as a function of burnup – 2.0% Gd_2O_3, 15.0% U-235, changing Gd-155 and Gd-157 enrichments.

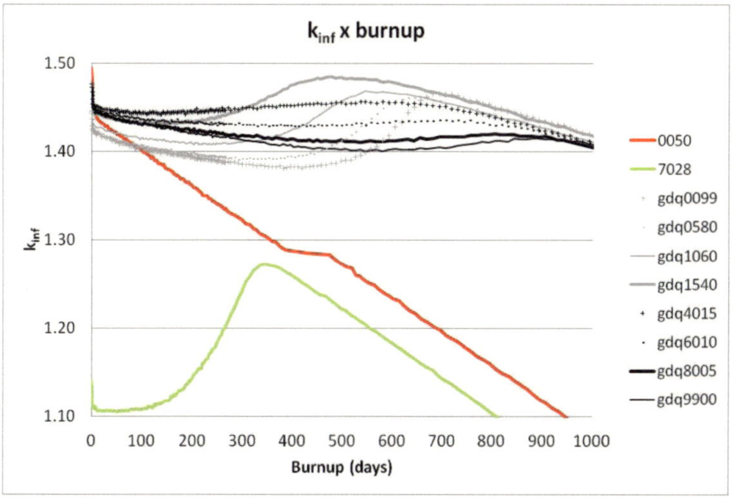

Figure 12. Zoom of Figure 11 (beginning of burnup interval).

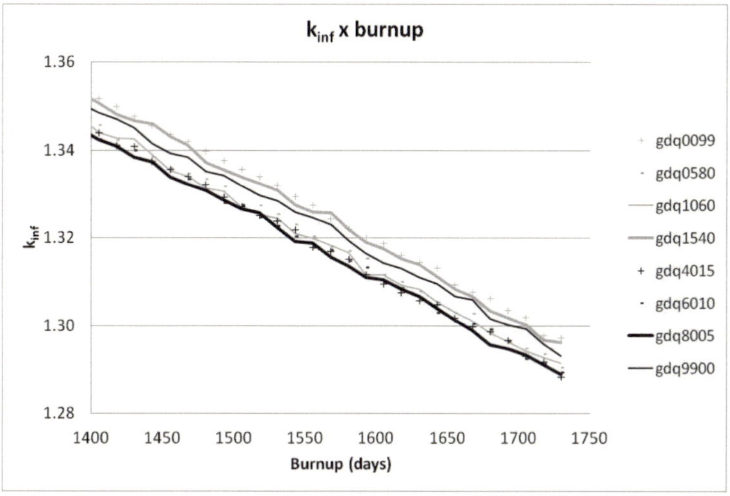

Figure 13. Zoom of Figure 11 (end of burnup interval).

3.2. Evolution of the infinite neutron multiplication factor as a function of the burnup – Fixed 15% U235, 100% Gd155 and Gd157 enrichments; Gadolinia contents change

Thus far, the studies performed have shown that a fuel composition with 2% Gd_2O_3 and 98% UO_2, enriched at 15% U-235 and at 100% Gd-155 is able to keep high reactivities at very long burnups. Furthermore, the reactivity peak of this composition occurs at zero burnup and it is a bit lower than the peak for the reference fuel. Nevertheless, this 2% Gadolinia content was settled based in [6]. Therefore, it is necessary to evaluate changes in such parameter.

The Figure 14 presents the evolution of the reactivity for compositions with 100% Gd-155 and 15.0% U-235 enrichments and Gadolinia contents from 1.0% to 4.0%, as listed in Table 4. Figure 15 shows the same results for a composition with 100% Gd-157. It is observed from these figures that when the gadolinium content is reduced to 1% the reactivity peaks surpass the maximum value achieved by the reference fuel 0050. However, this reactivity peak is not exceeded by fuel compositions with Gadolinia content within 1.5% to 4.0%. Moreover, it is very evident the advantage of using enrichments of 15% U-235 and 100% Gd-155 (instead of 100% Gd-157) due to the absence of reactivity peaks after the fresh fuel condition.

Strictly from the neutronic point of view, fuel compositions having the Gadolinia content within 1.5% to 4.0%, high Gd-155 enrichment (~100%) and percentage of U-235 around 15% can be technologically very attractive to keep the reactivity at high and stable levels, even for very long burnups. Nevertheless, this study is incomplete for neither has thermal-hydraulic nor thermo-mechanical analyses of these fuels. The physical model may also be improved towards a whole reactor core instead of a simple lattice. It is also absent any economic analysis regarding uranium and gadolinium at these levels of enrichment. These problems can be addressed in future works.

Finally, Figure 16 is presented to show clearly the evolution of fuel compositions having 100% Gd-155 and 2.0% Gd_2O_3 in different U-235 enrichments.

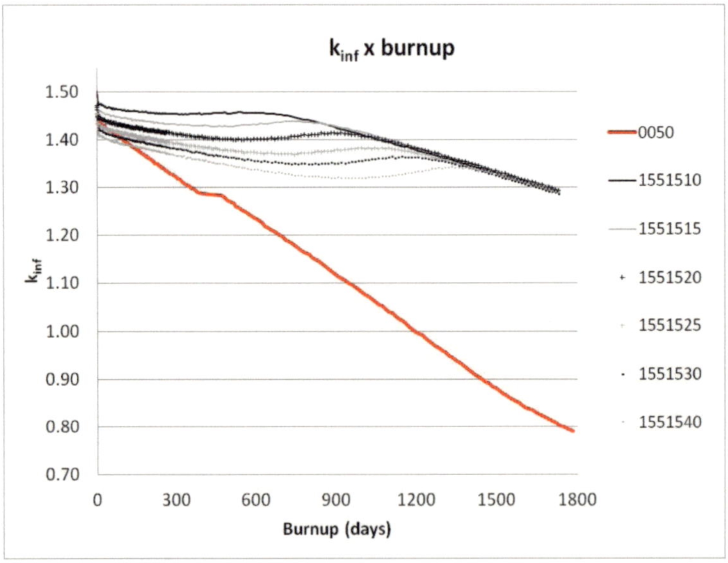

Figure 14. Infinite neutron multiplication factor as a function of burnup – 100% Gd-155, 15% U-235 and changing content of Gd_2O_3.

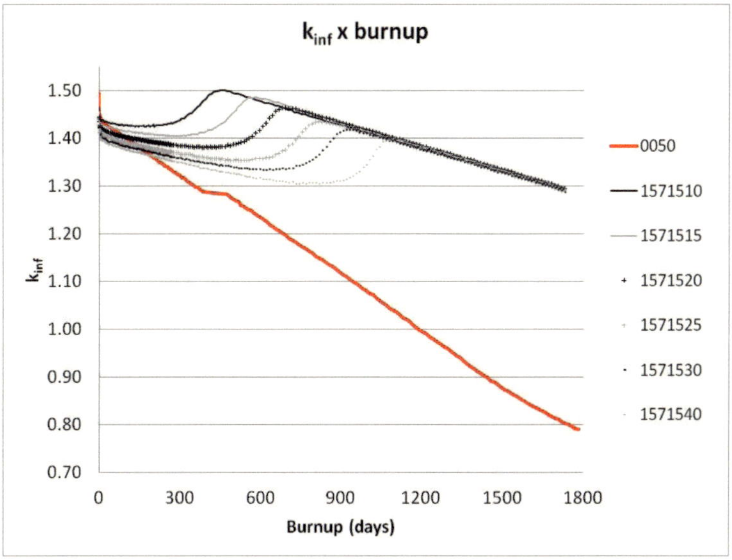

Figure 15. Infinite neutron multiplication factor as a function of burnup – 100% Gd-157, 15% U-235 and changing content of Gd_2O_3.

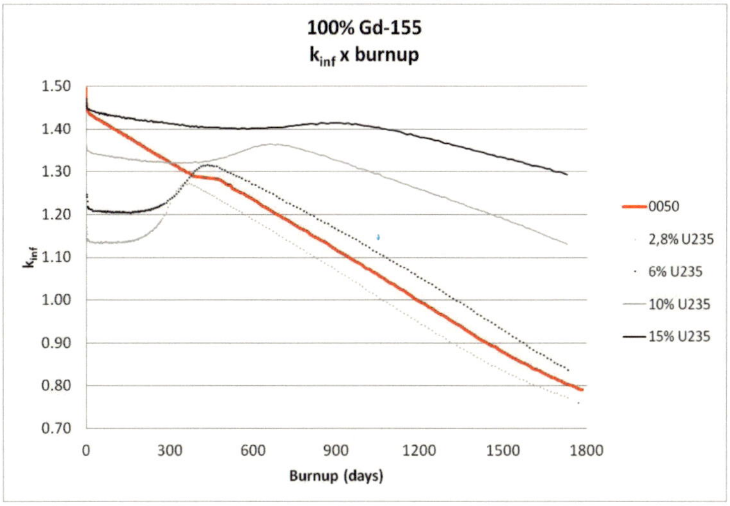

Figure 16. Infinite neutron multiplication factor as a function of burnup – fuel compositions having 100% Gd-155, 2.0% Gd_2O_3 and different U-235 enrichments.

4. Conclusions

Monte Carlo Burnup codes system Monteburns was used to simulate 3x3 lattices of rods containing several fuel compositions with different enrichments and contents of uranium and gadolinium oxides. The greater is the content of Gadolinia in the fuel the smaller is the reactivity of the fresh fuel. In addition, for fuels containing Gadolinia the reactivity peak is observed after some days of irradiation and not at zero burnup condition. These peaks moves to higher burnups as the Gadolinia content increases. The rise of the uranium-235 enrichment in the fuel can compensate the reduction of reactitivity due to the use of Gadolinia as burnable poison and even lead to higher burnups and longer cycles. It was found that fuel compositions having the Gadolinia content within 1.5% to 4.0%, 100% Gd-155 enrichment and 15% U-235 enrichment may keep high and stable levels of reactivity for very long burnups. In addition, the reactivity peak occurs at zero burnup, just like the fuels without any burnable poison, and the Gadolinia content is within a range that can be able to avoid the undesired effect in the fuel thermal conductivity.

Thus far, strictly from the neutronic point of view, fuel compositions having the Gadolinia content within 1.5% to 4.0%, high Gd-155 enrichment (~100%) and U-235 enrichment around 15% can be technologically very attractive. Nevertheless, additional analyses of thermal-hydraulic and thermo-mechanical properties for these fuels are needed. These problems can be addressed in future works as well as an economic analysis regarding uranium and gadolinium at these levels of enrichment and the physical model can be improved towards a whole reactor core model.

Author details

Hugo M. Dalle*, João Roberto L. de Mattos and Marcio S. Dias

*Address all correspondence to: dallehm@cdtn.br

Brazilian Nuclear Energy Commission, Nuclear Technology Development Centre, Campus da UFMG, Belo Horizonte, Minas Gerais, Brazil

References

[1] Poston D. L., Trellue H. R., *User's manual, version 2.0 for MONTEBURNS, version 1.0.* LANL, report LA-UR-99-4999, USA, 1999.

[2] NEI magazine, *Fuel Review: Design Data,* Nuclear Engineering International, September 2004, pages 26-35.

[3] Dalle H. M., Bianchini M., Gomes P. C., *A Temperature Dependent ENDF/B-VI.8 ACE Library for* UO_2*,* ThO_2*, ZIRC4, SS AISI-348, H2O, B4C and Ag-In-Cd*. Proceedings of the 2009 International Nuclear Atlantic Conference, INAC2009, Brazil, 2009.

[4] Dalle H. M, *Monte Carlo Burnup Simulation of the Takahama-3 Benchmark Experiment*. Proceedings of the 2009 International Nuclear Atlantic Conference, INAC2009, Brazil, 2009.

[5] X-5 Monte Carlo Team, *MCNP – A General Monte Carlo N-Particle Transport Code, Version 5*, Los Alamos National Laboratory, USA, 2005.

[6] Yilmaz S., Ivanov K., Levine S., Mahgerefteh M. *Development of Enriched Gd-155 and Gd-157 Burnable Poison Designs for a PWR Core*. Annals of Nuclear Energy 33 (2006), pages 439-445.

Chapter 5

Stability of γ-UMo Nuclear Fuel Alloys by Thermal Analysis

Fábio Branco Vaz de Oliveira and
Delvonei Alves de Andrade

Additional information is available at the end of the chapter

http://dx.doi.org/10.5772/53340

1. Introduction

The development of a nuclear fuel for research reactor applications must take into account the non-proliferation of nuclear grade materials for weapons requirements, according to the Reduced Enrichment of Research and Test Reactors (RERTR) program. To minimize a potential problem with the use of high enriched fuels, i.e., the production of Pu and its possible utilization in nuclear weapons, several reactors have been converted their fuels from high to low enrichment. It is now acceptable a level of less than 20% of U^{235}.

However, to compensate for the resultant loss of power, fuel densities must be raised and, thus, a new class of fuels, mainly the metallic ones, is under development. Those fuels are replacing the current oxides and silicides, and the most promising candidate has emerged from the binary system U-Mo, gamma phase stabilized, according firstly to the confirmation of its good behavior under fabrication and irradiation conditions, observed in several "in pile" tests performed over the years [1]. Compositions of 7 and 10 wt. % molybdenum are considered to be the most promising for the gamma-stabilized uranium fuels.

γUMo fuels can be used in two main configurations in fuel plates. If mixed as metallic powder in aluminum powder matrix, they are called dispersion fuels. The other way, a foil of γUMo is co-rolled with the aluminium cladding to achieve metallic bonding. In this configuration, they are called monolithic fuels. Since in the fabrication process, mainly in the rolling of the fuel plates, temperature is needed to ensure metallurgical stability, formation of new compounds can be observed [2]. For example, it was observed that, for the U-Mo-Al system, compounds of (γUMo)Alx, ($2 \leq x \leq 4$) have some deleterious results over the fuel plate's performance, due to the production of new uranium phases. Most of the new phases have low thermal conductivity, causing changes in the temperature profiles of the plate, leading to a non-uniform and excessive swelling, among other unwanted effects.

Works have been carried out with the objective of producing more stable and nonreactive fuel phases, as well as more chemically stable matrices. From the results of several experiments carried out over the years, called RERTR experiments, studying UMoX-Al system reactions in diffusion-pairs systems, it was determined that ternary additions of X = Zr, Ti or Si to the fuel γUMo particles and the addition of Si to the matrix are the most promising candidates for the minimization of the reactions and the formation of the interaction layer. Park et al. [3, 4] showed that under certain experimental conditions, an increase in the content of Zr, Si and Ti minimized the extent of the interaction layer for atomized fuels. However, such ternary elements could promote the γ-UMo decomposition into alpha-U, a more reactive phase, as a side deleterious effect. Most recently [5], it is shown that, for the usual compositions of 7, 10 and 12 % wt. Mo, the interaction layer is actually formed by 2 to 7 sub layers. The number of sub layers is a function of the molybdenum content.

In terms of the fabrication of γUMo alloys and their use as fuels, chemical resistance with the aluminium matrix is only one of the phenomena that must be taken into account, but also the temperatures at which the oxidation reactions starts. The presence of air in the processes of powder fabrication [6], like hydration-dehydration, atomization or machining, will change fuel's particle chemical and physical properties, so does its behavior under irradiation. The definition of the temperature of oxidation as one of the temperature limits of fabrication appears advantageous, to prevent oxidation. Together with the oxidation limit, the temperature with the start of reaction with aluminum is also used as a process temperature limit, mainly in the rolling of the fuel plates. They are also interesting in the verification of possible oxidation reactions occurring during the aluminum experiments.

The main objective of this work is to contribute to the study of the chemical stability of the high density and low enriched γ-UMo hypoeutectoid alloys fuels, currently being considered to replace the low density ones, such as U_3O_8 and the high density silicides USi, U_3Si_2 and U_3Si. For this to be accomplished, it is presented and discussed an alternative approach, to the study of the reactions of the formation of the interaction U-Mo-Al layers by means of solid state diffusion-pairs, based on thermal analysis.

It is given here an emphasis on the results of the technique of differential thermal analysis (DTA) and on the choice of a convenient parameter for the analysis of stability and definition of a stability criterion. This technique was chosen to perform the studies of the reactions, due to its simplicity and the production of results with high accuracy.

2. Experimental procedure

Samples of uranium alloys and aluminum were assembled in alumina crucibles, in the reaction chamber of the thermal analyzer furnace. Thermal cycles, comprising heating and cooling ramps, were utilized to the observation of the melting and solidification points of aluminum and any reaction products, when formed. After a vacuum level of 10^{-3} mbar, temperature was raised at 10°C/min from the room up to 1000°C, and decreased at the same rate. Temperatures and energies of reactions, showed as peaks in the HF curves, were extracted directly from the DTA curves and, when necessary, calculated. A blank consisting of pure aluminum was obtained, to serve as a comparison basis between the reactive and the non-reactive systems.

Analysis of the results was carried out based on the following considerations, given by the examination of the blank curve, and extensive to all the experiments. Differences between the melting points of aluminum and γUMo alloys are of the order of 600°C. Thus, during the heating ramp of the thermal cycle, the first peak shown in the curves was related to the melting of aluminum. Since the contact area between the solid fuel and liquid aluminum, produced during the thermal cycles, are larger than the presented by the aluminum when it is kept in its porous form, such as in the compacts, an increase in the probability of the reaction between Al and γUMo is expected. If a reaction occurs, the corresponding peaks must to appear, indicating release or absorption of heat, after the melting peak of aluminum. Since the mass of the system could be considered as a conserved quantity, due to the slight differences observed between the initial and final masses during the experiments, reaction indicates the formation of phases other than the original ones, usually having different thermal properties, when compared to the reactants. In previous papers, it was determined that, when a ternary UMoAl compound is formed, its stability is strongly dependent on its composition, some of them originating amorphous phases, with bad performance under irradiation.

To check for the changes in the composition of the original or reactant system, a careful examination of the cooling curve was also carried out. On the cooling ramp, the area under the peak corresponding to the aluminum solidification depends on the extent of the reaction of aluminum consumption in the heating step of the cycle. If no aluminum is consumed, indicating that no reaction occurred, it is expected the 1:1 proportion between the cooling (solidification) and heating (melting) areas under the peaks, as shown by the curve of the pure aluminum.

Some possible criteria for the stability of the alloys under aluminum, based on the experimental observations above, were devised, through a comparison between the energies given by the areas enclosed by both peaks, in the melting and solidification parts of the thermal cycle heat flow curves, obtained for each alloy and normalized for the reactant masses. Low cooling / heating ratios indicate low aluminum available for solidification, due to reaction and formation of a stable compound. To verify the possible oxidation of the samples, a comparison between the visual states of the samples, from the reactions with aluminum and oxygen, was made after the end of the thermal cycles. Also, it was made a comparison between the curves for aluminum and oxidation experiments.

Experiments for the stability studies in oxidizing atmosphere were carried out with the same thermal cycles used in the aluminum compatibility experiments. Two classes of samples were analyzed, the as cast and the thermally treated ones. Samples of the γUMo alloys (5 to 10% wt. Mo) were produced for the oxidation-DTA experiments. Synthetic air (80% nitrogen and 20% oxygen) was introduced into the furnace chamber to guarantee the continuous supply of oxygen during experiments.

The curves of heat flow (HF), mass and temperature, as a function of time, were obtained, and the strategy adopted was to simply register the points on the heat flow curves when the reaction with air started, as given by the DTA equipment. In addition, to compare the effects on stability, promoted by the homogenization treatments, a curve comparing the differences between the starting and maximum points for each composition was constructed, enabling us to derive some conclusions regarding gamma stability under oxygen. Normalization of the curves were carried out to the determination of the areas, the curves shown above are the original ones, given during the experiments.

Based on the experimental data, the next step was the construction of the stability curves. They are curves relating the area under some of the events defining a phenomena, for example, solidification and melting of aluminum peaks and possible reactions for the γUMo-Al systems, and temperatures of the starting of the reaction, in the case of the γUMo-O_2 systems, as a function of the composition x of the gamma γ-U_xMo alloys. The determination of the extremes (maxima or minima) of the curves, and the analysis of their surroundings, indicates a range of optimal composition, for the improvement of the γ-U_xMo alloys fuel properties in terms of fabrication and during reactor operation.

Analysis presented here was based on the curves given by the DTA equipment, and the phenomenology relating to the presence of the peaks, due to oxidation reactions, impurities, exothermic or endothermic reactions, changes of the physical states of the system, and the effects of other contaminants are being carried out.

3. Results and discussions

3.1. Aluminum

The first graph shows the results of the blank experiment, carried out for the same heating and cooling cycles, having as "reactant" only pure aluminum. It was observed the ratio between the areas under the solidification and melting events was 1:1.12, almost 1:1, indicating the conservation of the aluminum mass in the system, respecting our experimental errors, during the thermal cycle, and also indicating no interference in the experimental results due to aluminum mass loss.

Figures 2 to 5 shows the experimental heat flow curves obtained for each alloy composition, together with the thermal cycle, the curve in red. Points depicted in Figure 2 are called events associated with the phenomena, related to the stability analysis. They are shown in Tables 1 to 3, and described below, the symbol * refers to data for the as cast sample. The events are: A: first peak, before the start of the aluminum melting; B: start of the melting; C: endothermic heat flow peak for the melting; D: minimum, ending the melting process; E: start of a reaction between uranium alloy and aluminum; F: exothermic peak of reaction; G: end of the reaction; H: another transformation (reaction) peak; I: start of the physical transformation of some new compound formed; J: exothermic peak of the transformation; L: end of the transformation; M: solidification of aluminum starts; N: exothermic peak of the solidification, and O: end of the solidification.

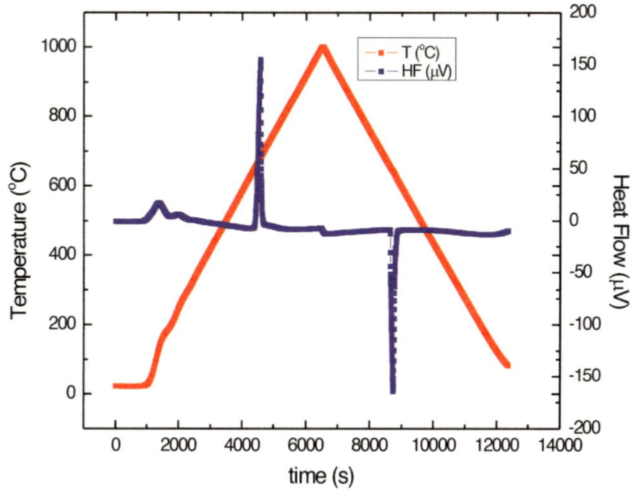

Figure 1. Heat flow and thermal cycle curves for pure aluminum [7].

To describe a specific phenomenon, it is suggested the following, based on the events taken as points in the HF curves, and according to the above description. A phenomenon like melting, solidification, etc., defines a set of coordinates given by the events. For example, the pure aluminium melting phenomenon is defined by the events B, C and D. From the data of table 1, their coordinates are B = (4429; 654.54; 0.38), C = (4580; 678.87; 154.78) and D = (4663; 694.12; -0.21).

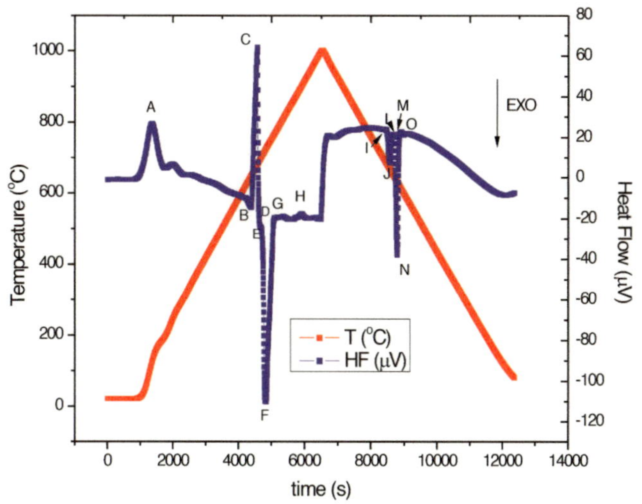

Figure 2. Heat flow curve for γU5Mo, with the events and related phenomena [7].

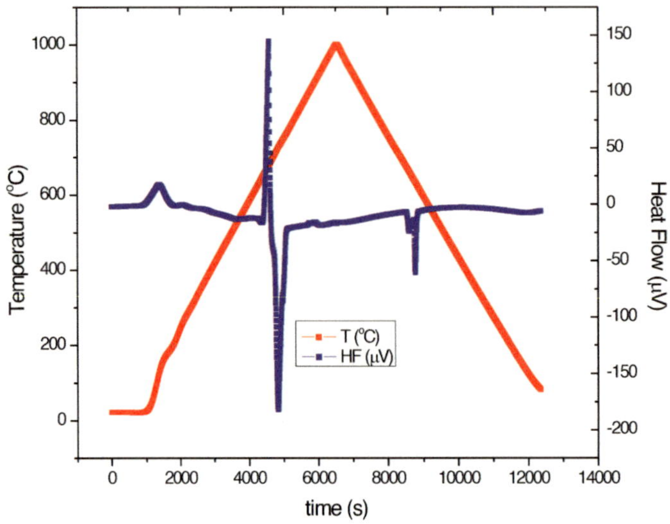

Figure 3. Heat flow for γU6Mo, homogenized 1000°C [7].

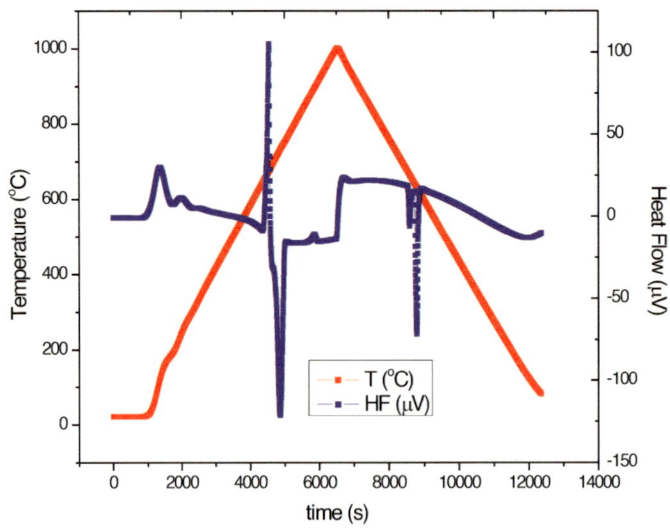

Figure 4. Heat flow for γU6Mo, as cast [7].

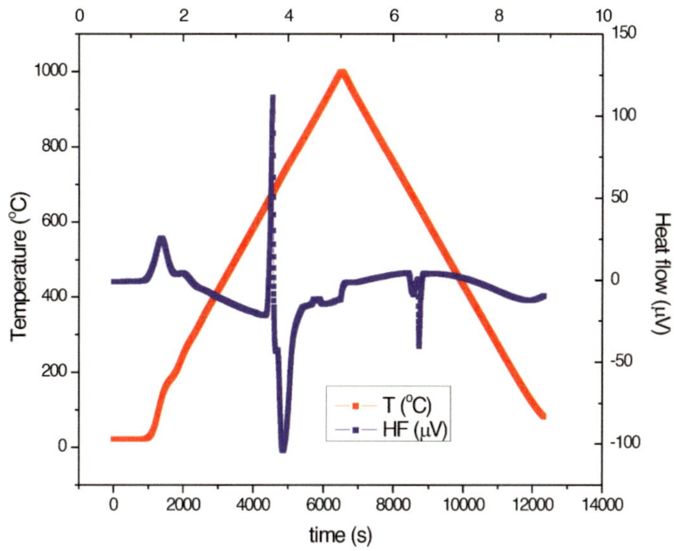

Figure 5. Heat flow for γU7Mo, homogenized 1000°C [7].

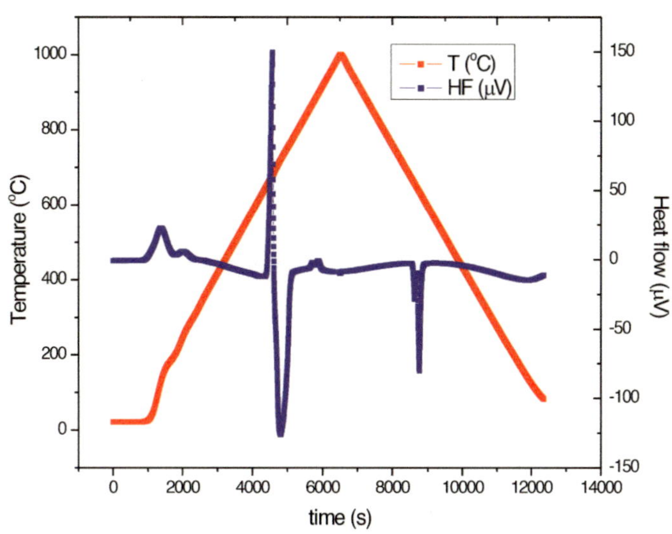

Figure 6. Heat flow for γU8Mo, homogenized 1000°C [7].

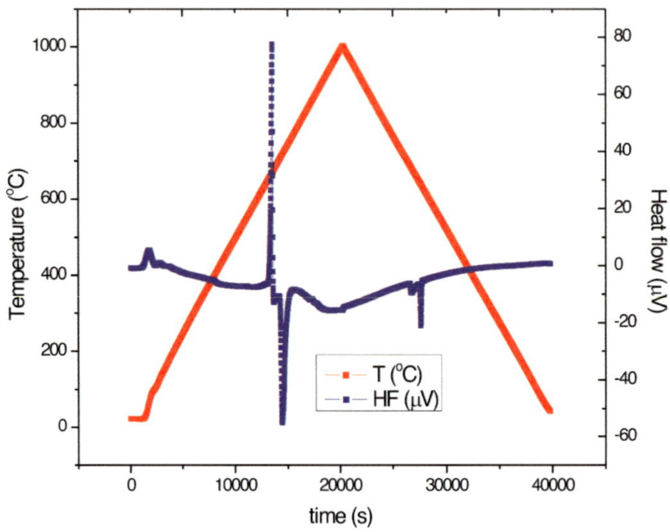

Figure 7. Heat flow curve for γU9Mo, homogenized 1000°C [7].

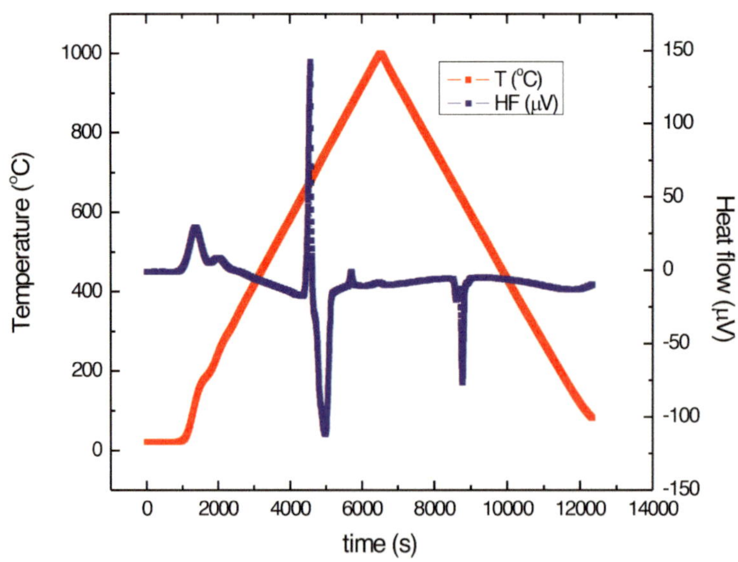

Figure 8. Heat flow curve for γU10Mo, homogenized 1000°C [7].

First assessment of the values of temperature, heat flow and times, for each of the points, are given in the Tables 1 to 3. Examination shows the evolution of peaks during heating and cooling, after the melting and before the solidification peaks, indicating regions for the formation of the reaction products. Visual examination of the samples showed no changes in their volumes, against the observed and expected volume changes in the oxidation experiments.

Event	γU5Mo	γU6Mo*	γU6Mo	γU7Mo	γU8Mo	γU10Mo	Al°
A	1358	1343	1365	1370	1385	1358	1370
B	4380	4355	4365	4350	4428	4353	4429
C	4570	4548	4575	4558	4573	4565	4580
D	-----	-----	-----	4665	-----	-----	4663
E	4690	4653	4690	4695	4613	4653	-----
F	4848	4858	4833	4868	4808	4970	-----
G	5074	5040	5075	5438	5125	5158	-----
H	5903	5855	5895	5735	5873	5700	-----
I	8498	8525	8493	8405	8525	8470	-----
J	8570	8578	8583	8568	8625	8590	-----
L	8700	8725	8658	8695	8705	8703	-----
M	8700	8725	8658	8695	8705	8703	8638
N	8790	8778	8763	8730	8760	8758	8738
O	8871	8918	8853	8870	8918	8835	8898

Table 1. Times (s) for the experimental events [7].

It was stated [8] that reactions of γUMo alloys with aluminum started typically at the temperature of 645°C. In Figure 1, it was indicated a temperature of 641°C for the melting point of the aluminum, and, as can be viewed in Figures 2 to 8, it is suggested that chemical reaction started soon after the melting phenomena. Thus, it is confirmed the validity of the fundamental assumptions of our approach, to the detection of possible reactions in the U-Mo-Al system. The formation of a layer of liquid aluminum in contact with the γUMo particles promotes an effective contact between both surfaces, enhancing the probability of reaction.

The entries of the Table 1 are the times of the occurrence of each event for each of the DTA experiments shown in the figures above. They were shown because of their importance in the calculus of the total heat released or absorbed in the system, which defines our stability parameter.

The entries of the Table 2 are the respective temperatures, and the entries of the Table 3 are the respective heat flows. In the following discussions and data analysis, it could be helpful to made the association X→ (t; T; Φ), where X = event, given by the A, B, C, etc., and the triple (t; T; Φ) giving its coordinates in the three-dimensional space comprised by the experiments. If no reaction occurs, peaks D and E are coincident, no F peak will appear. In this case, D defines aluminum transformation, and E defines a reaction in the system.

From the data listed in Tables 1, 2 and 3, it is possible to calculate the areas under the peaks, to the determination of the total heat released or absorbed on the system. These results are listed in the Table 4.

The areas were calculated by the determination of the areas below a peak of a set of events. For an example, a triangle basis gives the total time of the phenomena, and the heights are the maximum heat flows. For the purposes of the stability analysis, values of the areas were taken as positive. However, negative signs are still shown, to emphasize the thermodynamical nature of the phenomena, endothermic or exothermic.

Event	γU5Mo	γU6Mo*	γU6Mo	γU7Mo	γU8Mo	γU10Mo	Al°
A	123.32	117.6	124.31	125.88	130.64	123.31	126.24
B	647.54	643.77	644.14	643.22	654.69	643.19	654.54
C	678.15	674.42	678.86	674.5	676.85	676.48	678.87
D	-----	-----	-----	694.12	-----	-----	694.12
E	699.35	693.2	697.85	699.13	684.16	692.71	-----
F	726.92	728.75	723.76	729.2	719.69	747.24	-----
G	762.68	757.62	762.38	822.39	720.77	777.12	-----
H	901.02	893.09	898.95	871.4	894.95	866.97	-----
I	678.93	674.78	678.22	690.58	672.98	680.48	-----
J	667.3	666.39	663.75	664.36	656.88	661.06	-----
L	646.0	642.24	651.53	643.49	643.76	642.79	-----
M	646.0	642.24	651.53	643.49	643.76	642.79	654.62
N	631.96	634.48	634.48	638.15	635.48	634.52	612.32
O	615.99	611.08	619.74	615.07	609.19	621.53	640.27

Table 2. Temperatures (°C) for the experimental events [7].

For the reaction phenomena H, clearly shown in the figures with the exception of γU8Mo and γU10Mo alloys, it was difficult to define a complete phenomenon in terms of events. The region around this point is being analyzed, and the results will be presented in a future work, together with a more accurate interpretation of the peaks, based on thermodynamical and chemical consideration.

The studies of the reactions in the system U-Mo-Al are important in nuclear technology, since the formation of a stable layer of an insulating material is deleterious to the fuel's performance. Earlier works, like the previously mentioned in the introduction, determined the composition of this layer as a ternary of $(UMo)Al_3$ and $(UMo)Al_4$. However, most recent and accurate works [5] determined that the layer is a combination of at least 2 to 7 sublayers, each having specific composition. According to Perez et al [5], for an UMo alloy with 7 wt% Mo, it was determined for the sublayer near aluminum of UMo_2Al_{20}, UAl_4 and UAl. For the second sublayer, UMo_2Al_{20} and UAl_4, and up to the pure γ-U7Mo, the layers are composed by UMo_2Al_{20}, UAl_4 and UAl_3, UMo_2Al_{20} and UAl_3, $U_6Mo_4Al_{43}$, UMo_2Al_{20} and UAl_3, $U_6Mo_4Al_{43}$ and UAl_3, and finally $U_6Mo_4Al_{43}$ and γ-U. For an alloy of γ-U10Mo, 3 sublayers appeared, UMo_2Al_{20} plus UAl_4, $U_6Mo_4Al_{43}$, UMo_2Al_{20} and UAl_3, and the nearest from the fuel phase the composition is $U_6Mo_4Al_{43}$, UAl_3 and γ-U. This set of materials have different properties under irradiation, for example, $U_6Mo_4Al_{43}$ transforms to an amorphous structure at less than 1 damage per atom (dpa) and, at 100 dpa, presenting a high density of voids, which is related to possible sites for bubble nucleation in their structure. The presence of the ternary phases from the U-Mo-Al is responsible for the poor irradiation performance of the fuels, under irradiation conditions.

Evento	γU5Mo	γU6Mo*	γU6Mo	γU7Mo	γU8Mo	γU10Mo	Al°
A	26.71	29.,92	17.41	25.,55	22.71	29.19	16.86
B	-14.56	-8.59	-16.31	-21.6	-0.04	-17.31	0.38
C	64.31	104.87	144.76	111.51	149.6	142.04	154.78
D	-----	-----	-----	-43.44	-----	-----	-0.21
E	-24.28	-31.35	-39.38	-42.68	-1.24	-27.23	-----
F	-110.15	-121.39	-181.82	-103.98	-126.05	-111.52	-----
G	-21.14	-15.42	-24.06	-16.48	-11.01	-17.14	-----
H	-17.64	-10.9	-15.78	-11.8	-0.9	-1.19	-----
I	20.65	17.98	-6.66	4.4	-2.66	-5.68	-----
J	7.34	-5.76	-23.96	-8.86	-28.34	-20	-----
L	21.46	15.08	-14.8	0.64	-9.65	-12.58	-----
M	21.46	15.08	-14.8	0.64	-9.65	-12.58	-8.63
N	-37.81	-71.68	-60.61	-40.36	-80.25	-76.43	-164.3
O	22,.8	16.99	-6.72	4.31	-3.08	-10.48	-9.42

Table 3. Heat flows (μV) for the experimental events [7]

To solve this problem, it is studied in the RERTR experiments ternary additions to the γ-UMo alloys and also additions in the matrix, and the elements considered as promising are Zr, Ti, and Si. It is a convenient approach to keep the layer stable, since fuel properties and performance can be estimated with accuracy. γ-U7Mo and γ-U10Mo are the most studied for this purpose.

Event / m_i / E_i	γU5Mo	γU6Mo*	γU6Mo	γU7Mo	γU8Mo	γU10Mo	Al°
m_t	149.144	148.909	168.991	173.298	175.507	155.642	
m_c	75.836	73.94	89.5	97.437	91.654	77.095	
m_{Al}	73.308	74.692	79.491	75.86	83.853	78.546	82.010
BCD(E)							
Et	174.04	249.83	331.96	261.11	158.37	316.71	
Ec	342.27	503.14	626.79	464.40	303.25	639.38	
E_{Al}	354.07	498.07	701.75	596.49	331.47	627.57	441.89
Area	***25.9***	***37.2***	***56.1***	***45.3***	***27.8***	***49.3***	***36.2***
D(E)FG							
E_t	225.13	254.72	341.96	318.98	349.87	289.87	
E_c	442.76	512.98	645.68	567.33	669.96	585.21	
E_{Al}	458.03	507.82	722.89	728.70	732.28	574.39	
Area	***-33.6***	***-37.9***	***-57.8***	***-55.3***	***-61.4***	***-45.1***	---
H (Area)	***-17.64***	***-10.9***	***-15.78***	***-11.8***	***-0.9***	***-1.19***	
IJL(M)							
E_t	74.38	30.21	13.27	19.04	22.76	16.27	
E_c	36.55	60.83	25.05	33.87	43.58	32.85	
E_{Al}	37.80	60.22	28.05	43.50	47.63	32.25	
Area	***-2.7***	***-4.5***	***-2.2***	***-3.3***	***-3.9***	***-2.5***	---
L(M)NO							
E_t	68.71	113.69	57.52	42.79	89.66	54.35	
E_c	135.13	228.97	108.61	76.12	171.69	109.73	
E_{Al}	139.79	226.66	121.60	97.77	187.67	107.71	492.35
Area	***-10.3***	***-16.9***	***-9.7***	***-7.5***	***-15.7***	***-8.5***	***-40.4***

Table 4. Areas (mV.s), reduced for the samples masses [7].

However, it was shown by our results that there is a strong tendency for reactions in all the compositions, evidenced by the events and phenomena from the curves. One of the ways to support the above arguments is due to the evaluation of the differences in the areas above the solidification and melting peaks of the aluminum, shown in Figures 2 to 8. From the blank experiment, it was obtained an area ratio of 1.12, which was not kept constant in the U-Mo-Al system experiments. Thus, some amount of aluminum was consumed, probably by the reactions D(E)FG and H, as is shown in the heating ramps. During cooling, free aluminum solidifies, producing a less intense N peak. The difference allows us to estimate masses for the products and other quantitative parameters, together with the analysis of the IJL(M) phenomena.

In the first column of the Table 4, Event corresponds to the set of coordinates of a phenomenon, defined by the letters A, B, C, etc.; m_i is the masses considered for the analysis, for the calculations, i = c for the fuel, i = Al for the aluminum, and i = t for the total mass, the sum of of m_i and m_c. They appear in the first three lines of data. A row starting with E_i, i = c, Al or t, presents the normalization results for the respective masses. Since we are analyzing the behavior of the aluminum peaks and with aluminum, rows of interest are labeled with E_{Al}.

Data listed in Table 4 were obtained from Tables 1 to 3, and used to build the stability curve, presented on Figure 6. It can be noted that γU9Mo alloy data is not listed. It was detected a possible misinterpretation of the data during the software data process for this specific alloy. So, it is not included in the stability analysis graphic. A detailed verification is in progress, however will not be presented in the present article.

Phenomena like BCDE, DEFG, H, IJLM, LMNO and others that can be derived from the complete analysis of the heat flow or gain of mass curves, can all be studied for stability considerations. To exemplify, next figure is constructed considering the exothermic reaction peaks resulting from the DEFG phenomena, for each alloy.

From Figure 9 it is possible to observe that there is a tendency for the production of more exothermic reactions with aluminum when we consider compositions of 7 to 8%wt Mo. In terms of stability, it indicates that, if a reaction of the formation of the interaction layer is going to occur during fabrication or during reactor operation, compounds formed are probably more stable than those formed in the extremes of the composition.

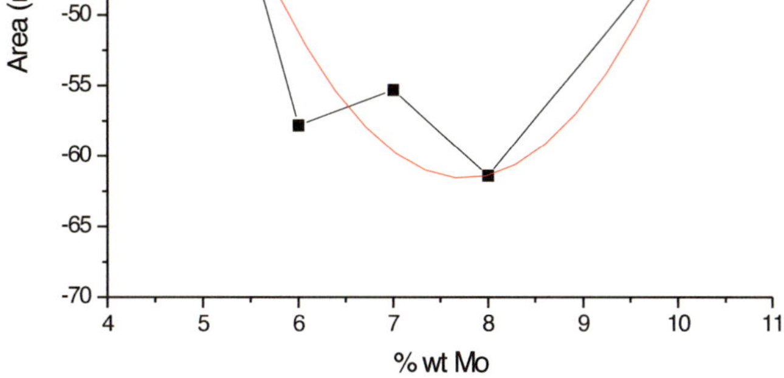

Figure 9. Results from the phenomenon DEFG.

Since the interaction layer formation is a phenomenon with high probability of occurrence, due to the temperatures involved during fabrication, irradiation conditions, etc., formation of such stable compounds are favoured.

Also from Figure 9, it is indicated that more stable compounds are formed in the region between 7 to 8% wt. Mo. The second-degree polynomial fit indicates the composition of 7.701% as a minimum, thus as the most stable one. This result is confirmed by the analysis of the results of the oxidation experiments, explained in the following item.

3.2. Oxygen or oxidizing atmosphere

The results of the differential thermal analysis for the γUMo alloys are presented below, Figures 10 to 21, showing the curves of mass gain, thermal cycle and heat flow. For each system, it was determined the temperatures where oxidation occurs, with the total mass gain after the experiments.

Since, during the fabrication of the fuel plates, we are interested in the temperature where chemical reactions start, first assessment of stability criteria was defined in terms of the maximum temperatures and the starting temperature of oxidation. Since a parabolic behavior is obtained, if we analyze the temperature differences between the as cast and thermally treated samples, it was stated as a possible second criterion, which takes into account the differences in microstructure for each class of samples.

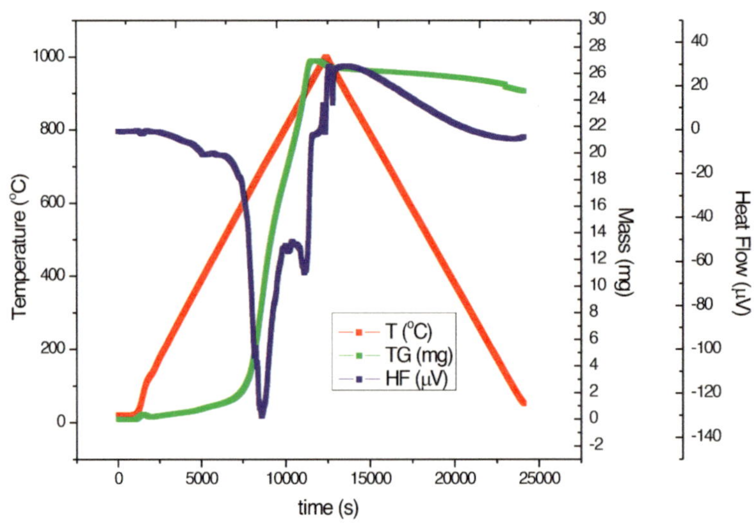

Figure 10. Curves of heat flow and mass gain, γU5Mo, homogenized 1000°C [7].

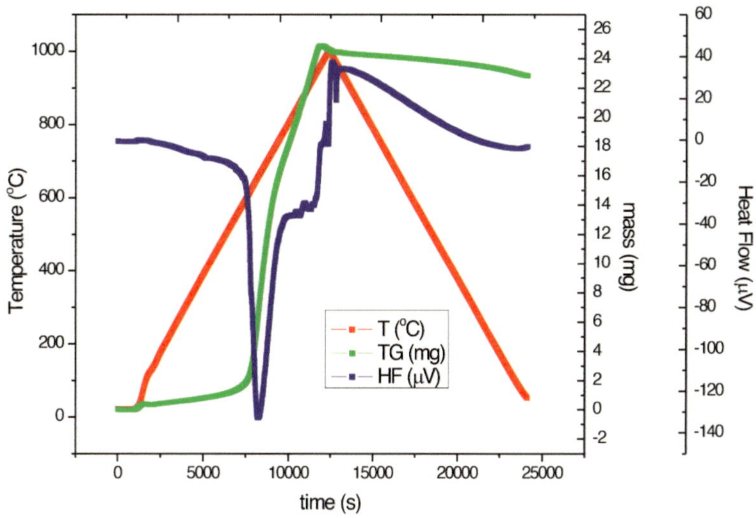

Figure 11. Curves of heat flow and mass gain, γU5Mo, as cast [7].

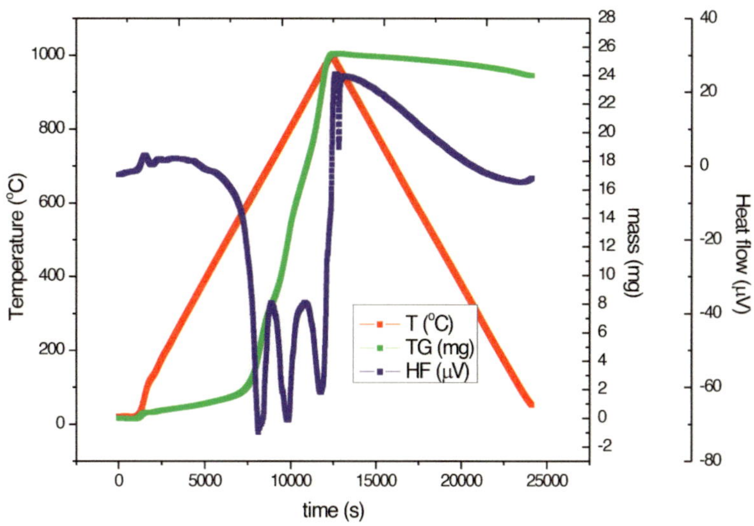

Figure 12. Curves of heat flow and mass gain, γU6Mo, homogenized 1000°C [7].

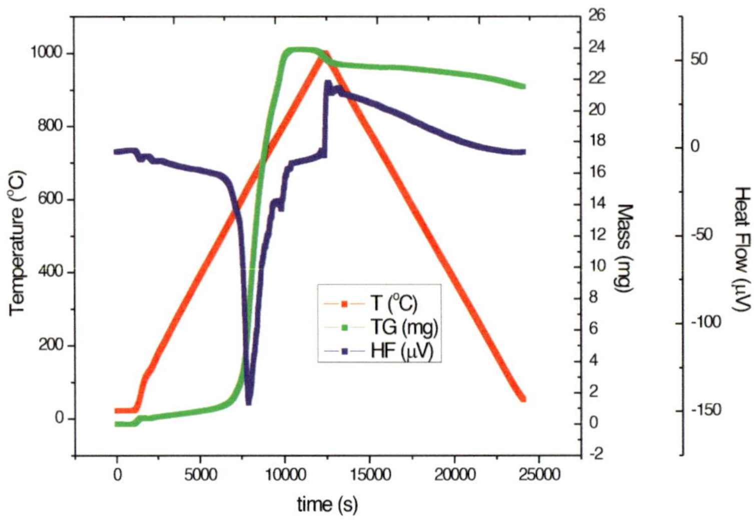

Figure 13. Curves of heat flow and mass gain, γU6Mo, as cast [7].

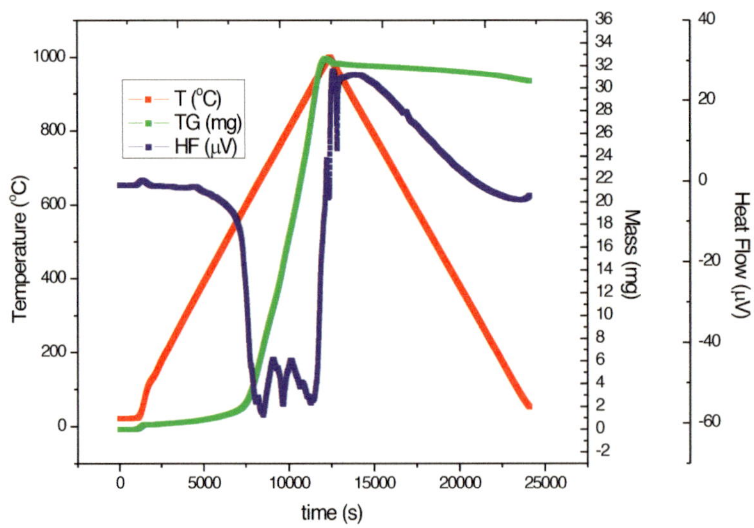

Figure 14. Curves of heat flow and mass gain, γU7Mo, homogenized 1000°C [7].

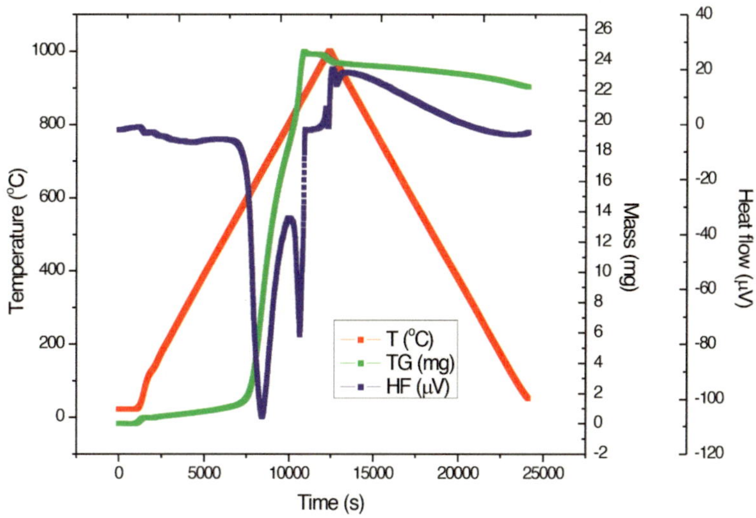

Figure 15. Curves of heat flow and mass gain, γU7Mo, as cast [7].

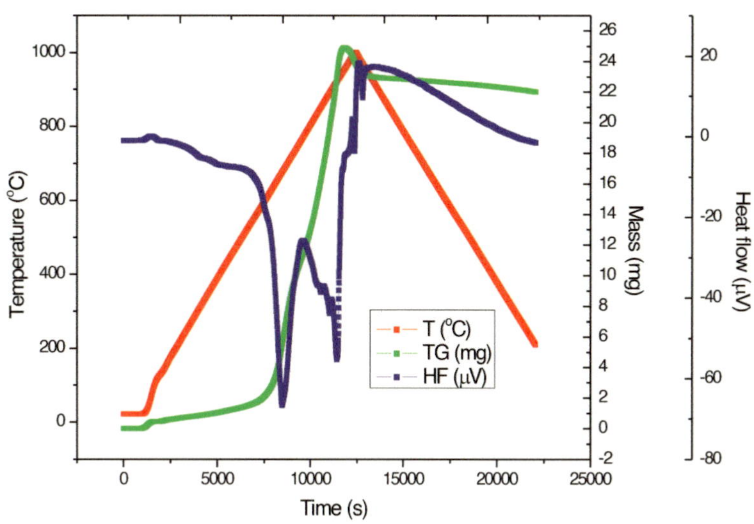

Figure 16. Curves of heat flow and mass gain, γU8Mo, homogenized 1000°C [7].

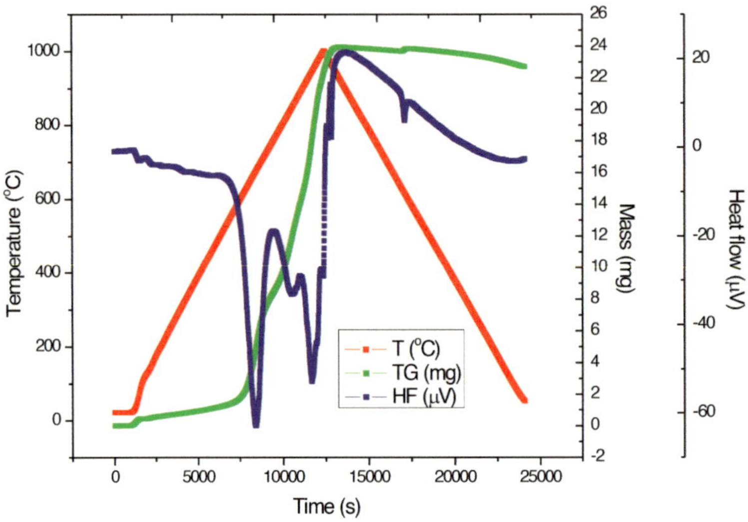

Figure 17. Curves of heat flow and mass gain, γU8Mo, as cast [7].

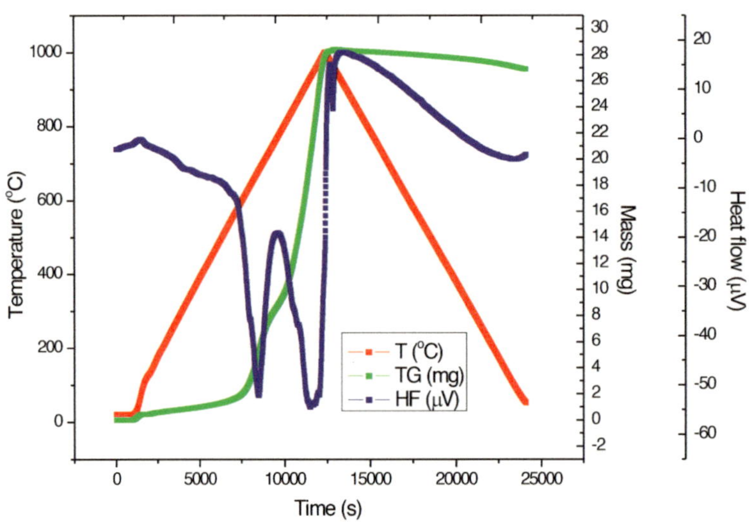

Figure 18. Curves of heat flow and mass gain, γU9Mo, homogenized 1000°C [7].

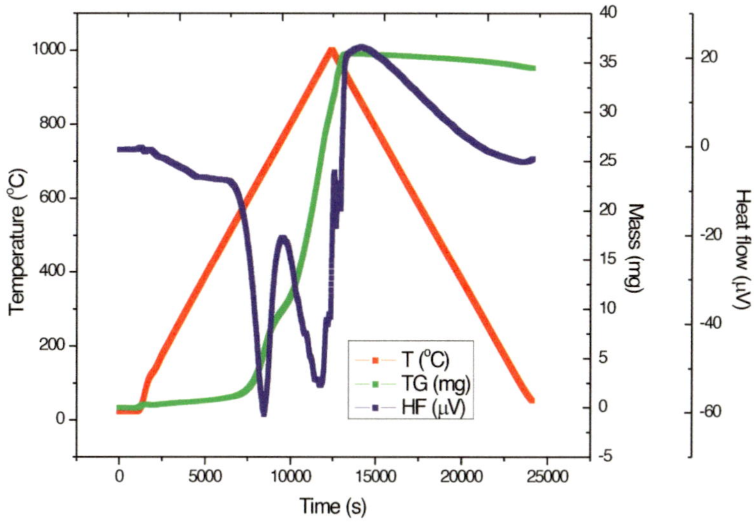

Figure 19. Curves of heat flow and mass gain, γU9Mo, as cast [7].

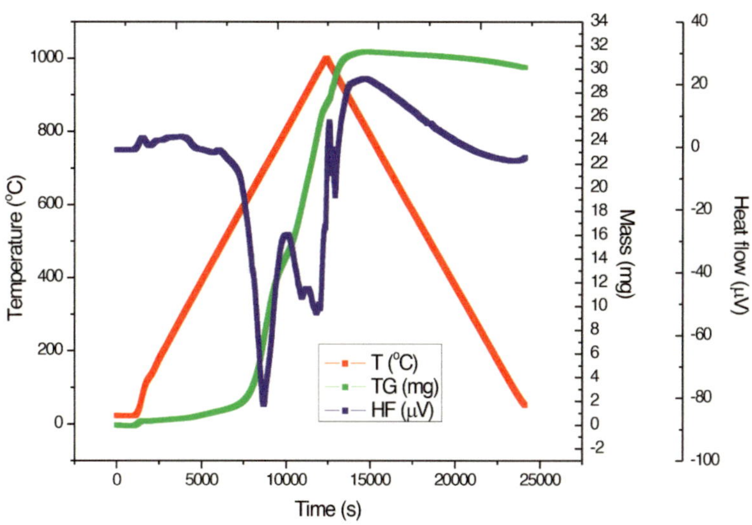

Figure 20. Curves of heat flow and mass gain, γU10Mo, homogenized 1000°C [7].

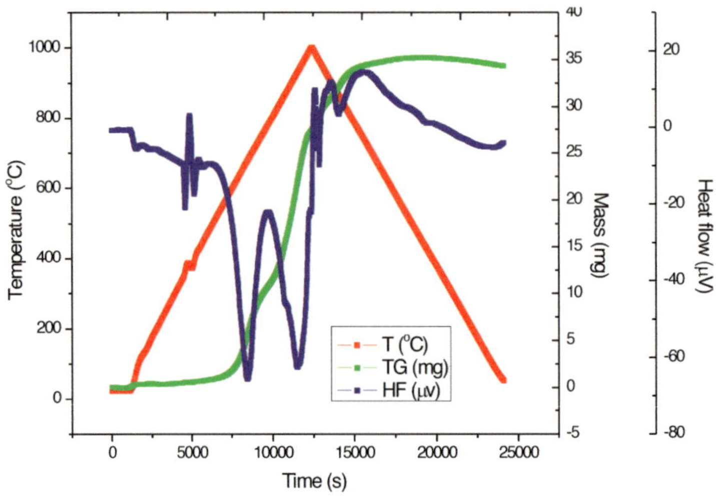

Figure 21. Curves of heat flow and mass gain, γU10Mo, as cast [7].

The temperatures of the start and maximum of oxidation are presented in Table 5 and shown graphically in the Figure 22. The choice for the first peak is mandatory for fabrication purposes, since it indicates a maximum of temperature to avoid oxidation of the fuel in an oxygen-rich environment.

The results shows that thermally treated alloys are more resistant to the oxidation than the as cast ones, and this resistance slightly increases with the increase in Mo content, showing a discontinuity for the case of the 6%wt, Figure 22. The abnormal behavior of the γU6Mo alloy is possibly due to several factors, like its more refined grain structure, some impurity that acted as inoculants, or even to a more pronounced Mo segregation, which possibly was not reduced with thermal treatments. This abnormal behavior was also observed during irradiation, in the first RERTR tests [1], and some authors suggest that at this percentage another kind of mechanism of alpha precipitation, instead of the gamma to alpha conversion, occurs.

	as cast		thermally treated	
Alloy composition	**Start temperature (°C)**	**Maximum temperature (°C)**	**Start temperature (°C)**	**Maximum temperature (°C)**
γU5Mo	549.70	660.31	581.35	687.55
γU6Mo	517.80	630.67	536.26	646.75
γU7Mo	542.21	672.36	553.29	680.34
γU8Mo	545.10	672.57	558.84	676.30
γU9Mo	535.25	675.82	565.35	679.73
γU10Mo	526.03	675.03	577.15	694.16

Table 5. Main temperature events, for the γUMo compositions tested [7]

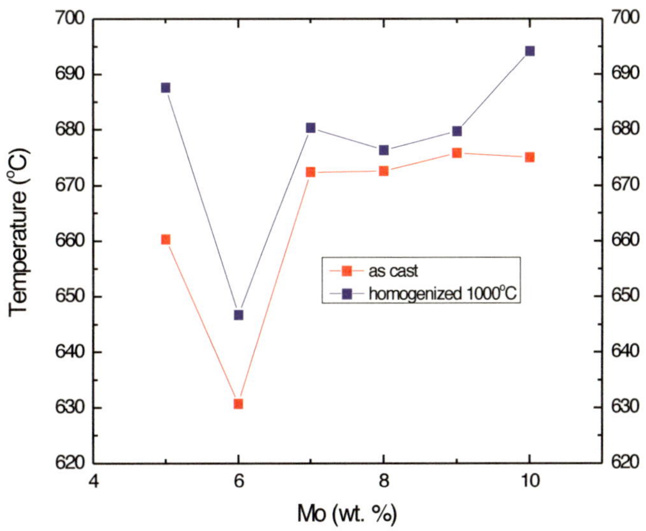

Figure 22. Temperatures of the first DTA peaks, γ-UMo alloys.

Instead of defining a peak or starting temperature as a stability criterion, it was found more accurate to state one which is related to the alloy's microstructure. This stability criterion is based on the determination of the curves of the differences in oxidation temperatures between the as cast and thermally treated samples, for each of the Mo contents, at the maximum and start temperatures. The respective graphics are shown below.

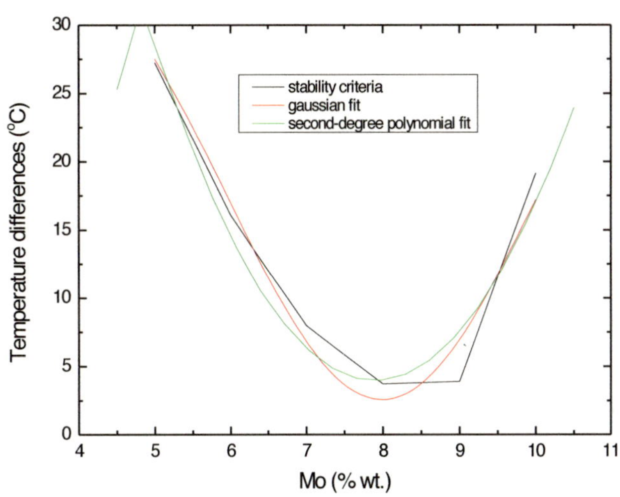

Figure 23. Oxidation temperature differences at maximum [7].

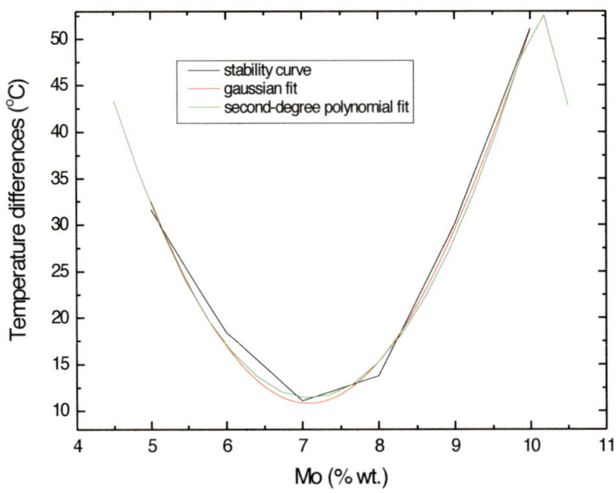

Figure 24. Oxidation temperature differences at start [7].

Comparing the microstructures of the as cast and thermally treated alloys, it is observed that high amounts of alpha phase are presented in the low Mo (5 to 6% wt) alloys, decreasing with the increase in the Mo content. On the other side, porosity increases with the Mo additions, as shown in the curves for alloy's density as a function of composition, see [7]. Since diffusion of oxygen is enhanced by the presence of porosities, and the reactive characteristic of the samples are characterized by the presence of the alpha phase, it is expected for the stability curve to have an extreme. Oxygen diffusion through the samples could be improved when the Mo content is at the extremes of composition. In the intermediate composition ranges there is neither enough alpha nor porosity to react with oxygen, leading to the behavior shown in Figures 23 and 24.

The point of minimum in both curves indicates whether the effect of the thermal treatments is minimal on these compositions. For the reactions with air, the point of minimum is at the value of 7.079 (54) %wt of molybdenum, for the curves of start of oxidation temperatures, while for the curve of maxima, the point of minimum was at 7.990 (147) %wt of molybdenum. For the ease of visualization, a second degree polynomial fit was applied to the curves. The minimum obtained is at 7.094 and 7.894 %wt of molybdenum, and the equations are:

$$\Delta T = 250.4587 - 67.403\%Mo + 4.7507\left(\%Mo\right)^2 \tag{1}$$

and:

$$\Delta T = 187.5971 - 46.5249\%Mo + 2.4968\left(\%Mo\right)^2 \tag{2}$$

Through the use of the differential operator [d / d(%*Mo*)] over the $\Delta T = f\ (\%Mo)$ curve, the minima mentioned above were obtained after $[d\Delta T/ d(\%Mo)] = 0$, since we are searching for % of Mo addition values for the optimal stabilization of the UMo alloys.

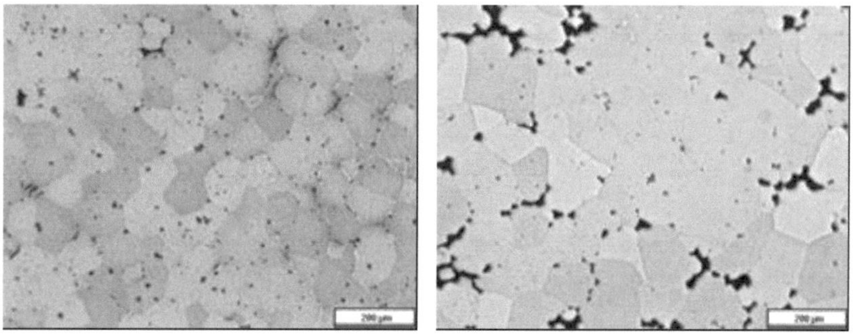

Figure 25. Micrographies of γUMo alloys, 5% wt. Mo (left) and 10%wt. Mo (right) [7].

In other words, the behavior of the temperature differences for the maximum and starting oxidation for a given %wt molybdenum concentration parameter, is explained by the fact that a thermal treatment applied to a stable alloy will not produce a remarkable effect in terms of changes in the oxidation process, if the alloy has a stable structure. So, the differences in the oxidation temperatures are at a minimum value for such compositions.

Also, the analysis performed for this set of experiments confirms the results of the experiments with aluminum, enabling the use of the temperature differences as a parameter for the evaluation of the stability of the alloys.

4. Conclusions

One of the main problems with γUMo for the use as fuels in research reactors is its reaction with the Al matrix. Additions of ternary elements have been considered to keep the fuel integrity in terms of its chemical and physical properties, during fabrication and irradiation. In the present work, the choice was for the analysis of the behavior of the binary γUMo alloys, in conditions near of the possible chemical environments found in the fabrication and under irradiation. Instead of the study of the effect of ternary additions in the stabilization of the alloys, it is also convenient to study the stability of the binary alloys, in order to set an optimal interval of Mo additions. It is suggested here the use of a differential thermal analysis to provide, based on the definition of some stability criteria, an alternative way to assess the alloys who promotes optimal properties.

The Gaussian or parabolic behavior can also be related to the alloy's stability. As explained above, hypoeutectoid γUMo alloys can be viewed, according to its compositions, as having remarkable different properties. When related to its behavior in the presence of the alpha phase, whose content decreases with the addition of Mo, or in terms of density, since porosity tends to increase with an increase in Mo percentage, stability results are very different. Thus it is natural to suppose that there will be an extreme in the stability parameter, adopted here as the difference in oxidation temperatures between the as cast and thermally treated samples for the same %wt. Mo. The minimum is shown in compositions of 7 to 8%wt. Mo, where the systems achieve optimal conditions in terms of stability, when thermal treatments produces minimal effects in the performance of the fuels under oxidation.

On the other hand, their behavior in the presence of aluminum shows an extreme near the same compositions of the oxidation experiments, also leading to the conclusion that between seven and eight percent of Mo we can obtain more stable alloys. The thermodynamics and possible mechanism of stabilization, and the parabolic and Gaussian behavior of the results, are objects of our current studies.

The results agreed with the choice of the composition of 7% wt. Mo for the fabrication of γUMo fuels and its use in research reactors. The data presented here is supported by previous results, regarding UMo fuel reactions with aluminium. Thus, it was shown that the DTA experiments carried out in this work, together with the definition of a suitable stability parameter, can be used as an alternative way to the analysis of the stability of the γ-UMo systems, for use as fuels in nuclear research reactors.

Author details

Fábio Branco Vaz de Oliveira and Delvonei Alves de Andrade

Nuclear and Engineering Center, Nuclear and Energy Research Institute, Brazilian Nuclear and Energy National Commission, IPEN-CNEN/SP, Cidade Universitária, São Paulo, SP, Brazil

References

[1] Meyer MK, et al. Irradiation Behavior of Uranium-Molybdenum Dispersion Fuel: Fuel Performance Data from RERTR-1 and RERTR-2, Proceedings of the XXII RERTR Meeting, Budapest, Hungary, 1999.

[2] Varela CLK, et al. Identification of Phases in the Interaction Layer Between U-Mo-Zr / Al and U-Mo-Zr / Al-Si. Proceedings of the XXIX RERTR Meeting, Prague, Czech Republic, 2007.

[3] Park JM, et al. Interdiffusion Behaviors of U-Mo-Zr / Al-Si. Proceedings of the XXVIII RERTR Meeting, Cape Town, South Africa, 2006.

[4] Park JM, et al. Interdiffusion Behaviors of U-Mo-Ti / Al-Si, Proceedings of the XXIX RERTR Meeting, Prague, Czech Republic, 2007.

[5] Perez E., Keiser Jr. DD, Sohn YH. Metallurgical and Materials Transactions A, vol. 42A, October 2011, pp. 3071-3083.

[6] Pasqualini EE. Advances and Perspectives in U-Mo Monolithic and Dispersed Fuels. Proceedings of the XXVIII International Meeting on Reduced Enrichment of Research Reactors, Cape Town, South Africa, 2006.

[7] Oliveira FBV. Development of a High Density Fuel Based on Uranium-Molybdenum Alloy with High Compatibility in High Temperatures. PhD thesis, University of São Paulo, 2008.

[8] Kim YS, et al. Interaction-Layer Growth Correlation for (U-Mo) / Al and Si added (U-Mo) / al Dispersion Fuels. Proceedings of the XXVIII RERTR Meeting, Cape Town, South Africa, 2006.

[9] Oliveira FBV, et al. Thermal and Chemical Stability of Some Hypoeutetoid γUMo Alloys. Proccedings of the 12th Research Reactor Fuel Management RRFM, Hamburg, Germany, 2008.

Chapter 6

Probabilistic Safety Assessment Applied to Research Reactors

Antonio César Ferreira Guimarães and
Maria de Lourdes Moreira

Additional information is available at the end of the chapter

http://dx.doi.org/10.5772/53012

1. Introduction

This chapter presents a Probabilistic Safety Assessment (PSA) and an uncertainty modeling review of a fuzzy approach applied to the Greek Research Reactor (GRR - 1) of the National Center for Scientific Research "Demokritos" [1]. The work was performed as part of the Probabilistic Safety Analysis (PSA) for the Research Reactor [2] in view of the development of new research reactors for radioisotopes production. As it occurs in any reliability study, statistically non-significant events report add a significant uncertainty level in the failure rates and basic events probabilities used on the Fault Tree Analysis (FTA) and in the probabilities of the EndState sequence in the Event Tree (ET) analysis. In order to model this uncertainty, a fuzzy approach was employed to reliability analysis of the GRR -1 Loss of Coolant Accident (LOCA) as Initiator Event (IE). As a case example, a guillotine rupture of the largest (10") pipe connected to the bottom of the reactor during full power operation is assumed as the initiator event. The final results have revealed that the proposed approach may be successfully applied to modeling of uncertainties in safety studies.

As part of the licensing process for nuclear power plants can be highlighted three important points and that are part of the official document of the IAEA [3], which can act as recommendations by the agency and should be adopted by anyone who is involved with nuclear reactors projects, they are:

Licensing principles should be established in the regulatory and legal framework. Examples of licensing principles can be presented as follows: The analysis approach to safety should be clearly defined, including the use of deterministic and *probabilistic methodologies* and analytical tools [3].

1. The following should be verified by the licensee to ensure that safety requirements are met: Design basis analyses and beyond design basis analyses, *fault tree analyses*, and *probabilistic safety assessments*, as appropriate [3].

2. There are several examples of documents and one important to be submitted to the Regulator is a preliminary safety analysis report before authorization to begin construction, which may include information on site evaluation, the design basis, nuclear and radiation safety, deterministic analyses and complementary probabilistic safety assessment [3].

It is notorious the use of probabilistic methods in the three recommendations mentioned in the agency's document, and thus our work seeks to make a small contribution to safety assessment studies for nuclear power plant and research reactors.

Probabilistic Safety Assessment (PSA) is a classical methodology that describes each accident's sequence through events trees (ETs), which combined success and failure of the performance or no safety system in an accident or a transient sequence. These initiating events result in sequences of the actions and system demands which may be modeled by the ETs. The accident consequences of the NPP status depend on the plant safety systems performance. The evaluation of the safety systems performance needs the component information, operational data, human error probability and physical phenomenon influence to each accident scenario. An overview of probabilistic safety assessment (PSA) methodologies used in the nuclear power licensing process and safety studies was introduced by Keller and Modarres [4].

Regardless of all innovations which were performed by this new generation of reactors, the possibility of accidents and faults of the security systems still remains. Therefore, committed studies should have accomplished in order to analyze the reliability of this plant concerning the DBAs, which take into account the possibilities of damage for the reactor core in the most different accident's scenarios. To develop the PSA study involving an important technique known as event tree approach, it is necessary the study of another technique known as fault tree analysis [5], that is the determination of a top event characterized by the failure of the present system in the ET. The Fault Tree analysis is being used in the last fifteen years to evaluate top event in probabilistic safety assessment (PSA level I & II) studies of Pressurized Water Reactor (PWR) nuclear power plants (NPP) and Research Reactors (RR).

The International Atomic Agency has published guides for PSA in research reactors [6] [7], in order to facilitate the application of PSA in that content. In this situation it is very common to employ some generic database, which are not applicable, because the results do not show the real situation of the system function for the future recommendations of project modifications or operational procedure, as the study concludes. In most cases, the Monte-Carlo methods have been used for uncertainty analysis and then, from the obtained results some other important aspects are addressed, like critical component and contribution of the uncertainty of each component for system general uncertainty [8].

Considering also the use of generic data, which was referred in last paragraph, the quantification and propagation of the uncertainty in this study is a very difficult assignment. In this context it is difficult to quantify reliability due to the large number of uncertainties associated

http://dx.doi.org/10.5772/53012

with the proper functioning of the front line system of research reactors. Therefore, an appropriate methodology should be proposed. The fuzzy methodology [9] and its engineering applications [10] [11] has been successively applied to uncertainty modeling in fault tree analysis. In this work the fuzzy set theory has been used in order to be accomplished. Considering the above mentioned, the main objective of the work presented here, was the development of an efficient fuzzy approach to be applied on reliability analysis of the Greek - I Research Reactor large break Loss of Coolant Accident (LOCA).

2. Description of fuzzy methodology

When the unreliability of each component has a point estimate, the top event unreliability will also be a point estimate. In this work, the component failure probabilities are considered as triangular fuzzy sets to incorporate the uncertainties of each relevant parameter. The membership function, $\mu_X(x)$, of a triangular fuzzy set is defined as:

$$\mu_X(x) = \max\left[0, 1 - \left|(x - x_1)/(x_2 - x_1)\right|\right], x_1 \leq x \leq x_2, \tag{1}$$

$$\mu_X(x) = 1, x = x_2, \tag{2}$$

$$\mu_X(x) = \max 0, 1 - \left|(x_3 - x)/(x_3 - x_2)\right|], x_2 \leq x \leq x_3, \tag{3}$$

$\mu_X(x) = 0$ otherwise,

with

$$\mu_X(x_2) = 1 \tag{4}$$

and

$$\mu_X(x_1) = \mu_X(x_3) = 0, \tag{5}$$

and $[x_1, x_3]$ are lower bound and upper bound of triangular fuzzy sets. These values may be obtained from the point median value and the error factor (EF) of the failure probability [12]. The lower bound, middle value, and the upper bound are defined as:

$$x_1 = q_p / EF \tag{6}$$

$$x_2 = q_p \tag{7}$$

$$x_3 = q_p.EF \tag{8}$$

where, EF = 5 if $0.01 < q_p$, EF= 3 if $0.001 < q_p < 0.01$, EF=10 if $q_p < 0.001$ and q_p is the point median value of the failure probability. The fuzzy evaluation of the failure probability of the top event in a fault tree is carried out using α-cut method. The top event can be represented by an N x 2 array, where N is the number of alfacuts.

2.1. Importance measures

The identification of critical component is essential for the safety analysis of any relevant system. Many measures are available in probabilistic approach like risk achievement worth, Birnbaum importance, Fussel Vesely importance and so on. Two different importance measures are introduced and they are (1) FIM - Fuzzy Importance Measure and (2) FUIM - Fuzzy Uncertainty Importance Measure.

2.1.1. Fuzzy Importance Measure (FIM)

The evaluation of the contribution of different basic events is essential to identify the critical components in the system. The top event failure probability by making the component 'i' fully unavailable (q=1) is:

$$Q_{q_i} = 1 = f\left(q_1, q_2, \ldots, q_{i-1}, 1, q_{i+1}, \ldots, q_n\right) \tag{9}$$

and for component 'i' fully available is:

$$Q_{q_i} = 0 = f\left(q_1, q_2, \ldots, q_{i-1}, 0, q_{i+1}, \ldots, q_n\right) \tag{10}$$

The fuzzy importance measure (FIM) is defined as:

$$FIM = ED[Q_{qi} = 1, Q_{qi} = 0] \tag{11}$$

where, ED [Q_{q_i}=1, Q_{q_i}=0,], is the Euclidean distance between two fuzzy sets A and B. ED is defined as:

$$ED\left[A, B\right] = \Sigma\left(\left(a_L - b_L\right)^2 + \left(a_U - b_U\right)^2\right)^{0.5} \alpha_i,\ i = 1,2,\ldots,N \tag{12}$$

http://dx.doi.org/10.5772/53012

where a_L, b_L and a_U, b_U are the lower and upper values of fuzzy set A and B respectively at each α-level.

2.1.2. Fuzzy Uncertainty Importance Measure (FUIM)

For the importance measure known as fuzzy uncertainty importance measure is proposed to identify the components which contribute maximum uncertainty to the uncertainty of the top event, and is defined as:

$$FUIM = ED\left[Q, Q_i \right] \tag{13}$$

where Q = top event failure probability, Qi = top event probability when error factor for component 'i' is unity ($EF_i = 1$), i.e. the parameter of the basic event has a point value or crisp value.

3. GRR — 1 Research reactor system description

3.1. Plant familiarization and information gathering

Reactor GRR-1 is a typical 5 MW pool-type reactor with MTR-type fuel elements [1] [2], cooled and moderated with demineralized light water. In line with the international Reduced Enrichment for Research and Test Reactor (RERTR) programme, the core has been recently fuelled with Low Enriched Uranium (LEU) elements of U3Si2-Al type. The fuel enrichment is 19.75% and the fissile loading is 12.34 g of 235U per plate. The equilibrium LEU core contains 28 standard fuel elements and 5 control fuel elements, arranged on a 6x9 element grid plate.

Each standard fuel element consists of 18 flat plates. The control fuel element is of the same size as the standard element but consists of only 10 plates, thus providing an inner gap for the insertion of the control blades. The control material is composed of Ag (80%), Cd (5%) and In (15%). The core is reflected by Beryllium on two opposite faces and is surrounded by a practically infinite thickness of pool water. One graphite thermal column is adjacent to one side of the core. In the middle of the core there is a flux trap. The core is suspended in a 9-m deep water pool of a volume of approximately 300 m^3. The fuel elements are cooled by circulating the water of the pool at a rate of 450 m^3/h. The water flows downward through the core, passes through a decay tank and then pumped back to the pool through the heat exchangers. A weighted flapper valve attached to the bottom of the core exit plenum enables natural circulation through the core in the absence of forced flow circulation. Core inlet temperature, i.e. pool water, is not permitted to exceed 45^0C. Pool temperature depends on reactor power, as well as on external temperature, because the latter affects heat dissipation in the cooling towers. In practice, core inlet temperature has been observed to vary in the range between 20^0C and 44^0C. Also quite homogeneous temperature conditions prevail in the pool, considering that similar measurements are routinely recorded from thermocouples located at distant positions in the pool (see Figure 1).

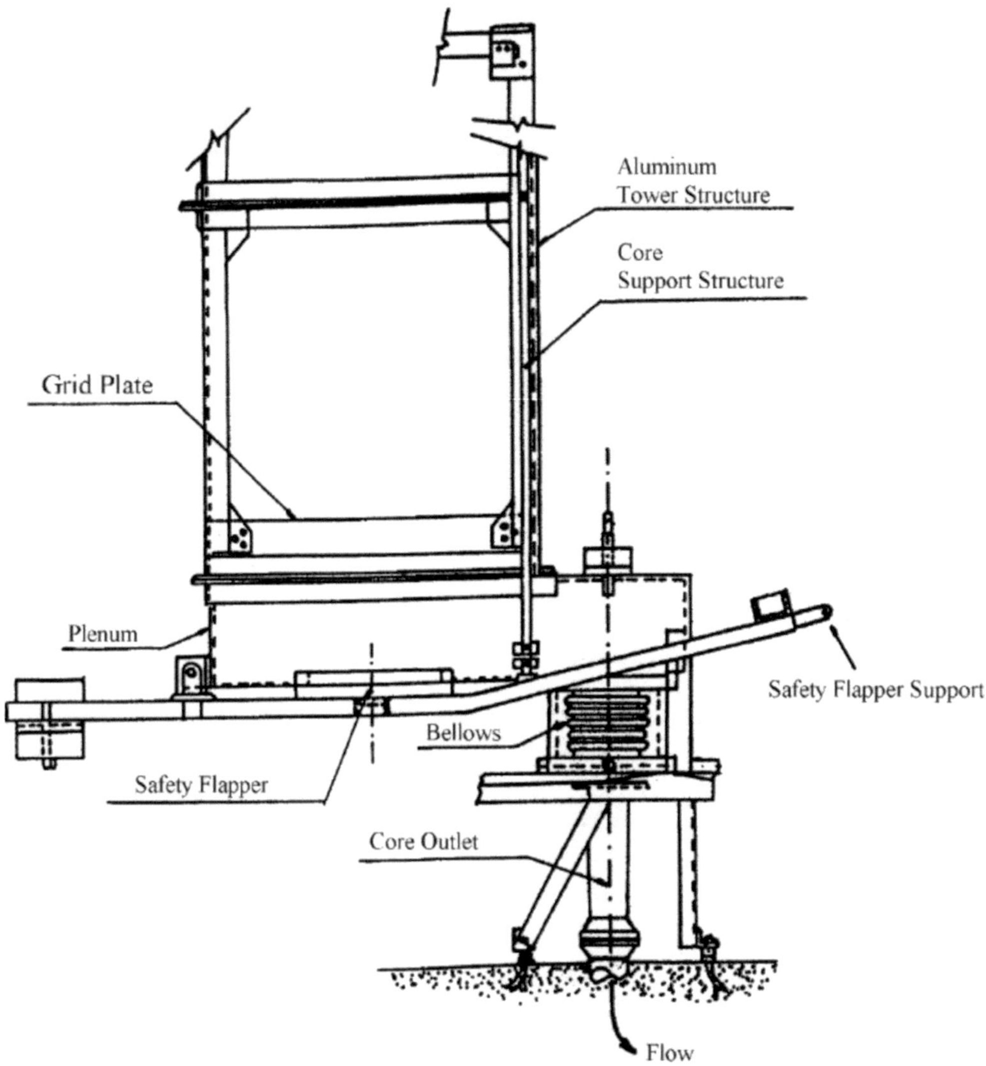

Figure 1. Coolant plenum and safety flapper (source Aneziris et al., 2004)

Determination and selection of plant operating states.

The following plant operation states have been considered.

1. Nominal full power operation (5MW)
2. Reduced power operation
3. Start-up operation
4. Reactor subcritical, reactor pool available.

Nominal full power operation is a plant operating state bracketing all others from the safety point of view. This is due to the fact that the reactor pool constitutes a large heat sink that is always available, regardless of the operating state of the reactor

Initiating event selection

An initiating event is an event that creates a disturbance in the plant and has the potential to lead to core damage, depending on the successful operation of the various mitigating systems in the plant. *Loss of Coolant Accidents* (LOCA) are all events that directly cause loss of integrity of the primary coolant pressure boundary. *Transient* initiators are those that could create the need for a reactor power reduction or shutdown and subsequent removal of decay heat.

3.2. Safety functions

Five basic safety functions incorporate the design of the Greek Research Reactor.

Functions that aiming at preventing core damage to occur following an initiating event:

1. Control reactivity
2. Remove core decay heat and stored heat
3. Maintain primary reactor coolant inventory
4. Protect containment integrity (isolation, overpressure)
5. Scrub radioactive materials from containment atmosphere

For each safety function will be presented the corresponding front-line system of the research reactor:

1. *Control reactivity* -**Reactor Protection System** (RPS): Automatic and Manual
2. *Remove core decay heat and stored heat:* **Primary Heat Removal System, Reactor Pool (Natural Convection) and Emergency Core Cooling System.**
3. *Maintain primary reactor coolant inventory* - **Reactor Pool Isolation**
4. *Protect containment integrity (isolation, overpressure):***Containment Isolation** and Emergency **Ventilation System**.
5. *Scrub radioactive materials from containment atmosphere:* Emergency Ventilation System

These front-line systems are described in detail the Safety Analysis of the Research Reactor.

Reactor Protection System (RPS) - The safety system consists of two independent safety channels, the magnet power supply, and the safety circuit with scrams, reverses interlocks and alarms.

Primary Heat Removal System - This system performs the basic safety function of heat removal from the reactor core both under power operation, as well as, following shutdown.

Reactor pool – Natural Convection - The reactor pool presents a major heat sink capable of independently absorbing the heat generated in the core in most of the cases. Natural convection is made possible through the opening of a weighted flapper valve sealing the core exit plenum.

Emergency Core Cooling System - In the event of a LOCA accident resulting in loss of the primary water and core uncovery the Emergency Core Cooling System (ECCS) can spray the reactor core through a 5cm diameter pipe with water coming from a 250 m3 storage tank located 30 m higher than the surface of the reactor pool. The water tank can be continuously filled by the city water.

Containment Isolation - In the event of an emergency, the normal ventilation system of the containment stops and the containment is isolated through the automatic closure of all existing openings. At the same time the Emergency Ventilation system starts operating.

Emergency Ventilation System - Following a manual scram, the pumps of the ventilation system stop and the emergency ventilation starts automatically, removing the possibly contaminated air in a rate of 1500 m3/h.

Electric Power Supply System - This is the only support system for the front line systems described above. The system consists of the main power which is received from the utility plus the following sources:

Non Break Unit: Stand-by Unit, Central Stand-by Unit and Diesel Motor

3.3. LOCA between others initiator events

Five initiating events have been identified in the Greek research reactor, which are the following: Loss of coolant (LOCA), Loss of Flow (LOFA), Excess reactivity, Loss of offsite power (LOOP), Flow Blockage. The associated thermal-hydraulic analysis is given in detail in [13]. Loss of flow might occur in the three ways: either owing to failure of both pumps, or to the failure of the safety flapper or to the butterfly value failure. Finally the following seven initiators are presented:

a. Loss of Coolant (LOCA)

b. Loss of Flow owing to pump failure

c. Loss of Flow owing to flapper failure

d. Loss of Flow owing to butterfly value failure

e. Excess reactivity

f. Loss of offsite power (LOOP)

g. Flow Blockage

4. Large LOCA as application methodology

As an application of our methodology was chosen fuzzy LOCA initiating event. Event Tree (ET) models the possible response of the reactor to loss of coolant. ET (see Figure 2) comprises the following events:

1. *LOCA (Initiate Event - IE)* - It is assumed that during full power operation there is a guillotine rupture of the largest (10″) pipe connected to the bottom of the reactor.

2. *Availability of reactor protection system* - Following LOCA the reactor protection system, both automatic and manual systems should shut down the reactor. Success of this event results in scram and hence in interruption of the fission chain reaction.

3. *Pool isolation* - Following LOCA the pool should be isolated from the cooling system. This occurs if the butterfly valves close, either manually or automatically, within 16 min following the accident. Successful isolation of the pool from the location of the break results in the core being immerged in the pool.

4.1. Event and fault tree of the LOCA

Since the probability of failure P is generally less than 0.1, the probability of success (1-P1) is always close to 1. Thus, the probability associated with the upper (success) branches in the tree is assumed to be 1 [14], see Fig. 2.

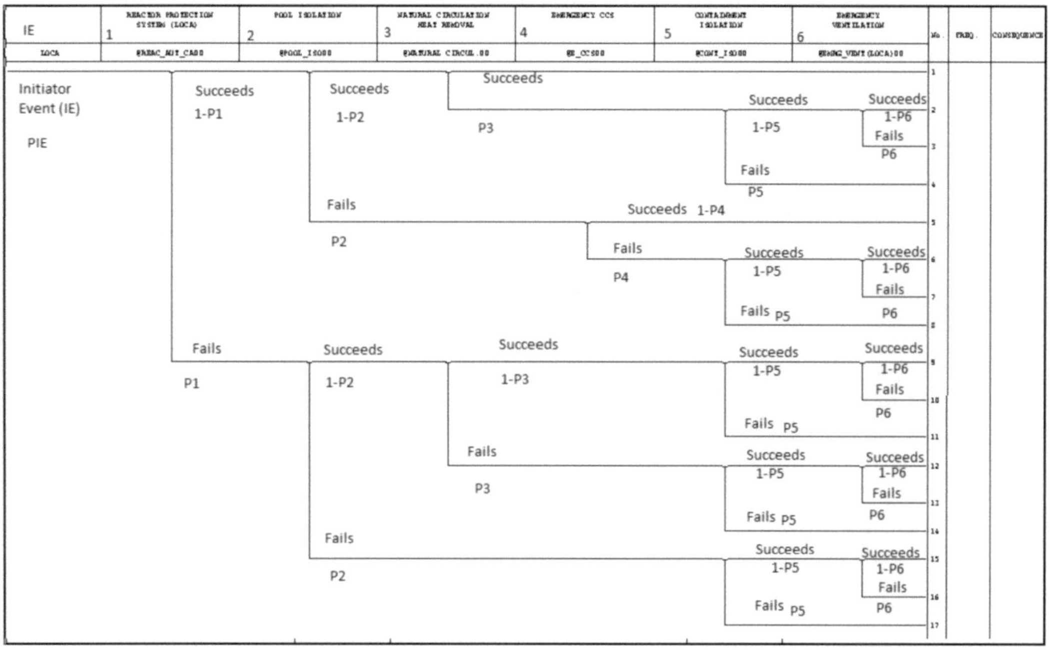

Figure 2. Event Tree (ET) for the initiating event LOCA

The probability of occurrence of events in a sequence is the product of conditional probabilities of the individual events in that chain. In this study, was considered that, the successive events in a sequence are *independent*, then the probability of a sequence is the product of unconditional probabilities of the individual events (so each front-line system has P failures as identical) [15].

The final results in terms of probabilities for all sequences, No. 1 - 17 (Eqs. 14 - 30), in the event tree, can be determined multiplying each value of probabilities in the branch, following the procedure described as:

$$\text{No. 1 PIE} \tag{14}$$

$$\text{No. 2 - PIE x } \left(1\text{-P1}\right) \text{ x}\left(1\text{-P2}\right) \text{ x P3 x } \left(1\text{-P5}\right) \text{ x } \left(1\text{-P6}\right) \tag{15}$$

$$\text{No. 3 - PIE x } \left(1\text{-P1}\right) \text{ x}\left(1\text{-P2}\right) \text{ x P3 x } \left(1\text{-P5}\right) \text{ x P6} \tag{16}$$

$$\text{No. 4 - PIE x } \left(1\text{-P1}\right) \text{ x}\left(1\text{-P2}\right) \text{ x P3 x P5} \tag{17}$$

$$\text{No. 5 - PIE x } \left(1\text{-P1}\right) \text{ x P2 x } \left(1\text{-P4}\right) \tag{18}$$

$$\text{No. 6 - PIE x } \left(1\text{-P1}\right) \text{ x P2 x P4 x } \left(1\text{-P5}\right) \text{ x } \left(1\text{-P6}\right) \tag{19}$$

$$\text{No. 7 - PIE x } \left(1\text{-P1}\right) \text{ x P2 x P4 x } \left(1\text{-P5}\right) \text{ x P6} \tag{20}$$

$$\text{No. 8 - PIE x } \left(1\text{-P1}\right) \text{ x P2 x P4 x P5} \tag{21}$$

$$\text{No. 9 - PIE x P1 x } \left(1\text{-P2}\right) \text{ x } \left(1\text{-P3}\right) \text{ x } \left(1\text{-P5}\right) \text{ x } \left(1\text{-P6}\right) \tag{22}$$

$$\text{No. 10 - PIE x P1 x } \left(1\text{-P2}\right) \text{ x } \left(1\text{-P3}\right) \text{ x } \left(1\text{-P5}\right) \text{ x P6} \tag{23}$$

$$\text{No. 11 - PIE x P1 x } \left(1\text{-P2}\right) \text{ x } \left(1\text{-P3}\right) \text{ x P5} \tag{24}$$

$$\text{No. 12 - PIE x P1 x } \left(1\text{-P2}\right) \text{ x P3 x } \left(1\text{-P5}\right) \text{ x } \left(1\text{-P6}\right) \tag{25}$$

$$\text{No. 13 - PIE x P1 x } \left(1\text{-P2}\right) \text{ x P3 x } \left(1\text{-P5}\right) \text{ x P6} \tag{26}$$

$$\text{No. 14 - PIE x P1 x }\left(\text{1-P2}\right)\text{ x P3 x P5} \tag{27}$$

$$\text{No. 15 - PIE x P1 x P2 x }\left(\text{1-P5}\right)\text{ x }\left(\text{1-P6}\right) \tag{28}$$

$$\text{No. 16 - PIE x P1 x P2 x }\left(\text{1-P5}\right)\text{ x P6} \tag{29}$$

$$\text{No. 17 - PIE x P1 x P2 x P5} \tag{30}$$

All probabilities of failure of each system (P1 - P6) are calculated using the Fault Tree methodology (see figs. 3 - 6). The final upper value of the FT is named of top event, and expressed by the probability calculated using the Minimal Cut Set (MCS). Fault Tree analysis is a technique by which many events that interact to produce other events can be related using simple logical relationships (AND, OR, etc.,); these relationships allow a methodical structure building that represents the system. Symbols called GATES (AND, OR,..), are used to graphically arranging the events into a tree structure, during the synthesis of the tree (represented in the Figs 3 - 6 by simbols 1, 2, 3,..., etc.).

A new approach was proposed in this study using fuzzy logic to try uncertainties, using the procedure described for FT and ET to calculate Top Event Probability and EndState Probability. Basic events in the FT are treated as fuzzy numbers, with lower bound and upper bound. The top event determined in each FT now represents a Fuzzy Top event, with lower and upper bound, to treat the uncertainty in the FT analysis. Following these procedures, the probabilities calculated for ET have fuzzy numbers to make the product, and then the addition, subtraction, and product of the two or more fuzzy numbers are done following the recommendations described in [16]. In this approach triangular membership functions are used as fuzzy numbers (triplet a_1 a_2 a_3).

The addition of triangular fuzzy number A =(a1, a2, a3) and B =(b1, b2, b3) is defined as:

$$\text{A+B= (a1+b1 , a2+b2 , a3+b3)} \tag{31}$$

Thus the addition of two triangular fuzzy numbers is again a triangular fuzzy number.

Similarly subtraction of two triangular fuzzy numbers is also a triangular fuzzy number and it can be given by the following expressions:

$$\text{A-B= (a1-b1 , a2-b2 , a3-b).} \tag{32}$$

The multiplication of two fuzzy numbers $A = (a1, a2, a3)$ and $B = (b1, b2, b3)$ denoted as $A*B$ can be defined as:

$$P=a1b1, Q=a2b2, R= a3b3. \quad (33)$$

It is evident that the resulting fuzzy number *A* **B* is not a triangular fuzzy number. But in most of the cases, computation with resulting fuzzy numbers becomes very tedious. Thus it is necessary to avoid the second and higher degree terms to make them computationally easy and therefore the product of two fuzzy numbers is reduced to a triangular fuzzy number *(P, Q, R)* or *(a1b1,a2b2, a3b3).*

Applying this concept in the expressions defined from ET, the final result of each sequence is also a triangular fuzzy number, reflecting the uncertainty.

Frequencies of initiating event (IE) appearing in Event Trees ET are estimated according to values in IAEA [7].

4.2. System fault tree with top event "Reactor protection system failure", in case of LOCA

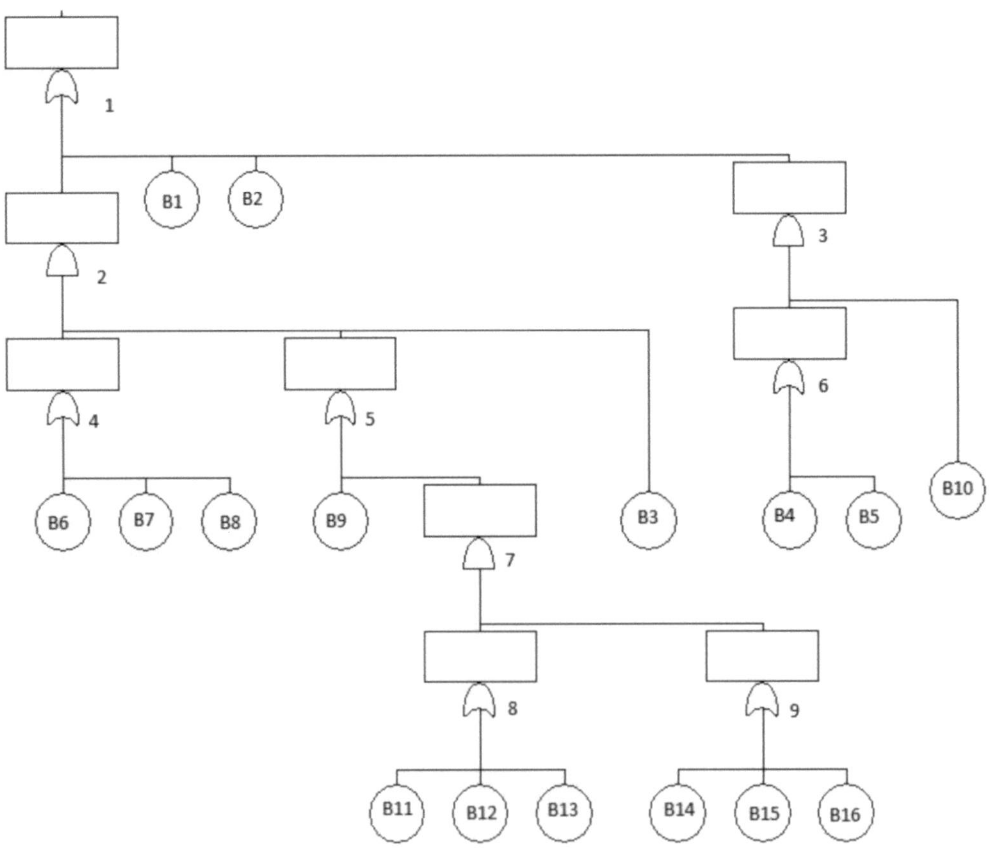

Figure 3. System Fault Tree with top event "Reactor protection system failure", in case of LOCA

4.3. System fault tree with top event "No pool isolation"

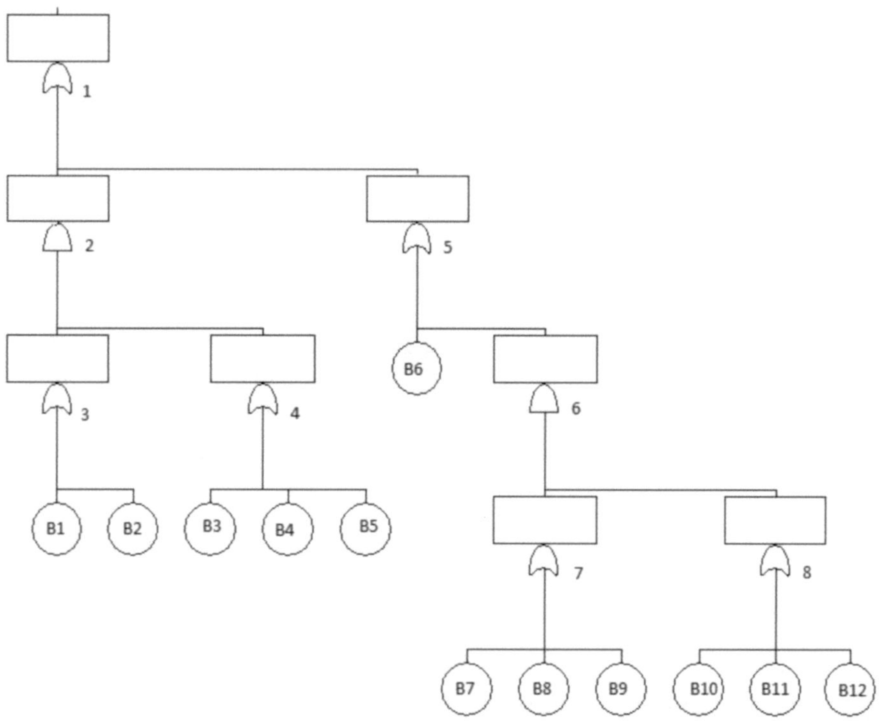

Figure 4. System Fault Tree with top event "No pool isolation"

4.4. System fault tree with top event "Natural circulation heat removal failure"

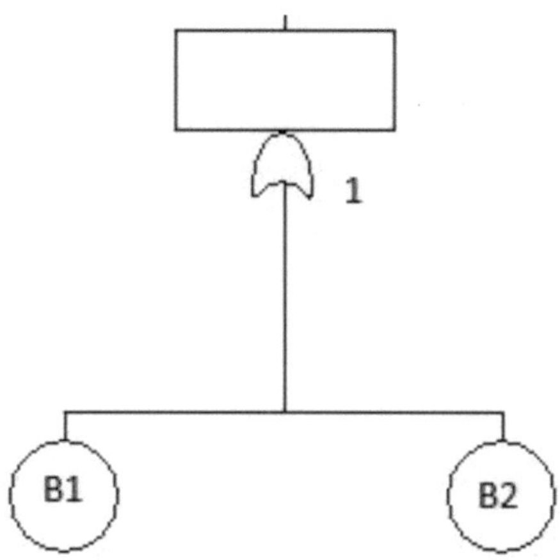

Figure 5. System Fault Tree with top event "Natural circulation heat removal failure"

4.5. System fault tree with top event "Emergency core cooling system failure"

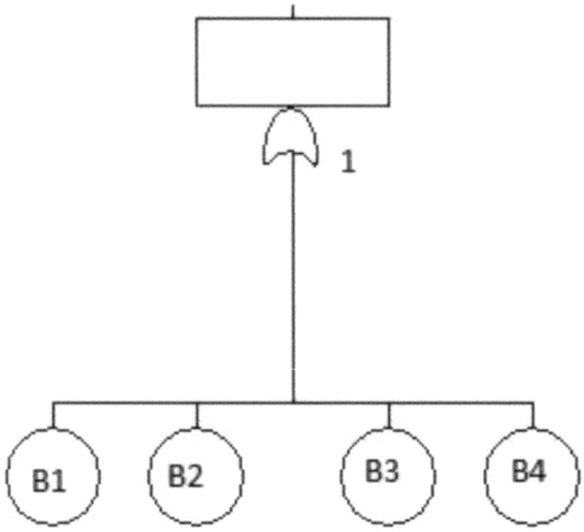

Figure 6. System Fault Tree with top event "Emergency core cooling system failure"

4.6. System fault tree with top event "Containment system failure"

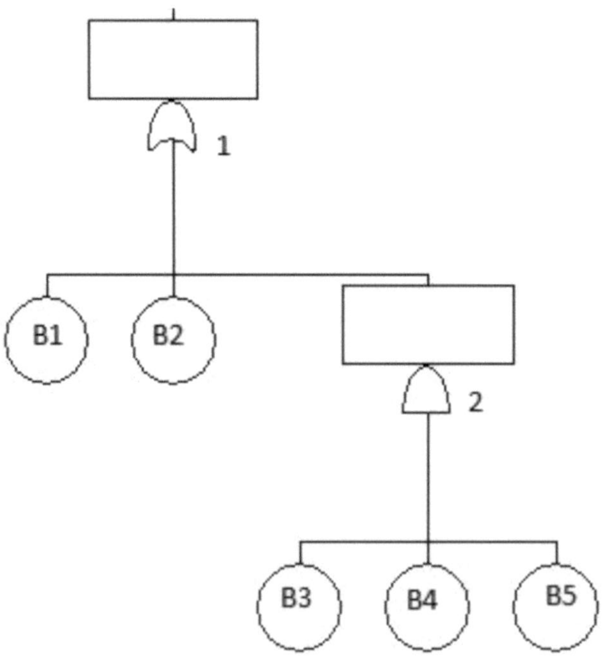

Figure 7. System Fault Tree with top event "Containment system failure"

4.7. System fault tree with top event "No Emergency ventilation", in case of LOCA

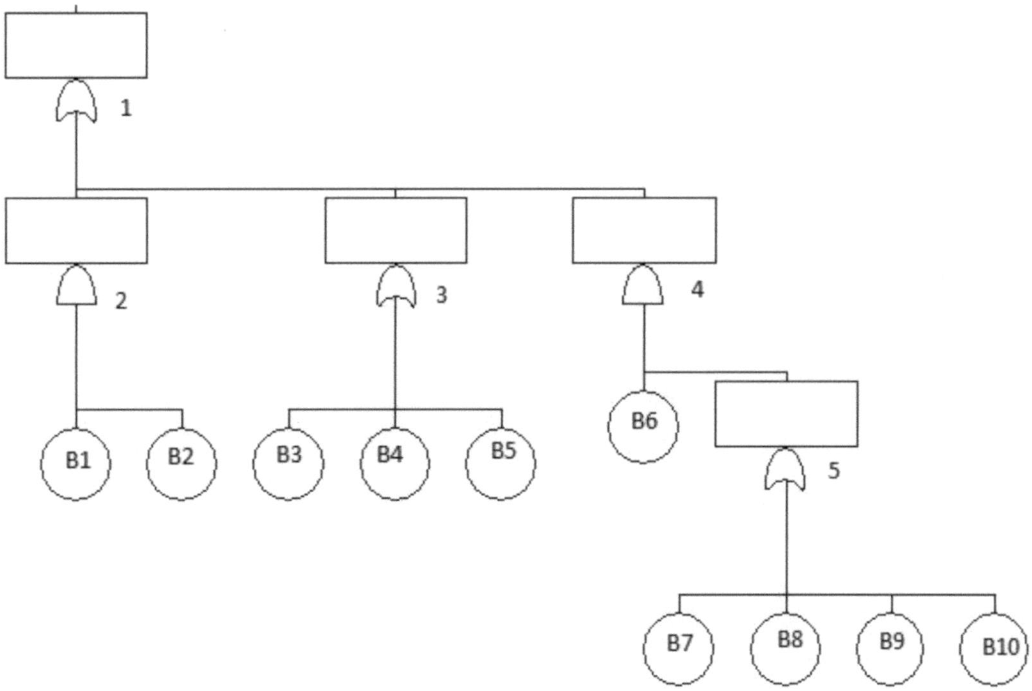

Figure 8. System Fault Tree with top event "No Emergency ventilation", in case of LOCA

5. Results

The studies presented in this chapter, considered a large LOCA GRR - 1 research reactor as a case example. A preliminary study pointed out the classical probabilistic safety analysis. We have used the systems fault tree approach to determine the top event probability in each system, i.e., Reactor Protection System (RPS), Pool Isolation (PI), Natural Circulation Heat Removal (NCHR), Emergency CCS (ECCS), Containment Isolation (CI), and Emergency Ventilation (EV). Applying the values of the probabilities assigned to each basic event in each front line system (see Table 1 - 6), the FuzzyFTA computer coding was used and it has been calculated the fuzzy top event probability to each system, considered in the event trees presented (Figs. 3 - 8). In the Table 7, we can see the upper and lower bound (where alfa-cuts = 0) and middle value (alfa-cuts = 1) of the fuzzy top event calculated to each front line system.

The results obtained by use of FuzzyFTA are presented in picture format. Figure 9 presents the result of the fuzzy top event to the RPS. The results for PI, NCHR, ECCS, CI and EV are not presented here due to graphical similarities. In Figure 10 and 11, is presented the ranking of components with respect to its fuzzy importance measure (FIM) and the fuzzy uncertainty importance measure (FUIM) of the component in relation to general uncertainty of the system for RPS. The results FIM and FUIM for PI, NCHR, ECCS, CI and EV are not presented here for graphical similarities.

The results obtained for each front line system, to probability of fuzzy top event, are used to calculate the probability of the EndState frequency by using the expressions given previously for the calculation using ET (section 4.1, Eqs 14 - 30). The probability of Event Initiator (PIE) for LOCA is 1.2 E-4 / year [1] [2]

In reference [17] was developed a guideline for estimating the lower and upper bound of the estimated failure rate. So the value of IE probability for purposes of calculating is (1.2E-05 1.2E-04 1.2E-03). For all operations of subtraction (1-P) in ET, where P is now a fuzzy number, for probabilities of failure of each system, the value "1" becomes (1 1 1).

Substituting the values of probability and performing fuzzy operations of subtraction, and multiplication can find the final values shown in Table 8 for fuzzy probabilities of each EndState sequences.

Component failure rates and the corresponding unavailabilities for front-line and support systems are given in [2]. The source of the failure rates is the IAEA database [18].

Basic Event	Component Failure Identification	Probability
1	B1 - Failure of all 5 rods	2.28E-06
2	B2 - Failure of electromagnets to disengage	1.92E-06
3	B3 - NOR no eletric power	0.99999
4	B4 - Gate slow scram fails	6.97E-4
5	B5 - Relay T3 stuck closed	6.97E-4
6	B6 - Sensor fails to give signal	2.35E-2
7	B7 - Relay T1 fails to open	6.97E-4
8	B8 - No eletric power	2.779E-06
9	B9 - Human error	0.01
10	B10 - NOR no eletric power	0.99999
11	B11 - No eletric power	2.779E-6
12	B12 - Sensor T2 fails	2.35E-2
13	B13 - Relay T2 fails to open	6.97E-4
14	B14 - Sensor fails to give signal	2.35E-2
15	B15 - Relay T1 fails to open	6.97E-4
16	B16 - No eletric power	2.779E-6

Table 1. Description of Basic Events and Probabilities for RPS

http://dx.doi.org/10.5772/53012

Basic Event	Component Failure Identification	Probability
1	B1 - operador fails to close manual butterfly valves	1.00E-02
2	B2 - manual butterfly valves fail in open position	3.60E-06
3	B3 - operator fails to give signal to pneumatic valves	1.00E-02
4	B4 - no eletric power	2.779E-06
5	B5 - pneumatic valves fail stuck in open position	8.20E-05
6	B6 - Human error	1.00E-02
7	B7 - No Eletric Power	2.779E-06
8	B8 - Sensor T2 fails	2.35E-02
9	B9 - Relay T2 fails to open	6.97E-04
10	B10 - Sensor fails to give signal	2.35E-02
11	B11 - Relay T1 fails to open	6.97E-04
12	B12 - No eletric Power	2.779E-06

Table 2. System Fault Tree with top event "No pool isolation"

Basic Event	Component Failure Identification	Probability
1	B1 - flapper fails to open (stuck)	1.44E-05
2	B2 - wrong weight	1.00E-02

Table 3. System Fault Tree with top event "Natural circulation heat removal failure"

Basic Event	Component Failure Identification	Probability
1	B1 - hole in hose	1.20E-04
2	B2 - operator fails to connect hose	1.00E-02
3	B3 - water valve fails stuck closed	3.60E-06
4	B4 - no water in tank	1.20E-04

Table 4. System Fault Tree with top event "Emergency core cooling system failure"

Basic Event	Component Failure Identification	Probability
1	B1 - gates of ventil, system fail to remain closed	1.01E-04
2	B2 - doors fail to remain closed	1.44E-05
3	B3 - air pump #2 fails to stop	3.24E-04
4	B4 - air pump #1 fails to stop	3.24E-4
5	B5 - air pump #3 fails to stop	3.2E-04

Table 5. System Fault Tree with top event "Containment system failure"

Basic Event	Component Failure Identification	Probability
1	B1 - No sinal	5.82E-04
2	B2 - Human error	1.00E-02
3	B3 - Filters fail	3.44E-04
4	B4 - Air pump F failed	2.27E-03
5	B5 - Two valves (h) fails to open	1.68E-03
6	B6 - Loss of offsite power	1.00E-04
7	B7 - AC Generator fails	2.79E-03
8	B8 - Diesel motor fails	8.20E-03
9	B9 - Switches fail stuck	3.20E-03
10	B10 - Human error	1.00E-02

Table 6. System Fault Tree with top event "No Emergency ventilation"

Front Line Systems / Fuzzy top event	LOWER BOUND Alfa-cuts = 0	MIDDLE VALUE Alfa-cuts = 1	UPPER BOUND Alfa-cuts = 0
(1)REACTOR PROTECTION SYSTEM (LOCA)	0.00138	0.0353	0.724
(2) POOL ISOLATION	0.00203	0.0107	0.0671
(3) NATURAL CIRCULATION HEAT REMOVAL	2.00E-3	1.00E-2	5.00E-2
(4) EMERGENCY CCS	0.00201	0.0101	0.0511
(5) CONTAINMENT ISOLATION	1.01E-5	1.15E-4	1.01E-3
(6) EMERGENCY VENTILATION	0.00135	0.0043	0.0155

Table 7. Upper and Lower Bound of the fuzzy top event

http://dx.doi.org/10.5772/53012

Fuzzy Top Event for Full and MCS FTA for Alfa-Levels

Alfa-Level	LowerBoundFull	UpperBoundFull	LowerBoundMCS	UpperBoundMCS
0	3.02E-5	0.11	0.00138	0.724
0.1	8.03E-5	0.0891	0.00274	0.64
0.2	0.000153	0.0709	0.00455	0.554
0.3	0.00025	0.0553	0.00682	0.469
0.4	0.00037	0.042	0.00954	0.386
0.5	0.000516	0.0309	0.0127	0.308
0.6	0.000688	0.0217	0.0163	0.236
0.7	0.000887	0.0143	0.0204	0.171
0.8	0.00111	0.00862	0.0249	0.116
0.9	0.00137	0.00444	0.0299	0.0701
1	0.00165	0.00165	0.0353	0.0353

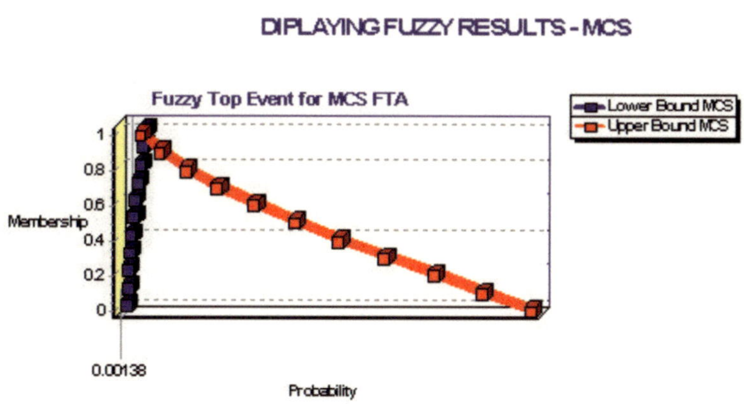

Figure 9. In the case of RPS, show the lower bound and upper bound ranking to fuzzy top event considering the Fault Tree full and with Minimal Cut Set (MCS). Only the MCS results are represented in graphic form.

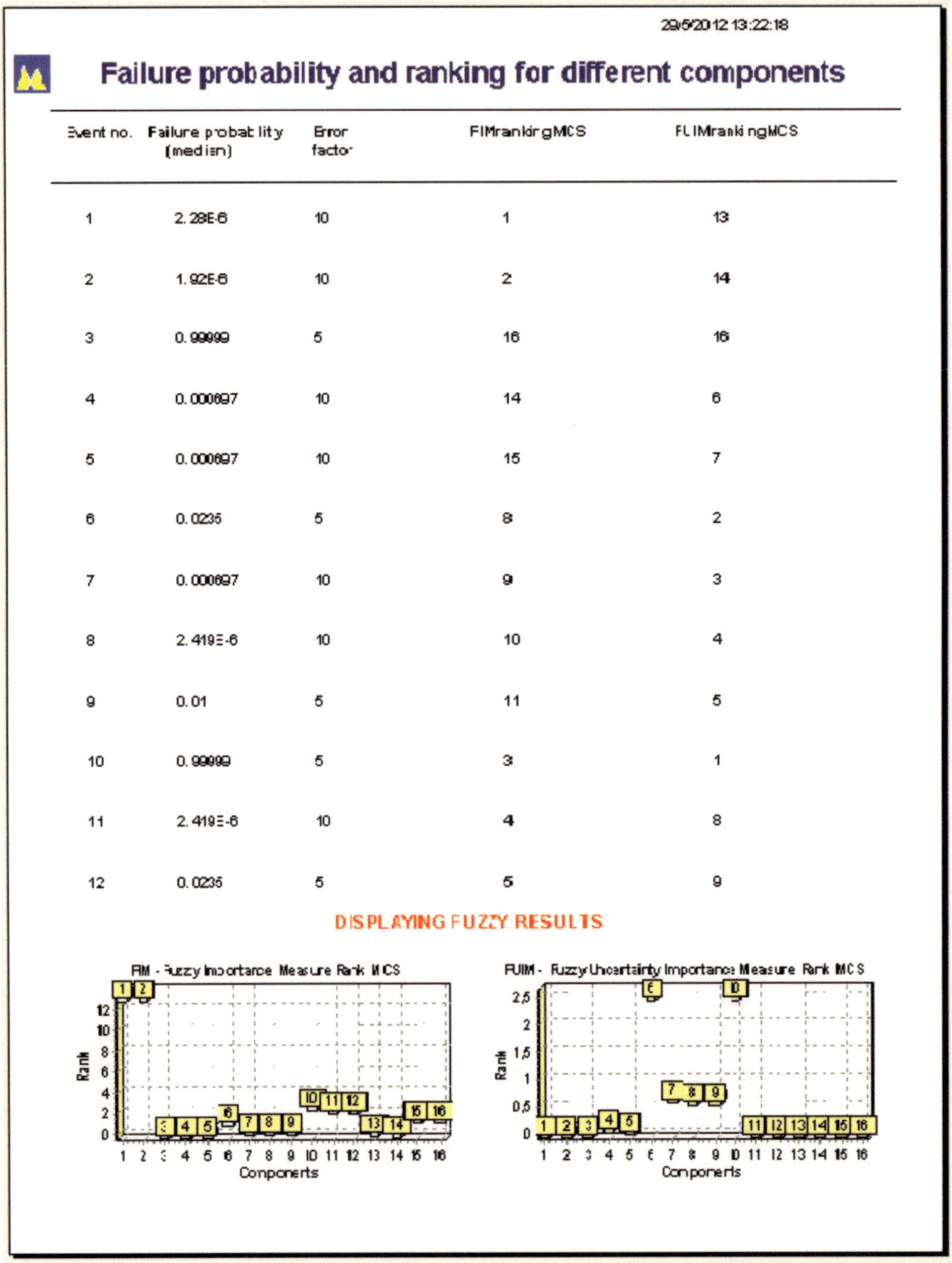

29/5/2012 13:22:18

Failure probability and ranking for different components

Event no.	Failure probability (median)	Error factor	FIMrankingMCS	FUIMrankingMCS
1	2.28E-6	10	1	13
2	1.92E-6	10	2	14
3	0.99999	5	16	16
4	0.000697	10	14	6
5	0.000697	10	15	7
6	0.0235	5	8	2
7	0.000697	10	9	3
8	2.419E-6	10	10	4
9	0.01	5	11	5
10	0.99999	5	3	1
11	2.419E-6	10	4	8
12	0.0235	5	5	9

Figure 10. RPS results of FIM and FUIM ranking (importance and uncertainty of the each basic event to general system) and graphic representation of these results, using FT with MCS.

http://dx.doi.org/10.5772/53012

29/5/2012 13:25:28

Event no.	Failure probability (median)	Error factor	FIMrankingMCS	FUIMrankingMCS
13	0.000697	10	12	11
14	0.0235	5	13	12
15	0.000697	10	6	10
16	2.419E-6	10	7	15

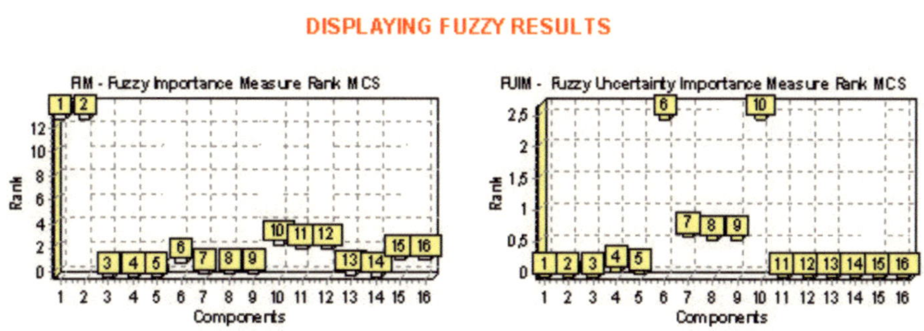

Figure 11. RPS continuation

No.	End State Fuzzy Frequency
1	(1 1 1)
2	(2.38E-08 1.14E-06 1.52E-05)
3	(3.22E-11 4.92E-09 2.40E-07)
4	(2.41E-11 1.20E-10 6.00E-08)
5	(2.43E-08 1.28E-06 8.05E-05)
6	(4.89E-04 1.29E-06 4.11E-6)
7	(6.61E-14 5.57E-11 6.37E-08)
8	(4.96E-16 4.94E-15 4.15E-09)
9	(1.65E-08 4.23E-06 8.68E-04)
10	(2.23E-11 1.82E-08 1.34E-05)
11	(1.39E-13 4.87E-10 8.77E-07)
12	(3.31E-11 4.23E-08 3.62E-05)
13	(4.47E-14 1.82E-10 7.75E-07)
14	(3.31E-16 4.87E-12 4.34E-08)
15	(4.96E-02 4.53E-08 6.99E-05)
16	(4.53E-14 1.94E-10 9.03 E-07)
17	(2.80E-16 4.05E-10 6.71E-08)

Table 8. Frequency of Release - EndState of the Event Tree

6. Conclusion

Recent studies about severe accidents in conventional research reactors pointed to the very low core melt frequency from a initiator event Loss of Coolant (LOCA) of **1.4E-6** by year, for Frequency of core damage and **1.16E-2** for Conditional probability of core damage (/year), given initiator.

The lack of event data record or the use of generic data might have led to high uncertainty level in crisp core melt frequency. The results achieved showed this tendency and the need to apply an uncertainty modeling approach.

The proposed methodology, likewise, was able to generate more realistic and statistically significant numbers. A fuzzy approach is able to estimate consistent values and thresholds for safety assessment as well as to model the high uncertainty level inherent to front-line systems. The data range (lower and upper bounds) showed on Table 8, permit us to conclude that the front line systems introduced by research reactor, in the event tree (ET) have significantly elevated the safety plant level.

The case study presented here has confirmed the great advantage of applying this methodology to the LOCA initiator event in current research reactors and future reactor projects for radioisotopes production.

The uncertainty evaluation presented here allow us to propose the use of this methodology as an alternative approach to be applied in probabilistic safety assessments, particularly in cases where relevant operational data records are not available such as innovative design.

With the assistance of the FuzzyFTA program it was possible to determine the top event of each fuzzy fault tree for each front line system, thus including the calculation of the forecast uncertainty due to the presence of uncertainty in the basic events. It was also possible, using fuzzy concept, to determine measures fuzzy (**FIM**) which allows the components' rank number, determining the most important components which have greater or lesser relevance to the system as a whole, well as the measure of uncertainty of each component through its ranking (**FUIM**), and its importance to overall system uncertainty. Furthermore, in case of the RPS the two components more important for the RPS system are components 1 and 2. For the overall system uncertainties the two components that are the components most contribute 6 and 10. Thus, it can be the same reasoning applies for the other front-line system using the results obtained by using the FuzzyFTA program for each system.

Using approximations in operations of fuzzy numbers it can be calculated the fuzzy uncertainty associated with probability for each EndState sequence, as if the results were as triangle function. The results thus obtained allow calculating the risk associated with the event initiator, with a degree of uncertainty. Thus, in calculating the final risk (probability of the EndState sequence x consequence), its reduction may be aided by the use of the values of uncertainties in the probabilities in the EndStates of the sequences, together with changes at the level of projects to reduce the general uncertainties of the each front-line system of the event tree and by the end the associated risk.

Improvements can be made in new projects of research reactors for radioisotopes production using this concept to increase the reliability of the project.

The unavailability of the system also plays an important factor when the nuclear project is involved with the production of radioisotopes.

Acknowledgements

This research has been supported by Brazilian Nuclear Energy Commission (CNEN), Brazilian Council for Scientific and Technological Development (CNPq), Research Support Foundation of State of the Rio de Janeiro (FAPERJ) and Financier of Studies and Project (FINEP).

Author details

Antonio César Ferreira Guimarães* and Maria de Lourdes Moreira

*Address all correspondence to: tony@ien.gov.br

Nuclear Engineering Institute, Rio de Janeiro, RJ, Brazil

References

[1] Aneziris, O. N, Housiadas, C, Papazoglou, I. A, & Stakakis, M. (2001). Probabilistic Safety Analysis of the Greek Research Reactor, Demo 01/2, NCSR _Demokritos, Athens, Greece. http://www.ipta.-demokritos.gr/chousiadas/christosweb/PSA-Dem.pdf

[2] Aneziris, O. N, Housiadas, C, Stakakis, M, & Papazoglou, I. A. Probabilistic safety analysis of a Greek Research Reactor, Annals of Nuclear Energy (2004). , 31(2004), 481-516.

[3] International Atomic Energy AgencySpecific Safety Guide, Licensing Process for Nuclear Installations, (2010). Vienna.(SSG-12)

[4] Keller, M, & Modarres, M. (2005). A Historical overview of probabilistic risk assessment development and its use in the nuclear power industry: a tribute to the late Professor Norman Carl Rassmussen, Reliability Engineering & System Safety , 89(2005), 271-285.

[5] Nrc, U. S. Fault Tree Handbook (NUREG-0492), (1981).

[6] International Atomic Energy Agency(1987). Probabilistic Safety Assessment for Research Reactors. IAEA-TECDOC-400, Vienna.

[7] International Atomic Energy Agency(1989). Application of Probabilistic Safety Assessment to Research Reactors. IAEA-TECDOC-517, Vienna.

[8] Guimarães, A. C. F, & Ebecken, N. F. F. (1999). FuzzyFTA; a fuzzy fault tree system for uncertainty analysis, Annals of Nuclear Energy, , 26(6), 523-532.

[9] Zadeh, L. A. (1987). Fuzzy Sets and Applications: Selected Papers. Wiley: New York.

[10] Guimaraes, A. C. F, Lapa, C. M. F, & Cabral, D. C. Adaptive Fuzzy System for Degradation Study in Nuclear Power Plants Passive Components. Progress in Nuclear Energy, (2006). , 48, 655-663.

[11] Guimaraes, A. C. F. ; Lapa, C. M. F.. Hazard and Operability Study using Approximate Reasoning in Nuclear Power Plants. Nuclear Engineering and Design, USA, v. 236, n. 12, p. 1256-1263, 2006.

[12] Liang, G. S, & Wang, M. J. (1993). Fuzzy Fault-Tree Analysis Using Failure Possibility, Microelectronic Reliability, , 33(4), 583-597.

[13] Housiadas, C. Thermal-hydraulic calculations for the GRR-1 research reactor core conversion to low enriched uranium fuel", DEMO 99/5, NCSR "Demokritos", Athens, (1999).

[14] Reactor Safety StudyU.S. NRC Rep. WASH-1400, NUREG 75/014, October (1975).

[15] Mccormick, N. J. Reliability And Risk Analysis, Method and Nuclear Power Applications, ACADEMIC PRESS, Inc., (1981).

[16] Tyagi, S. K, Pandey, S, & Tyagi, D. R. Fuzzy set theoretic approach to fault tree analysis. International Journal of Engineering, Science and Technology, (2010). , 2(5), 276-283.

[17] Tompmiller, D. A, Eckel, J. S, & Kozinsky, E. J. Human reliability data bank for nuclear power plant operations, A review of existing human reliability data banks

http://dx.doi.org/10.5772/53012

(General Physics Corporation and Sandia National Laboratories NUREG (CR-2744). Washington DC: U.S. Nuclear Regulatory Commission, (1982)., 11

[18] International Atomic Energy Agency(1997). Generic Component Reliability Data for Research Reactor PSA. IAEA-TECDOC-930, Vienna.

Chapter 7

Generation IV Nuclear Systems: State of the Art and Current Trends with Emphasis on Safety and Security Features

Juliana P. Duarte, José de Jesús Rivero Oliva and
Paulo Fernando F. Frutuoso e Melo

Additional information is available at the end of the chapter

http://dx.doi.org/10.5772/53140

1. Introduction

Fifty years ago, on June 26, 1954, in the town of Obninsk, near Moscow in the former USSR, the first nuclear power plant was connected to an electricity grid to provide power. This was the world's first nuclear power plant to generate electricity for a power grid, and produced around 5 MWe [1]. This first nuclear reactor was built twelve years after the occurrence of the first controlled fission reaction on December 2, 1942, at the Manhattan Engineering District, in Chicago, Illinois, US. In 1955 the USS Nautilus, the first nuclear propelled submarine, equipped with a pressurized water reactor (PWR), was launched. The race for nuclear technology spanned several countries and soon commercial reactors, called first generation nuclear reactors, were built in the US (Shippingport, a 60 MWe PWR, operated 1957-1982, Dresden, a boiling water reactor, BWR, operated 1960-1978, and Fermi I, a fast breeder reactor, operated 1957-1972) and the United Kingdom (Magnox, a pressurized, carbon dioxide cooled, graphite-moderated reactor using natural uranium).

A few years after the projects had developed many nuclear safety concepts were extended and then implemented in second-generation nuclear systems, consisting of reactors currently in operation, as the PWR, CANDU (Canadian Deuterium Uranium Reactor), BWR (Boiling Water Reactor), GCR (Gas-Cooled Reactor), and VVER (Water - Water Power Reactor), the latter developed by the former Soviet Union. At this time, other concepts were studied in parallel, such as liquid metal cooled reactors and the reactors with thorium and uranium molten salt, which did not propagate commercially and/or remained in experimental countertops. Operating or decommissioned power reactor designs can be found in [2].

With operating experience gained in recent decades, digital instrumentation development and lessons learned from the accidents at TMI (Three Mile Island), Chernobyl and recently Fukushima, Generation III and III+ reactor designs have incorporated improvements in thermal efficiency and included passive system safety and maintenance costs and capital reduction. There are several designs of these so called advanced reactors and some are being built in the U.S. and China.

In 2001, nine countries (Argentina, Republic of Korea, Brazil, Canada, Republic of South Africa, United Kingdom, France, United States and Japan) signed the founding document of Generation IV International Forum (GIF) in order to develop nuclear systems that can fulfill the increasing world electric power needs with high safety, economics, sustainability and proliferation resistance levels [3]. The Russian Federation, People´s Republic of China, Switzerland and Euratom have joined this group. Since then, GIF has selected the six most promising reactor system designs to be developed until 2030 and has created research groups on materials, fuel and fuel cycle, conceptual design, safety, thermal-hydraulics, computational methods validation and others in order to develop the necessary technology on an international cooperation basis.

According to IAEA [4], an accessible, affordable and sustainable energy source is fundamental to the development of modern society. Current scenarios predict a global demand for primary energy 1.5-3 times higher in 2050 as compared to today, and a 200% relative increase in the demand for electricity. Nuclear power is an important source that should be considered, because it is stable power on a large scale, with virtually no greenhouse gas emissions and low environmental impact as compared to fossil fuels, and can produce heat for chemical processes in industry and for hydrogen generation. In this context, GIF is seeking to develop more economical, sustainable and safe nuclear reactors, from their fuel cycles to decommissioning and waste treatment, and thus meet the world's energy needs.

This chapter is organized as follows. Section 2 discusses the goals for Generation IV reactor systems, section 3 discusses the current six Generation IV reactor system design description and research on the subject. Section 4 focuses on the first reports concerning reactor safety and risks [5], particularly the Integrated Safety Assessment Methodology (ISAM). Section 5 addresses economic aspects of this new reactor generation [6]. Section 6 concerns the discussion on proliferation resistance and physical protection [7]. Section 7 encompasses a set of general conclusions on the subject, addressing mainly the relevance of these nuclear system concepts.

2. Goals for generation IV nuclear energy systems

Before selecting the reactors that will be part of the Next Generation Nuclear Plant (NGNP), the founding countries of the Generation IV International Forum selected eight key objectives able to make these reactor designs vital in the near future. These eight goals based on concepts of sustainability, economic competitiveness, safety, physical protection and nuclear proliferation resistance are described below [3]:

http://dx.doi.org/10.5772/53140

1. Sustainability-1 – NGNP will provide sustainable energy generation that meets clean air objectives and provides long-term system availability and effective fuel utilization for worldwide energy production;
2. Sustainability-2 – NGNP will minimize and manage their nuclear waste and notably reduce the long-term administrative burden, thereby improving protection for the public health and the environment.
3. Economics-1 – NGNP will have a clear life-cycle cost advantage over other energy sources.
4. Economics-2 – NGNP will have a level of financial risk comparable to other energy projects.
5. Safety and Reliability-1 – NGNP operations will excel in safety and reliability.
6. Safety and Reliability-2 – NGNP will have a very low frequency and degree of reactor core damage.
7. Safety and Reliability-3 – NGNP will eliminate the need for offsite emergency response.
8. Proliferation resistance and Physical Protection – NGNP will increase the assurance that they are very unattractive and the least desirable route for diversion or theft of weapons-usable materials, and provide increased physical protection against acts of terrorism.

With these concepts in mind, projects already in operation, in test or just conceptual have been analyzed and six promising nuclear reactors have been selected to be designed and built in this endeavor.

3. The six reactor concepts under discussion

Based on GIF goals, six reactor designs have been selected to be developed and constructed by 2030. Such reactors must meet the safety and security, sustainability, economics and proliferation resistance criteria defined as essential for this new generation. That is, projects must: consider the entire fuel cycle to obtain a higher fuel burn-up and consequently less actinides in the final waste, reducing their lifetime in final repositories; increase the thermal efficiency with combined cycles and cogeneration; aim at the production of hydrogen, be intrinsically safe, with negative reactivity coefficients and achieve competitive costs since their construction.

GIF has divided the involved countries in working groups formed by laboratories, universities, and government agencies, according to the experience and interest of each to develop the projects of the following reactors: Very-high-temperature reactor (VHTR), Gas-cooled fast reactor (GFR), Supercritical-water-cooled reactor (SCWR), Sodium-cooled fast reactor (SFR), Lead-cooled fast reactor (LFR) and Molten salt reactor (MSR). For each project specific groups of materials, computational methods, fuel and fuel cycle, thermal-hydraulics, safety and operation and others have been created. The first four reactor concepts are completely defined and the remaining two are in progress.

A major challenge for almost every project is the development of new materials for the reactor primary systems that can tolerate temperatures up to 1000 °C without reducing safety margins. Another point is the need to develop and validate computer codes both in neutronics and thermal-hydraulics projects with little or no available operational experience. These issues and the obtained solutions will be discussed next.

3.1. Very-high-temperature reactor (VHTR)

The VHTR is one of the most promising projects of Generation IV reactors, given the experience in gas-cooled reactors that many countries developed in recent decades. The first reactors of this type were designed by the UK and France for the production of plutonium [8], were graphite-moderated and used CO_2 as a coolant, which limited the maximum temperature at 640 °C, when chemical reactions occur between gas and moderator. Later, the use of helium (although more costly) was justified by having better heat transfer properties, and permit increasing the core outlet temperature.

The current GIF design consists of a helium-cooled reactor with graphite as a moderator, TRISO fuel and coolant outlet temperature above 900 °C. There are two core concepts: the prismatic block-type and the pebble bed-type. The first type follows the line of the High Temperature Engineering Test Reactor (HTTR) developed and built by Japan initially with coolant exit temperature of 850 °C and then 950 °C in April 2004 [9]. The second is the result of the German program, which was later imported by the People's Republic of China and developed in the Republic of South Africa as the Pebble Bed Modular Reactor (PBMR). Both designs use TRISO fuel (see Figure 1), which consists of a spherical micro kernel of oxide or oxycarbide fuel and coating layers of porous pyrolytic carbon (buffer), inner dense pyrolytic carbon (IPyC), silicon carbide (SiC) and dense outer pyrolytic carbon (OPyC).

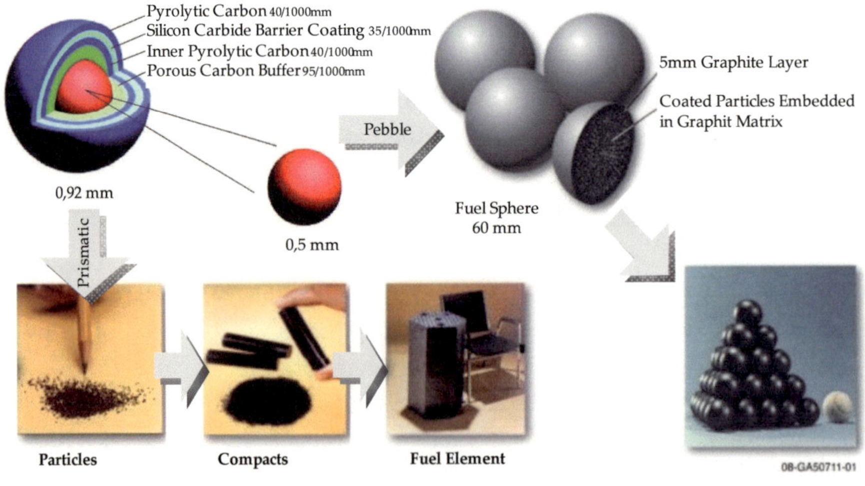

Figure 1. TRISO fuel for prismatic block-type and pebble bed-type cores [10,11].

The goal of achieving an exit temperature of 1000°C coolant will allow the VHTR to generate electricity with high efficiency and provide heat for hydrogen production and for refineries, petrochemical and metallurgical industries. For this purpose, research is being done to evaluate the use of other materials, such as uranium-oxicarbide UCO and ZrC, which increase the capacity of TRISO fuel burn-up, reduce the permeability of fission products and increase resistance to high temperatures in case of an accident (above 1600 °C) [12]. Although a once-through uranium fuel cycle is planned reactors have potential for this deep-burn of plutonium and minor actinides, as well as the use of thorium based fuel [13].

The experimental reactor HTTR in Japan and HTR-10 in China support the development of the VHTR, in particular providing important information on safety and operational characteristics. The research group on computational methods validation and benchmarking uses these reactors to validate computer codes in the areas of thermal hydraulics, thermal mechanics, core physics and chemical transport.

3.2. Gas-cooled fast reactor (GFR)

Another project selected by GIF of a gas-cooled reactor is the GFR, the fast-neutron-spectrum, helium-cooled and closed fuel cycle reactor. The selection of this reactor was based on its excellent potential for long-term sustainability, in terms of the use of uranium and of minimizing waste through reprocessing and fission of long-lived actinides [14,15]. In terms of non-proliferation, the objective of high burn-up together with actinide recycling results in spent fuel that is unattractive for handling. The use of gas as coolant means achieving high temperatures, so that this design also has the purpose of providing process heat and enabling hydrogen production.

Figure 2. Updated layout of ALLEGRO featuring two main heat exchange loops [15].

One of the proposed designs is a 1200 MWe reactor, operating at 7 MPa, coolant exit temperature of 850°C, an indirect combined cycle with He-N_2 gas mixture for intermediate gas cycle and natural convection as passive safety system [14]. At least two fuel concepts that meet the proposed design are being studied: a ceramic plate-type fuel assembly and a ceramic pin-type fuel assembly. In this latter, the fuel assembly is based on a hexagonal lattice of fuel-pins and the materials used are uranium and plutonium carbide as fuel, and silicon carbide as cladding [16].

An experimental reactor called ALLEGRO (Fig. 2) with 80 MWth will be the first built GFR with the objective of demonstrating the feasibility and qualifying technologies for fuel, fuel assembly and new safety systems. It is important to notice the interface between GFR and VHTR reactors, which use helium as a coolant and power conversion technology using gas turbines and cogeneration, as many efforts on component and materials development are being held together.

3.3. Supercritical-water-cooled reactor (SCWR)

The SCWR concept emerged at the University of Tokyo in 1989 and became a global concern after being selected by the Generation IV International Forum in 2002 [17]. The SCWR is a high-temperature, high-pressure, water-cooled reactor that operates above the thermodynamic critical point of water (374 °C, 22.1 MPa). This last feature eliminates coolant boiling, since both liquid and gaseous states can coexist. Thus, according to [18], the need for recirculation and jet pumps, pressurizer, steam generators, and steam separators and dryers in current LWRs is eliminated. Still, according to [18] the main mission of this reactor design is the generation of low-cost electricity.

A more advanced project is the Japanese 1620 MWe SCWR consisting of a pressure-vessel type, once-through reactor and a direct Rankine cycle system. Figure 3 is a simplified system where steam leaves the pressure vessel at 560 °C and 25 MPa, passes through moisture separator valves and control valves, high and low pressure turbines connected to the generator, condenser, low and high pressure pumps, a deaerator (to remove dissolved gases) to return to the pressure vessel. Because of supercritical water condition, no phase change occurs, which means that the coolant continuously passes from the liquid state to the gaseous state.

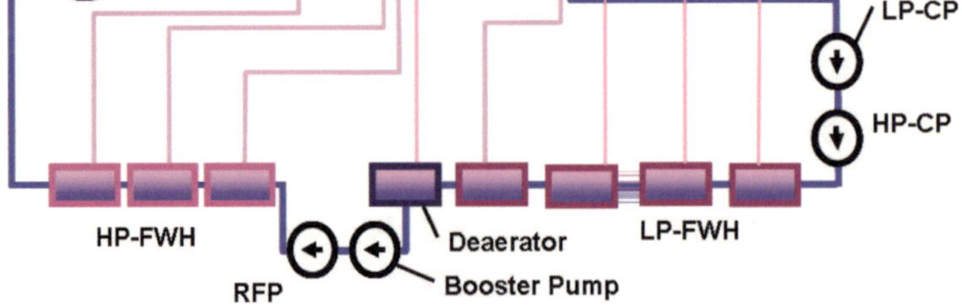

Figure 3. The Japanese Supercritical Water Reactor (JSCWR) [17].

Another concept proposed by Canada is generically called CANDU-SCWR [19], it is a pressured tube type reactor with fuel channels separating the light water coolant from the heavy water moderator. The supercritical steam is led directly to the high-pressure turbine, eliminating the need of steam generators as the JSCWR.

3.4. Sodium-cooled fast reactor (SFR)

The Sodium-cooled Fast Reactor is among the six candidate designs selected for their potential to meet Generation IV technology goals. It is a fast-spectrum, sodium-cooled reactor and closed fuel cycle for efficient management of actinides and conversion of fertile uranium Three options are under consideration (Figure 4): a large size loop type sodium-cooled reactor with uranium-plutonium oxide fuel (1500 MWe), a medium size pool-type system with 600 MWe and a small size modular type with plutonium-minor actinide-zirconium metal alloy fuel [20].

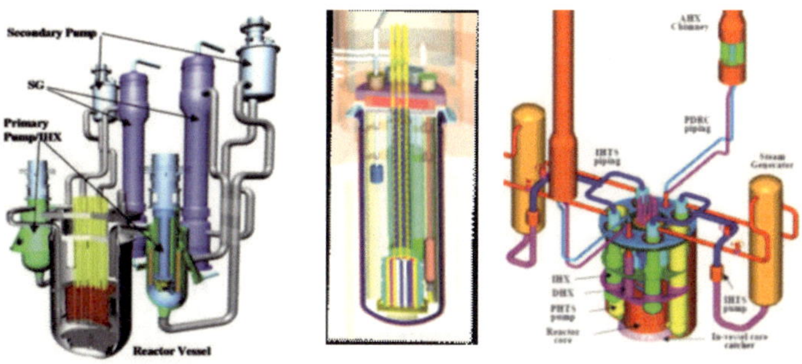

Figure 4. Loop, modular and pool configuration [20,21]

Heat capacity and high thermal conductivity of liquid-metal coolants provides large thermal inertia against system heating during loss-of-flow accidents. Furthermore, due the non-corrosive characteristic of sodium coolant, the reactor core and primary system components do not degrade even over very long residence times in the reactor, so that maintenance requirements are minimized. However sodium reacts chemically with air and water and requires a great coolant system seal.

3.5. Lead-cooled fast reactor (LFR)

The application of lead coolant in nuclear systems began in the 1970s in the ancient Soviet Union with systems cooled by Lead-Bismuth Eutectic (LBE), which were developed for nuclear-powered submarines [22]. Two of the currently proposed designs for international cooperation are the Small Secure Transportable Autonomous Reactor (SSTAR, Fig. 5) and the European Lead-cooled System (ELSY). Both concepts are lead-cooled, with passive decay heat removal and nitride fuels. A closed fuel cycle is expected for the efficient conversion of fertile uranium and actinide management for this reactor.

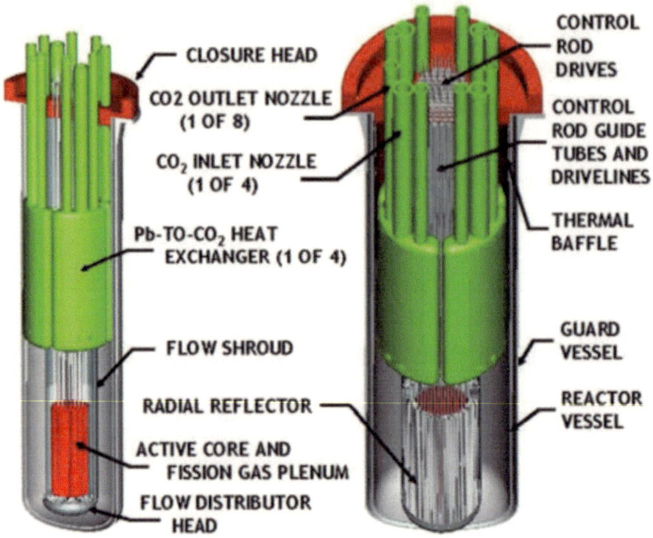

Figure 5. Small Secure Transportable Autonomous Reactor [22].

Lead flows in the reactor core cooling it by natural circulation and passes through Pb-to-CO_2 heat exchangers located between the vessel and the cylindrical shroud. It is noteworthy that lead has been chosen as coolant rather than LBE to drastically reduce the amount of alpha-emitting ^{210}Po isotope produced [22]. Many studies on the choice of fuel and fuel cycle have been proposed since 2004 when an international cooperation through GIF began. Features of this reactor, like enabling fissile self-sufficiency, autonomous load following, operation simplicity, reliability, transportability, as well as a high degree of passive safety, make it a unique proliferation resistant, safe and economical design.

3.6. Molten salt reactor (MSR)

Investigation of molten-salt reactors started in the late 1940s as part of the United States' program to develop a nuclear powered airplane [23]. Today, research on this reactor has been resumed due to its inclusion as one of the six Generation IV reactor types. MSRs have been initially considered as thermal-neutronic-spectrum graphited-moderated concepts, but since 2005 R&D has focused on the development of fast-spectrum MSR concepts. The Molten Salt Fast Reactor core is a cylinder of the same diameter and height, where nuclear reactions occur in the fuel salt. Two options have been considered for the fuel cycle: 233U-started MSFR and TRU(fuelled with transuranic elements)-started MSFR [24]. The salt processing scheme relies on both on-line and batch processes (Fig. 6) to satisfy the constraints for a smooth reactor operation while minimizing losses to waste stream [25]. A simplified schematic of the online processing can be seen in Figure 6, where fuel fission gas extraction and actinide separation occur.

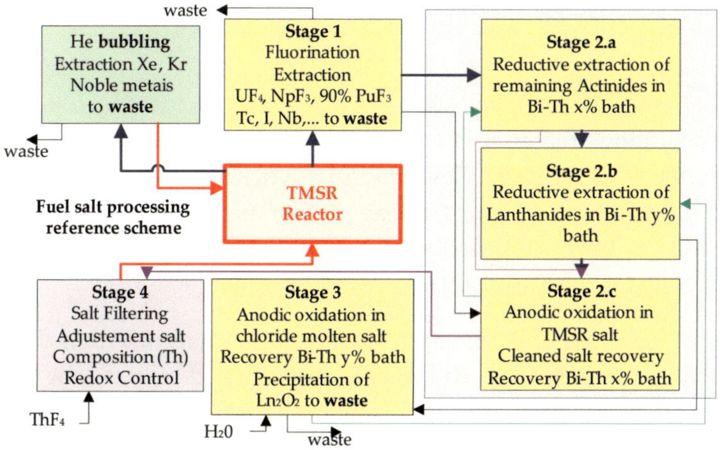

Figure 6. TMSR (MSFR) reference fuel salt processing [25].

A survey of the main features of the reactors discussed here is displayed in Table 1. It is important to remember that these projects are in development and there are different views in the same reactor line. Some reactors are already testing their designs with full construction and will be essential for future validation methodologies. If these six reactors achieve their ultimate goals, they will not only bring more advanced technology, but new concepts of safer, sustainable and economical nuclear reactors. Some of these aspects will be discussed in more detail in the following sections.

	GFR	SCWR	MSR	SFR	VHTR	LFR
Neutrons energy	Fast	Thermal	Thermal or fast	Fast	Thermal	Fast
Power (MWe)	1200	Up to 1500	1000	50 - 1500	~200	19.8 - 600
Fuel type	Plate or ceramic	UO_2 pellet	Molten salt	MOX or Metal alloy	TRISO	MOX or Nitrides
Coolant	Helium	Light water	Molten salt	Sodium	Helium	Lead
Moderador	-	Light water	-	-	Graphite	-
Outlet temperature	850°C	625°C	1000°C	~500°C	1000°C	~500°C
Pressure * (MPa)	7	25	~0.1	~0.1	5 - 7	~0.1
Fuel Cycle	Closed	once through	Closed	Closed	once through	Closed

* Primary system pressure.

Table 1. Survey of Generation IV Nuclear Reactors.

4. The integral safety assessment methodology (ISAM)

4.1. GIF safety philosophy

Since the very beginning, Safety and Reliability issues were extensively considered, representing three of the eight goals of Generation IV Nuclear Energy Systems [26]. Safety, together with waste management and nuclear proliferation risks, remains as one of the key problems of nuclear energy. Consequently, a new generation of Nuclear Power Plants (NPPs) has to face and solve these problems convincingly.

In spite of past nuclear accidents, including the most recent at Fukushima Daiichi, NPPs have an excellent safety record and still remain as a safe and high-power energy production technology without greenhouse gases emissions. Unfortunately, public acceptance continues to be an important impediment.

In this context, the Generation IV International Forum (GIF) safety objectives have been oriented to substantially upgrade safety and enhance public confidence in NPPs, by means of an increasing use of inherent safety features, a major reduction of core damage frequency and by eliminating the need for offsite emergency response in case of accidents [26]. It was immediately recognized the necessity of a standard methodology for Generation IV safety assessments. The methodology would allow a uniform safety evaluation of different NPP concepts with respect to Generation IV safety goals [26].

During the past 60 years safety assessment has evolved, from the initial deterministic analysis through conservative assumptions and calculations, to an increasing application of a best-estimated deterministic approach, in conjunction with probabilistic methods [27, 28]. Best-estimated and probabilistic assessments identify potential accidents scenarios that could be important contributors to risk [28]. As a consequence, Probabilistic Safety Analysis (PSA) is having a more prominent and fundamental role in the design process and licensing analysis, as part of an integrated approach, risk-informed, which also includes Deterministic Analysis and Defense in Depth Philosophy [29]. This new safety assessment conception, integrating probabilistic and deterministic analysis with an extensive application of the Defense in Depth principle, constitutes the basis for the Generation IV safety assessment methodology.

4.2. Phenomena identification and ranking tables (PIRT)

In 2004, the Idaho National Engineering and Environmental Laboratory proposed a Research and Development Program Plan, identifying the R&D needs for the next generation nuclear plant (NGNP) design methods [30]. Later, in 2008, the Idaho National Laboratory prepared a report for the US Department of Energy with a R&D Technical Program Plan for the NGNP methods [31]. Both reports were focused on the development of tools to assess the neutronic and thermal-hydraulic behavior of the Very High Temperature Reactor (VHTR) systems, using a Phenomena Identification and Ranking Tables (PIRT) informed R&D process.

The first two steps of the proposed methodology are 1) Scenario Identification, where operational and accident scenarios that require analysis are indentified, and 2) PIRT, where important phenomena are identified and ranked for each scenario. In the following stages of the methodology, analysis tools are evaluated to determine whether important phenomena can be calculated and operational and accident scenarios that require study are analyzed.

PIRT was developed in the late 1980s, for the qualification of deterministic safety codes for Light Water Reactors [32, 33]. It is a systematic way of gathering information from experts on a specific subject, and ranking the importance of the information, in order to meet some decision-making goal [34]. An important part of the process is to also identify the associated uncertainties, usually by scoring the knowledge bases for the phenomena. Example of successful PIRT applications in thermal-hydraulics, severe accidents, fuels, materials degradation, and nuclear analysis may be found in [34]. An extensive application of PIRT is reported in [33, 35, 36], for the determination of code applicability for the analysis of selected scenarios with uncertainty evaluations.

4.3. The risk-informed and performance-based approach

In 2007, the United States Nuclear Regulatory Commission (NRC) developed a risk-informed and performance-based regulatory structure for the licensing of future NPPs, with broader use of design specific risk information and applicable to any reactor technology [29]. Defense in Depth remains basic to this framework, providing safety margins to compensate for the uncertainties in the requirements for design, construction and operation.

In this model, PSA and the Licensing Basis Events (LBEs) deterministic calculations are closely linked. LBEs are selected developing a Frequency-Consequence (F-C) curve, which is used together with the plant specific PSA.

The performance-based approach is applied whenever possible, so that performance history is used to focus attention on the most important safety issues. Objective criteria are established for evaluating performance.

4.4. Defense in depth: The objective provisions tree (OPT)

IAEA has also proposed a new safety approach and a methodology to generate technology-neutral safety requirements for advanced and innovative reactors [37]. The document identified several areas requiring further development, such as the replacement of qualitative safety objectives for quantitative ones, the enhancement of Defense in Depth, further development of PSA, the application of methodologies early in the design process, the use of an iterative design process to demonstrate the adequacy of Defense in Depth and a comprehensive review of the existing safety approach.

The Objective Provisions Tree (OPT) methodology [38, 39] was suggested as a tool to systematically examine all possible options for provisions to prevent and/or control challenging mechanisms jeopardizing Defense in Depth. OPT is a systematic critical review of the Defense in Depth implementation. It identifies the required provisions that jointly ensure the prevention or control of a mechanism that represents a challenge for a safety function, which is part of a Defense in Depth Level. The existence of several challenges for each safety function and several mechanisms contributing to a given challenge leads to a tree structure. The set of provisions, jointly ensuring the prevention of one mechanism, constitutes a Line of Protection (LOP).

The proposed main pillars of the new IAEA safety approach are Quantitative Safety Goals (correlated with each level of Defense in Depth), Fundamental Safety Functions, and Defense in Depth (generalized, including probabilistic considerations). Quantitative Safety Goals are based on the condition that plant states with significant frequency of occurrence have only minor or no potential radiological consequences, according to the F-C curve, called Farmer´s Curve.

Defense in Depth essential characteristics were described as exhaustive, balanced and graduated. "Balanced" means that no family of initiating events should dominate the global frequency of plant damage states. "Graduated" means that a progressive defense excludes the possibility of a particular provision failure to generate a major increase in potential consequences, without any possibility of recovering the situation at an intermediate stage.

4.5. The risk and safety working group (RSWG)

A Risk and Safety Working Group (RSWG) was created in the frames of the Generation IV Forum to develop a safety approach for Generation IV Nuclear Systems. After its initial meeting in 2005, the first major work product, establishing the basis for the required safety approach, was released by the end of 2008 [40]. The document recommended the early application of a cohesive safety philosophy based on a concept of safety that is "built-in, not added-on". The identification of risks must be as exhaustive as possible. The remaining lack of exhaustiveness of the accident scenarios should be covered by the notion of enveloped situations and the implementation of the Defense in Depth principles. A re-examination of the safety approach was proposed, complementing the IAEA suggestions [37] with the following attributes:

- Risk-informed, combining both probabilistic and deterministic information.
- Understandable, traceable, and reproducible.
- Defensible. Whenever possible, known technology should be used.
- Flexible. New information and research results should be easily incorporated.
- Performance-based.

RSWG recommended a safety approach that manages simultaneously deterministic practices and probabilistic objectives, handling internal and external hazards. Besides the Defense in Depth characteristics previously mentioned (exhaustive, balanced and graduated) [37], RSWG added the followings: tolerant and forgiving. "Tolerant" means that no small deviation of a physical parameter outside the expected range can lead to severe consequences (ab-

http://dx.doi.org/10.5772/53140

sence of cliff-edge effects). "Forgiving" means the availability of a sufficient grace period and the possibility of repair during accident conditions.

RSWG recognized safety related technology gaps for the six reactor concepts selected by the Generation IV Forum in different technical areas such as updated safety approach, fuel, neutronics, thermal aerolics/hydraulics, materials & chemistry, fuel chemistry, passive safety and severe accident behavior. An effective mix of modeling, simulation, prototyping and demonstrations should be used to reduce the existing uncertainties and lack of knowledge.

4.6. The integral safety assessment methodology (ISAM)

During 2008 the RSWG began the development of the methodology for Generation IV safety assessments, stated in [26]. The methodology would integrate PSA and several other techniques, such as PIRT and OPT, with an extensive deterministic and phenomenological modeling [41]. In 2009 the RSWG focused its work on the development of a methodology that was denominated Integrated Safety Assessment Methodology (ISAM) [42].

ISAM consists of five distinct analytical tools which are structured around PSA:

- Qualitative Safety Requirements/Characteristic Review (QSR).
- Phenomena Identification and Ranking Table (PIRT),
- Objective Provision Tree (OPT).
- Deterministic and Phenomenological Analyses (DPA).
- Probabilistic Safety Analysis (PSA).

From its original conception, it was clearly established that ISAM is not intended to dictate requirements or compliance with safety goals to designers. Its intention is solely to provide useful insights into the nature of safety and risk of Generation IV systems for the attainment of Generation IV safety objectives. ISAM will allow evaluation of a particular Generation IV concept or design relative to various potentially applicable safety metrics or "figures of merit" (FOM) [42, 43].

During 2010 the RSWG focused its work on the finalization of ISAM methodology presented in 2009. ISAM was conceived as a methodology providing specific tools to examine relevant safety issues at different points in the design evolution, in an iterative fashion through the development cycle. It is considered well integrated, and when used as a whole, offers the flexibility to allow a graded approach to the analysis of technical issues of varying complexity and importance [5].

Finally, the document describing the Generation IV Integrated Safety Assessment Methodology (ISAM) was available in June, 2011 [5]. According to this report, a principal focus of RSWG is the development and demonstration of an integrated methodology that can be used to evaluate and document the safety of Generation IV nuclear systems.

An important remark is that the safety approach for Generation IV nuclear systems should differ from the one, usually applied to previous reactor generations, in which safety is generally "added on" by applying safety assessments to relatively mature designs and introducing the results in many cases as "backfits". ISAM is therefore intended to support achievement of safety that is "built-in" rather than "added on" by influencing the direction of the concept and design development from its earliest stages [5].

It is envisioned that ISAM will be a "tool kit" used in three different ways:

- As an integrated methodology, throughout the concept development and design phases, revealing insights capable of influencing the design evolution. ISAM can develop a more detailed understanding of safety-related design vulnerabilities, and resulting risk contributors. New safety provisions or design improvements can be identified, developed, and implemented relatively early.
- Applying selected elements of the methodology separately at various points throughout the design evolution to yield an objective understanding of safety-related issues (such as risk contributors, safety margins, effectiveness of safety provisions, uncertainties, etc.) important to decision makers.
- In the late stages of design maturity, for decision makers and regulators to measure the level of safety and risk associated with a given design relative to safety objectives or licensing criteria ("post facto" application).

ISAM is essentially a PSA-based safety assessment methodology, with the additional strength of other tools, tailored to answering specific types of questions at various stages of design development. The methodology is well integrated. Although individual analytical tools can be selected for separate and exclusive use, the full value of the integrated methodology is derived from using each tool, in an iterative fashion and in combination with the others, throughout the development cycle. Figure 7 shows the overall task flow of ISAM and indicates which tools are intended for each phase of Generation IV system technology development [5].

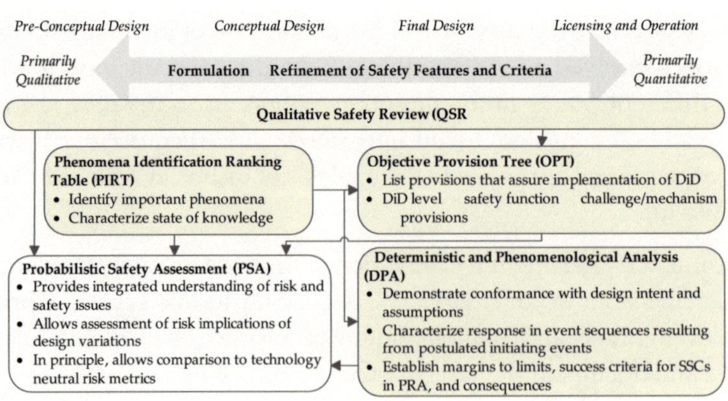

Figure 7. ISAM Task Flow [5].

The methodology comprises several stages, from the pre-conceptual design to licensing and operation, with the corresponding safety features and criteria moving from primarily qualitative to quantitative. In the early design stages, the main role corresponds to the qualitative safety analysis techniques QSR, PIRT and OPT, but Deterministic and Phenomenological Analysis (DPA) and Probabilistic Safety Assessment (PSA) are introduced earlier in comparison with the current practice for previous reactor generations. DPA and PSA are the key techniques to be applied during the last stages corresponding to final design, licensing and operation, where safety criteria are mainly quantitative.

4.7. The qualitative safety features review (QSR)

Only one of the tools integrated in ISAM is completely new, and was developed specially for Generation IV Reactor Systems: the Qualitative Safety features Review (QSR). QSR is intended to provide a systematic means of ensuring and documenting that the evolving Generation IV system concept of design incorporates the desirable safety-related attributes and characteristics that are identified and discussed in [40].

QSR is conducted using a template structure organized according to the first four levels of Defense in Depth (prevention, control, protection and management of severe accidents). The review is based on an exhaustive check list of safety good practices and recommendations applicable to Generation IV systems. Design options are evaluated, identifying their strength or weakness, and qualified as favorable, unfavorable or neutral in relation with the desirable characteristics.

An exhaustive and detailed check list is essential for the QSR quality. To generate the list, a top-down functional approach is conducted, as shown in Figure 8 [5]. Firstly, the recommendations from RSWG, IAEA standards, INSAG and INPRO guidelines and other references, organized according to the levels of Defense in Depth, are detailed as much as possible, for a technology neutral condition. Finally, the defined neutral characteristics are used to develop the set of specific recommendations for a particular technology. In this process, the characteristics and features are grouped in four classes, moving from general recommendations to detailed specific attributes [5]:

Class 1: Generic and Technology neutral.

Class 2: Detailed and Technology neutral.

Class 3: Detailed and Technology neutral but applicable to a given safety function.

Class 4: Detailed and applicable to a given safety function and specific technology.

RSWG has developed check lists covering desirable Classes 1 to 3 safety characteristics for the first four levels of Defense in Depth. Class 3 recommendations were determined for the safety function "Decay Heat Removal" [5]. It is considered that QSR applications will be important for the identification of Generation IV R&D needs.

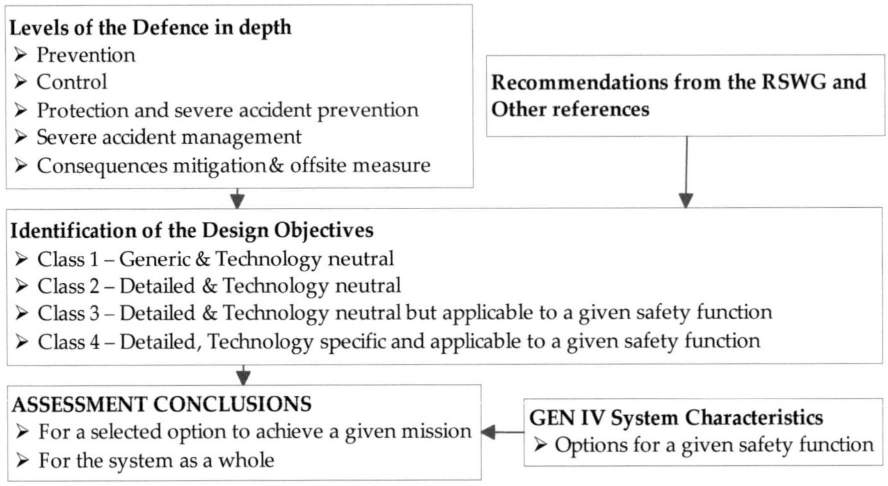

Figure 8. Top-down approach to establish safety recommendations derived from a QSR [5]

4.8. PIRT and OPT

In ISAM, PIRT and OPT are tools that complement each other. PIRT is used to identify phenomena impacting accident scenarios while OPT finds out the necessary Defense in Depth provisions to prevent, control or mitigate the corresponding phenomena. They can be iteratively applied from conceptual to mature designs. Both techniques are basically qualitative tools based on expert elicitation.

PIRT is more expert-dependant due to the complexity and diversity of possible phenomena involved in accident scenarios. The expert panel selects a FOM, for example, the fission products release, the maximum core coolant temperature, etc.; and the PIRT process is applied to each of the identified accident scenarios to determine and categorize the safety-related phenomena and the uncertainties in their knowledge. Usually a four level scale is used to rank the phenomena importance (high, medium, low or insignificant) and existing knowledge (full, satisfactory, partial or very limited). The ranked values of importance and uncertainty are useful to determine R&D effort priorities for accident scenarios. Safety issues located in the region of high importance and large uncertainty have the greatest priorities.

OPT is a systematic and structured top-down method for a fully characterization of the Defense in Depth architecture by identifying in detail the set of objective realistic countermeasures against a great diversity of safety-deteriorating mechanisms. Following a deductive approach, the model goes from each level of defense throughout the safety objectives and physical barriers; down to the safety functions and their challenges, until challenging mechanisms at the lowest tree levels are deduced and the associated sets of objective provisions in opposition to them, constituting LOPs, are determined. Figure 9 shows the tree structure of the OPT process [39].

http://dx.doi.org/10.5772/53140

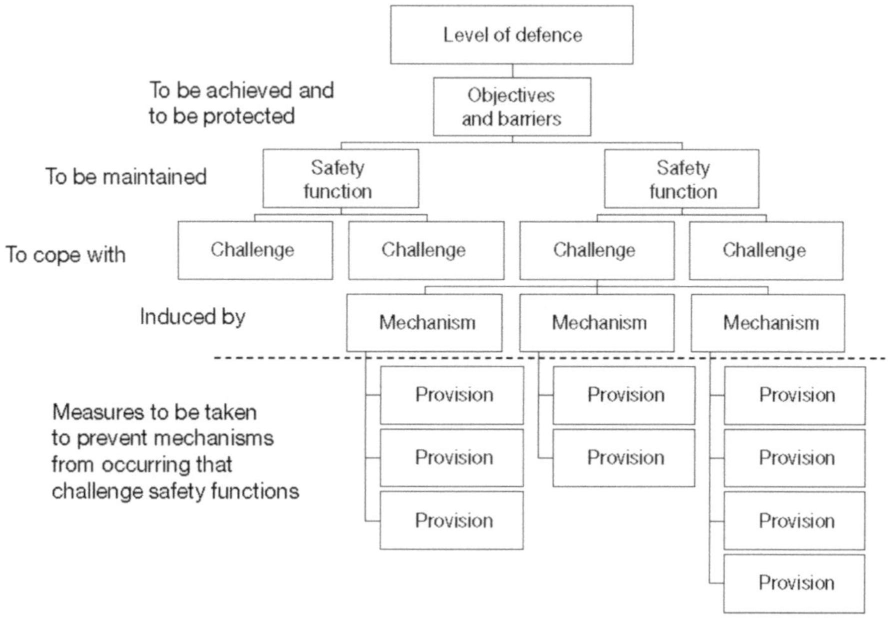

Figure 9. OPT process to determine Defense in Depth objective safety provisions [39]

The application of OPT is expected to contribute from early stages to the built-in safety approach proclaimed as part of Generation IV safety assessment philosophy. Its final objective is to ensure that Defense in Depth satisfies the desirable attributes of being exhaustive, balanced, progressive, tolerant and forgiving [5]. During the pre-conceptual and conceptual design phases OPT will serve as an important guidance to research efforts in order to achieve the mentioned Defense in Depth attributes by means of robust, reliable and simple design solutions. OPT can also identify degradation mechanisms representing feedbacks to PIRT.

4.9. DPA and PSA

PSA is the principal basis of ISAM, but DPA also constitutes a vital component, providing support to PSA, as well as to PIRT and OPT. Following the Risk-informed methodology, ISAM integrates PSA, DPA and Defense in Depth (evaluated using the OPT). The two mutually complemented methodologies, PSA and DPA, contribute to assess important phenomena identified in the PIRT process, and can also evaluate the effectiveness of LOPs deduced by means of OPT. DPA is indispensable to understand NPP safety issues, providing quantitative insights, important as PSA inputs. Best-estimated deterministic computer codes are preferred, incorporating sensitivity analyses to cover the existing uncertainties, depending on the design stage. One important challenge is the upgrade, development and validation of deterministic computer codes and their necessary input data to perform convincing safety assessments of some innovative reactor concepts.

RSWG supports the idea of applying PSA from the earliest practical stages of the design process, and continuing its application iteratively throughout the evolution of the design concept until its maturity, in the stages of final design, licensing and operation. The PSA scope should comprise both internal and external events. PSA is recognized as a fundamental tool to prioritize properly design and operational issues which are more significant to safety, contributing in this way to a proper balance between costs and safety effectiveness of Generation IV Nuclear Energy Systems. PSA advantages for a systematic understanding and evaluation of risk uncertainties is also remarked. It is expected that PSA will contribute to understand differences in the level of safety of diverse technical proposals and select the designs that better fulfill the selection safety criteria for a given Generation IV reactor concept.

The traditional PSA metrics of Core Damage Frequency (CDF) and Large Early Release Frequency (LERF) remain useful, although the first one does not apply to all the new reactor concepts. It has been recommended the generalization of CDF as "undesirable event with significant source term mobilization". The principal risk metric that should be used for comparing Generation IV concepts is the Farmer's Curve, oriented to the Generation IV Safety objective regarding the elimination of off-site emergency response [5]. To achieve its leading role, PSA must be validated as part of a rigorous quality assurance program, including a peer review conducted by independent experts.

4.10. ISAM: Experiences and perspectives

ISAM is a relatively new methodology which is still under development and adjustment. It is recognized that the methodology will need to be modified or updated based on the lessons and findings derived from the Fukushima accident [5, 44].

An example of a preliminary ISAM application to the Japanese Sodium-cooled Fast Reactor (JSFR) is described in [5]. A PIRT process was conducted for the reactor shutdown by the passive Self-Actuated Shutdown System (SASS) upon an Unprotected Loss of Flow (ULOF) accident. OPT results are presented for the safety function "Core heat removal" of Defense in Depth Level 3, indentifying several mechanisms for the challenge "Degraded or disruption of heat transfer path" and their corresponding LOPs. Applicability of DPA and PSA is also shown.

RSWG demonstration that ISAM can be used to evaluate and document the safety of Generation IV nuclear systems, supported by an extended use of the methodology in practical applications, is only beginning but has excellent perspectives of becoming a future reality. It will certainly depend on international efforts and progress in the materialization of Generation IV reactor concepts. RSWG has been asked for GIF to work toward the provision of increasingly detailed guidance for application of ISAM in the development of Generation IV systems [44].

5. Generation IV: Economical aspects [6]

The economic modeling working group (EMWG) was formed in 2004 for developing a cost estimating methodology to be used for assessing Generation IV International Forum (GIF) systems against its economic goals. Its creation followed the recommendations from the economics crosscut group of the Generation IV roadmap project that a standardized cost estimating protocol be developed to provide decision makers with a credible basis to assess, compare, and eventually select future nuclear energy systems, taking into account a robust evaluation of their economic viability. The methodology developed by the EMWG is based upon the economic goals of Generation IV nuclear energy systems, as adopted by GIF: to have a life cycle cost advantage over other energy sources, to have a level of financial risk comparable to other energy projects (i.e., to involve similar total capital investment and capital at risk).

This section briefly describes an economic model for Generation IV nuclear energy systems [6] and the accompanying software [45] in which the guidelines and models were implemented. These tools will integrate cost information prepared by Generation IV system development teams during the development and demonstration of their concept, thus assuring a standard format and comparability among concepts. This methodology will allow the Generation IV International Forum (GIF) Experts Group to give an overview to policy makers and system development teams on the status of available economic estimates for each system and the relative status of the different systems with respect to Generation IV economic goals. Figure 10 displays the structure of the integrated nuclear energy economic model. The following discussion is based on this figure.

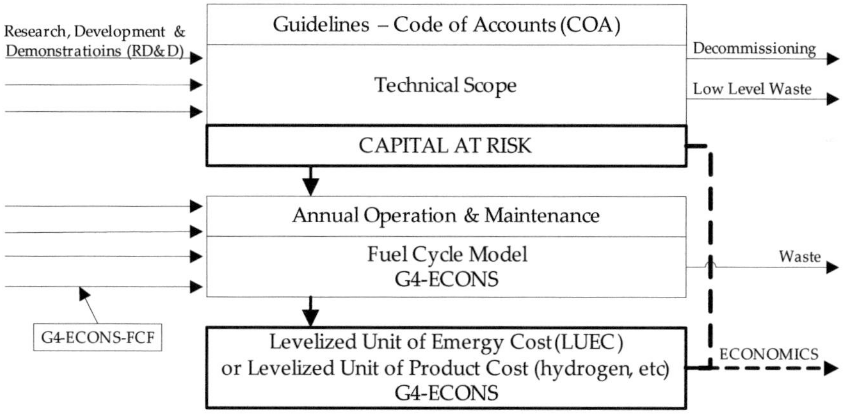

Figure 10. Structure of the integrated nuclear energy economic model [6].

The model is split in four parts: construction/production, fuel cycle, energy products, and modularization.

Cost estimates prepared by system design teams should report the overall direct and indirect costs for reactor system design and construction (base construction cost) and an estimate of the reactor annual operation and maintenance costs. The intent is that these costs be developed using the GIF COA described in [6], prepared by the methods outlined therein. The decision maker, however, needs more than just the overall costs in each life cycle category. Of particular interest are the cost per kilowatt of installed capacity and the cost of electricity generation (cost per kilowatt-hour) from such systems, including the contribution of capital and non-fuel operations.

Ref. [6] describes how interest during construction (IDC), contingencies, and other supplemental items are added to the base construction cost to obtain the total project capital cost. This total cost is amortized over the plant economic life so that the capital contribution to the levelized unit of energy cost (LUEC) can be calculated. Operation and maintenance (O&M) and decontamination and decommissioning (D&D) costs, along with electricity production information, yield the contributions of non-fuel costs to the overall cost of electricity. These algorithms have been derived from earlier Oak Ridge National Laboratory (ORNL) nuclear energy plant databases (NECDB) [46, 47], to calculate these costs.

Fuel cycle materials and services are purchased separately by the utility or the fuel subcontractor. For fuel cycles commercially deployed, there are mature industries worldwide that can provide these materials and services. Markets are competitive, and prices are driven by supply and demand. The fuel cycle model requires as inputs the amount of fuel needed for the initial core and subsequent equilibrium cores, along with the fissile enrichment of the uranium or plutonium, and, for uranium, the transaction tails assay assumed by the enrichment service provider. The EMWG model uses algorithms similar to those described in [48] to estimate the overall cost for each step and ultimately the unit cost contribution of fuel to electricity cost. Background material on the economic aspects of fuel cycle choices including information on nuclear materials and fuel cycle service unit costs for conventional reactor types that use commercially available fuels can be found in NEA reports [48, 49]. These documents include cost data on fuel reprocessing and high-level waste disposal for closed fuel cycles and spent fuel disposal for the once-through option.

Innovative fuel cycles or fuel cycle steps for which no industrial scale or commercial facilities currently exist, especially for fuel fabrication, reprocessing, and waste disposal are also addressed. For example, the Very-High-Temperature Reactor system will require high-temperature particle fuel and the SFR (Sodium-Cooled Fast Reactor) system might require innovative pyrometallurgical and pyrochemical facilities for fuel fabrication, reprocessing, and re-fabrication. For such systems, price data for fuel cycle services generally are not readily available. Therefore, a unit cost of fuel cycle services, such as $/kgHM (heavy metal) for fuel fabrication, should be calculated using a methodology similar to that used for LUEC calculation for the reactor system. The design team must supply data on the design and construction costs for the facilities, along with an estimate of their annual production rates and operation costs. Algorithms discussed in [6] can produce rough approximations of the unit costs.

The heat generated by some Generation IV systems has the potential for uses other than electricity generation, such as the production of hydrogen by thermal cracking of steam. There are also possible co-production models where the heat is used for both electricity production and process heat applications. The energy products model deals with these issues and is also discussed.

Cost issues and possible economic benefits that might result from modularization or factory production of all or part of a reactor system are also discussed [6].

The aim is to furnish a standardized cost estimating protocol to provide decision makers with a credible basis to assess, compare, and eventually select future nuclear energy systems taking into account a robust evaluation of their economic viability. To provide a credible, consistent basis for the estimated costs, early estimates of the evolving design concepts are expected to be based on conventional construction experience of built plants. This limitation is desirable from a consistency point of view because it can provide a reasonable starting point for consistent economic evaluation of different reactor concepts.

Of particular interest in this sense is the work by Hejzlar et al [50], related to the Gas Cooled Fast Reactor (GFR). They discuss challenges posed by the GFR when striving for the achievement of balance among the Generation IV goals. According to Carelli et al [51], the nuclear option has to face not only the public opinion sensibility, mainly related to plant safety and waste disposal issues, but also the economic evaluation from investors and utilities, particularly careful on that energy source and in deregulated markets. Smaller size nuclear reactors can represent a viable solution especially for developing countries, or countries with not-highly-infrastructured and interconnected grids, or even for developed countries when limitation on capital at risk applies. A description of Small-Medium size Reactor (SMR) economic features is presented, in a comparison with the state-of-the-art Large size Reactors. A preliminary evaluation of the capital and operation and maintenance (O&M) costs shows that the negative effects of the economies of scale can be balanced by the integral and modular design strategy of SMRs.

6. Proliferation resistance and physical protection [7]

Technical and institutional characteristics of Generation IV systems are used to evaluate system response and determine its resistance against proliferation threats and robustness against sabotage and terrorism. System response outcomes are expressed in terms of a set of measures.

The methodology is organized to permit evaluations to be performed at the earliest stages of system design and to become more detailed and more representative as design evolves. Uncertainty of results is incorporated into the evaluation.

The results are intended for three types of users: system designers, program policy makers, and external stakeholders. Program policy makers will be more likely to be interested in the high-level results that discriminate among choices, while system designers and safeguards experts will be more interested in results that directly relate to design options that will improve their performance (e.g., safeguards by design).

The proliferation resistance and physical protection Working Group has based its specification of the evaluation methodology scope on the definition of the Generation IV proliferation resistance and physical protection goal. The Generation IV Technology Roadmap (DOE,

2002b) [52] formally defined the following proliferation resistance and physical protection goal for future nuclear energy systems:

Generation IV nuclear energy systems will increase the assurance that they are a very unattractive and the least desirable route for diversion or theft of weapons-usable materials, and provide increased physical protection against acts of terrorism.

The definition of proliferation resistance adopted by the Working Group agrees with the definition established at the international workshop sponsored by the International Atomic Energy Agency (IAEA, 2002b)[53].

Formal definitions of proliferation resistance and physical protection have been established as presented next.

Proliferation resistance is that characteristic of a nuclear energy system that prevents the diversion or undeclared production of nuclear material and the misuse of technology by the host state seeking to acquire nuclear weapons or other nuclear explosive devices.

Physical protection (robustness) is that characteristic of a nuclear energy system that prevents the theft of materials suitable for nuclear explosives or radiation dispersal devices and the sabotage of facilities and transportation by sub-national entities or other non-host state adversaries.

The proliferation resistance and physical protection technology goal for Generation IV nuclear energy systems, when combined with the definitions of proliferation resistance and physical protection, is therefore as follows.

A Generation IV nuclear energy system is to be the least desirable route to proliferation by hindering the diversion of nuclear material from the system and hindering the misuse of the nuclear energy system and its technology in the production of nuclear weapons or other nuclear explosive devices.

A Generation IV nuclear energy system is to provide enhanced protection against theft of materials suitable for nuclear explosives or radiation dispersal devices and enhanced protection against sabotage of facilities and transportation. The proliferation resistance and physical protection methodology provides the means to evaluate Generation IV nuclear energy systems with respect to the following categories of proliferation resistance and physical protection threats:

Proliferation Resistance – Resistance to a host state's acquisition of nuclear weapons by:

- Concealed diversion of material from declared flows and inventories;
- Overt diversion of material from declared flows and inventories;
- Concealed material production or processing in declared facilities;
- Overt material production or processing in declared facilities;
- Concealed material production or processing by replication of declared equipment in clandestine facilities.

- Physical Protection (robustness)
- Theft of nuclear weapons-usable material or information from facilities or transportation;
- Theft of hazardous radioactive material from facilities or transportation for use in a dispersion weapon (radiation dispersal device or "dirty bomb");
- Sabotage at a nuclear facility or during transportation with the objective to release radioactive material to harm the public, damage facilities, or disrupt operations.

Figure 11 illustrates the most basic methodological approach. For a given system, analysts define a set of challenges, analyze system response to these challenges, and assess outcomes. The challenges to the nuclear energy system are the threats posed by potential proliferate states and by sub-national adversaries. The technical and institutional characteristics of Generation IV systems are used to evaluate the system response and determine its resistance to proliferation threats and robustness against sabotage and terrorism. The outcomes of the system response are expressed in terms of proliferation resistance and physical protection measures.

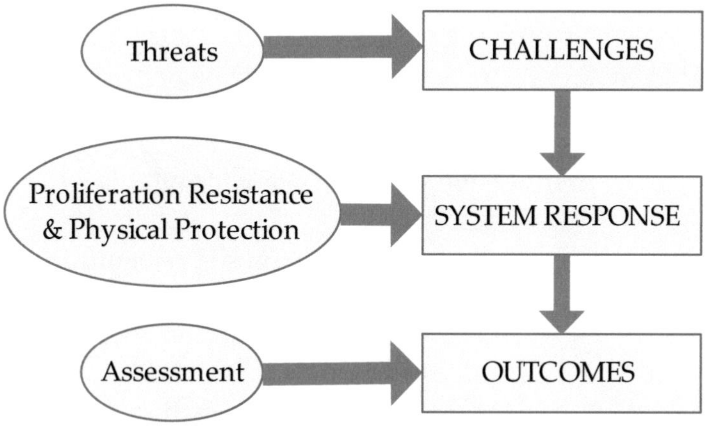

Figure 11. Basic framework for the proliferation resistance and physical protection evaluation methodology [7]

The evaluation methodology assumes that a nuclear energy system has been at least conceptualized or designed, including both intrinsic and extrinsic protective features of the system. Intrinsic features include the physical and engineering aspects of the system; extrinsic features include institutional aspects such as safeguards and external barriers. A major thrust of the proliferation resistance and physical protection evaluation is to elucidate the interactions between intrinsic and extrinsic features, study their interplay, and then guide the path toward an optimized design.

The structure for the proliferation resistance and physical protection evaluation can be applied to the entire fuel cycle or to portions of a nuclear energy system. The methodology is organized as a progressive approach to allow evaluations to become more detailed and more representative as system design evolves. Proliferation resistance and physical protection evaluations should be performed at the earliest stages of design when flow diagrams are first developed in order to systematically integrate proliferation resistance and physical protection robustness into the designs of Generation IV nuclear energy systems along with the other high-level technology goals of sustainability, safety and reliability, and economics. This approach provides early, useful feedback to designers, program policy makers, and external stakeholders from basic process selection (e.g., recycling process and type of fuel), to detailed layout of equipment and structures, to facility demonstration testing.

Figure 12 provides an expanded outline of the methodological approach. The first step is threat definition. For both proliferation resistance and physical protection, the threat definition describes the challenges that the system may face and includes characteristics of both the actor and the actor's strategy. For proliferation resistance, the actor is the host state for the nuclear energy system, and the threat definition includes both the proliferation objectives and the capabilities and strategy of the host state. For physical protection threats, the actor is a sub-national group or other non-host state adversary. The physical protection actors' characteristics are defined by their objective, which may be either theft or sabotage, and their capabilities and strategies.

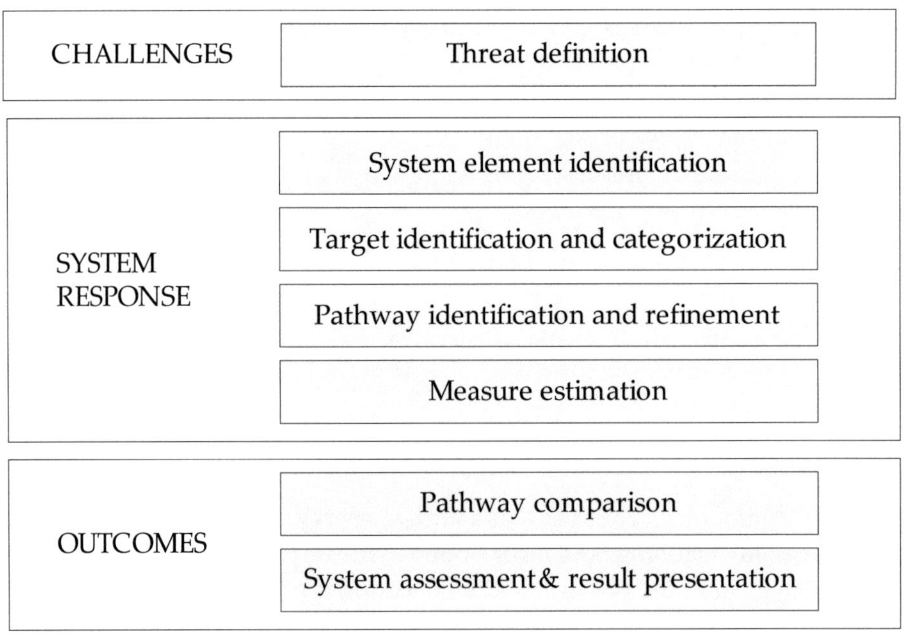

Figure 12. Framework for the proliferation resistance and physical protection evaluation methodology [7]

The proliferation resistance and physical protection methodology does not determine the probability that a given threat might or might not occur. Such evaluations may come from national threat evaluation organizations. The proliferation resistance and physical protection evaluation is based on design features of facilities as well as institutional considerations. Therefore, the selection of what potential threats to include is performed at the beginning of

http://dx.doi.org/10.5772/53140

a proliferation resistance and physical protection evaluation, preferably with input from a peer review group organized in coordination with the evaluation sponsors. The uncertainty in the system response to a given threat is then evaluated independently of the probability that the system would ever actually be challenged by the threat. In other words, proliferation resistance and physical protection evaluations are challenge dependent.

The detail with which threats can and should be defined depends on the level of detail of information available about the nuclear energy system design. In the earliest stages of conceptual design, where detailed information is likely limited, relatively stylized but reasonable threats must be selected. Conversely, when design has progressed to the point of actual construction, detailed and specific characterization of potential threats becomes possible.

When threats have been sufficiently detailed for the particular evaluation, analysts assess system response, which has four components:

1. System Element Identification. The nuclear energy system is decomposed into smaller elements or subsystems at a level amenable to further analysis. The elements can comprise a facility (in the systems engineering sense), part of a facility, a collection of facilities, or a transportation system within the identified nuclear energy system where acquisition (diversion) or processing (proliferation resistance) or theft/sabotage (physical protection) could take place.

2. Target Identification and Categorization. Target identification is conducted by systematically examining the nuclear energy system for the role that materials, equipment, and processes in each element could play in each of the strategies identified in the threat definition. Proliferation resistance targets are nuclear material, equipment, and processes to be protected from threats of diversion and misuse. Physical protection targets are nuclear material, equipment, or information to be protected from threats of theft and sabotage. Targets are categorized to create representative or bounding sets for further analysis.

3. Pathway Identification and Refinement. Pathways are potential sequences of events and actions followed by the actor to achieve objectives. For each target, individual pathways are divided into segments through a systematic process, and analyzed at a high level. Segments are then connected into full pathways and analyzed in detail. Selection of appropriate pathways will depend on the scenarios themselves, the state of design information, the quality and applicability of available information, and the analyst's preferences.

4. Estimation of Measures. The results of the system response are expressed in terms of proliferation resistance and physical protection measures. Measures are the high-level characteristics of a pathway that affect the likely decisions and actions of an actor and therefore are used to evaluate the actor's likely behavior and outcomes. For each measure, the results for each pathway segment are aggregated as appropriate to compare pathways and assess the system so that significant pathways can be identified and highlighted for further assessment and decision making

The measures for proliferation resistance are:

- Proliferation Technical Difficulty – The inherent difficulty, arising from the need for technical sophistication and materials handling capabilities, required to overcome the multiple barriers to proliferation.
- Proliferation Cost – The economic and staffing investment required to overcome the multiple technical barriers to proliferation, including the use of existing or new facilities.
- Proliferation Time – The minimum time required to overcome the multiple barriers to proliferation (i.e., the total time planned by the Host State for the project).
- Fissile Material Type – A categorization of material based on the degree to which its characteristics affect its utility for use in nuclear explosives.
- Detection Probability – The cumulative probability of detecting a proliferation segment or pathway.
- Detection Resource Efficiency – The efficiency in the use of staffing, equipment, and funding to apply international safeguards to a nuclear energy system.
- The measures for physical protection are:
- Probability of Adversary Success – The probability that an adversary will successfully complete the actions described by a pathway and generate a consequence.
- Consequences – The effects resulting from the successful completion of the adversary's action described by a pathway.
- Physical Protection Resources – The staffing, capabilities, and costs required to provide PP, such as background screening, detection, interruption, and neutralization, and the sensitivity of these resources to changes in the threat sophistication and capability.

By considering these measures, system designers can identify design options that will improve system proliferation resistance and physical protection performance. For example, designers can reduce or eliminate active safety equipment that requires frequent operator intervention.

The final steps in proliferation resistance and physical protection evaluations are to integrate the findings of the analysis and to interpret the results. Evaluation results should include best estimates for numerical and linguistic descriptors that characterize the results, distributions reflecting the uncertainty associated with those estimates, and appropriate displays to communicate uncertainties.

Further literature on the subject of this section comprises, for example, the paper by Bari et al [54], where the general methodology for proliferation resistance and physical protection is discussed and applied. An application to an example sodium fast reactor is discussed in terms of elicitation in [55]. An application concerning nuclear fuel cycles is discussed in [56]. A practical tool to assess proliferation resistance of nuclear energy systems is discussed in [57].

Proliferation resistance is discussed in [58] concerning the mobile fuel reactor, which is not one of Generation IV concepts in discussion, but interesting insights may be found therein. Penner et al [59] discuss new reactor designs and construction where a Generation IV design perspective is presented and proliferation resistance is set as an issue of utmost importance. Lennox et al [60] discuss the plutonium issue from the point of view of Generation IV designs.

The molten salt reactor is focused on proliferation issues in Ref. [61] also. Here proliferation considerations are discussed in face of the reactor operation because without the removal of plutonium and uranium from the fuel mixture, the reactivity starts to fluctuate and needs compensation. Uri and Engel [62] discuss non-proliferation attributes of molten salt reactors, as, for example, less plutonium stocks.

Myths of the proliferation resistance approach are focused in Ref. [63].

7. Conclusions

The discussion presented in this chapter clearly shows that much effort has been developed on a worldwide basis for conceiving the reactors that will be in use around 2030. Due to its beginning as a military weapon, nuclear energy is not an energy option that is accepted without strong restrictions in many countries. This resistance has been particularly aggravated immediately after the accidents in Three Mile Island, Chernobyl and the recent one in Fukushima. Many lessons learned from these accidents have been employed in the conception of Generation IV reactors, as many of them had already been implemented in Generation III reactors, like Westinghouse's AP1000.

The Generation IV philosophy for reactor development brings into light concerns about sustainability, economic viability, safety, and security translated into the concepts of proliferation resistance and physical protection. These are new concepts that are playing the dominante roles in reactor development for the future. It is also noteworthy that safety analysis is to be stressed, mainly the application of the risk-informed decision making approach for licensing purposes. Certainly, this integrated phylosophy will do much for turning nuclear energy systems much more acceptable by the final users.

Author details

Juliana P. Duarte[1], José de Jesús Rivero Oliva[1] and Paulo Fernando F. Frutuoso e Melo[2]

1 Department of Nuclear Engineering, Polytechnic School, Federal University of Rio de Janeiro, Brazil

2 Nuclear Engineering Graduate Program, COPPE, Federal University of Rio de Janeiro, Brazil

References

[1] IAEA - http://www.iaea.org/newscenter/news/2004/obninsk.html (accessed 03 September 2012).

[2] IAEA - Reference Data Series No 2. Nuclear Power Reactors in the World. Vienna; 2012.

[3] Generation IV International Forum. http://www.gen-4.org/ (accessed on 20 May 2012).

[4] IAEA - Nuclear Energy Series No. NP-T-1.8. Nuclear Energy Development in the 21st Century: Global Scenarios and Regional Trends. Vienna; 2010.

[5] GIF - Risk and Safety Working Group (RSWG). An Integrated Safety Assessment (ISAM) for Generation IV Nuclear Systems, June de 2011.

[6] GIF - The Economic Modeling Working Group Of the Generation IV International Forum. Cost Estimating Guidelines for Generation IV Nuclear Energy Systems. GIF/EMWG/2007/004. Revision 4.2, 26 September 2007.

[7] GIF - The Proliferation Resistance and Physical Protection Evaluation Methodology Working Group. Evaluation Methodology for Proliferation Resistance and Physical Protection of Generation IV Nuclear Energy Systems, September 2011.

[8] O'Conner TJ. Gas Reactors - A Review of the Past, an Overview of the Present and a View of the Future. Proceedings of GIF Symposium, 9-10 September 2009, Paris, France. 2009.

[9] Ueta S, Aihara J, Sawa K, Yasuda A, Honda M and Furihata M. Development of high temperature gas-cooled reactor (HTGR) fuel in Japan. Progress in Nuclear Energy 2011; 53 788-793.

[10] AIEA - Status report 70 - Pebble Bed Modular Reactor (PBMR). Available from http://aris.iaea.org/ARIS/reactors.cgi?requested_doc=report&doc_id=70&type_of_output=html (accessed 03 September 2012).

[11] Allen T, Busby J, Meyer M, Petti D. Materials challenges for nuclear systems. Materials Today. 2010; 13(12) 14-23.

[12] GIF. Annual Report 2010. 2010. Available from http://www.gen-4.org/PDFs/GIF_2010_ annual_report.pdf (accessed 20 May 2012).

[13] Brossard P, Abram TJ, Petti D, Lee YW and Fütterer MA. The VHTR Fuel and Fuel Cycle Project: Status of Ongoing Research and Results. Proceedings of GIF Symposium, 9-10 September 2009, Paris, France. 2009.

[14] Anzieu P, Stainsby R and Mikityuk K. Gas-cooled Fast Reactor (GFR): Overview and Perspectives. Proceedings of GIF Symposium, 9-10 September 2009, Paris, France. 2009.

[15] Stainsby R, Peers K, Mitchell C, Poette C, Mikityuk K, Somers J. Gas cooled fast reactor research in Europe. Nuclear Engineering and Design, 2011; 241 3481-3489.

[16] Martín-del-Campo C, Reyes-Ramírez R, François J-L, Reinking-Cejudo A. Contributions to the neutronic analysis of a gas-cooled fast reactor. Annals of Nuclear Energy 2011; 38 1406-1411.

[17] AIEA - Status report 71 - Japanese Supercritical Water-Cooled Reactor (JSCWR). Available from http://aris.iaea.org/ARIS/download.cgi?requested_doc=report&doc_id =71&type_of_output=pdf (accessed on 20 August 2012).

[18] Idaho National Laboratory. http://nuclear.inl.gov/deliverables/docs/appendix_2.pdf (accessed on May 2012)

[19] Khartabil H. SCWR: Overview. Proceedings of GIF Symposium, 9-10 September 2009, Paris, France. 2009.

[20] Ichimiya M, Singh BP, Rouault J, Hahn D, Glatz JP, Yang H. Overview of R&D Activities for the Development of a Generation IV Sodium-Cooled Fast Rector System. Proceedings of GIF Symposium, 9-10 September 2009, Paris, France. 2009.

[21] Chang YI, Grandy C, Lo Pinto P, Konomura M. Small Modular Fast Reactor Design Description. 2005. Available on http://www.ne.anl.gov/eda/Small_Modular_Fast_Reactor_ANL_SMFR_1.pdf (accessed on 25 August 2012).

[22] Cinotti L, Smith C F, Sekimoto H. Lead-cooled Fast Reactor (LFR): Overview and Perspectives. Proceedings of GIF Symposium, 9-10 September 2009, Paris, France. 2009.

[23] Rosenthal MW, Kasten PR, Briggs RB. Molten-Salt Reactors - History, Status and Potential. Oak Ridge National Laboratory, October 10, 1969.

[24] Delpech S, Merle-Lucotte E, Heuer D, Allibert M, Ghetta V, Le-Brun C, Doligez X, Picard G. Reactor physic and reprocessing scheme for innovative molten salt reactor system. Journal of Fluorine Chemistry 2009; 130 11-17.

[25] Renault C, Hron M, Konings R, Holcomb DE. The Molten Salt Reactor (MSR) in Generation IV: Overview and Perspectives. Proceedings of GIF Symposium, 9-10 September 2009, Paris, France. 2009.

[26] DOE. A technology roadmap for Generation-IV nuclear energy systems. GIF-002-00, U.S. DOE Nuclear Energy Research Advisory Committee and the Generation IV International Forum, 2002.

[27] Modarres M, Kim IS. Deterministic and Probabilistic Safety Analysis. Handbook of Nuclear Engineering. Cacucci, D. G.(Ed.). Volume III. Reactor Analysis. Springer; 2010, p1742-1812.

[28] D'Auria, F, Giannotti W, Cherubini M. Integrated Approach for Actual Safety Analysis. Nuclear Power – Operation, Safety and Environment. Tsevetkov, P. (Ed.) InTech; 2011, p29-46.

[29] U.S. NRC. Feasibility Study for a Risk-Informed and Performance-Based Regulatory Structure for Future Plant Licensing. Main Report. NUREG-1860. Vol. 1. 2007.

[30] Schultz, R. R. et al. Next Generation Nuclear Plant – Design Methods Development and Validation Research and Development Program Plan. Idaho National Engineering and Environmental Laboratory. INEEL/EXT-04-02293, 2004.

[31] Schultz, R. R. et al. Next Generation Nuclear Plant Methods Research and Development Technical Program Plan. Idaho National Laboratory. INL/EXT-06-11804, 2008.

[32] Boyack B. et al. Quantifying Reactor Safety Margins, Application of Code Scaling, Applicability, and Uncertainty Evaluation Methodology to Large-Break, Loss-of-Coolant Accident, NUREG/CR-5249, 1989.

[33] Vilim R. B. et al. Initial VHTR Accident Scenario Classification: Models and Data. Generation IV Nuclear Energy System Initiative. Argonne National Laboratory. ANL-GenIV-057, 2005.

[34] Diamond, D. J. Experience Using Phenomena Identification and Ranking Technique (PIRT) for Nuclear Analysis. Brookhaven National Laboratory. BNL-76750-2006-CP, 2006.

[35] Vilim R. B. et al. Prioritization of VHTR System Modeling Needs Based on Phenomena Identification, Ranking and Sensitivity Studies. ANL-GenIV-071, 2006.

[36] Best Estimate Safety Analysis for Nuclear Power Plants: Uncertainty Evaluation. Safety Report Series No. 52. IAEA. 2008.

[37] Proposal for a Technology-Neutral Safety Approach for New Reactors Design. IAEA-TECDOC-1570. 2007

[38] Considerations in the Development of Safety Requirements for Innovative Reactors: Application to Modular High Temperature Gas Cooled Reactors. IAEA-TECDOC-1566. 2003.

[39] IAEA - Safety Report Series No 46. Assessment of Defense in Depth for Nuclear Power Plants. Vienna, Austria, 2005.

[40] Generation IV International Forum. Basis for the Safety Approach for Design & Assessment of Generation IV Nuclear Systems. RSWG Report. GIF/RSWG/2007/002, 2008.

[41] GIF. Annual Report. 2008. Available from http://www.gen-4.org/PDFs/GIF_2008_Annual_Report.pdf (accessed on 22 July 2012).

[42] Gen IV International Forum. Annual Report, 2009. Available from http://www.gen-4.org/PDFs/GIF-2009-Annual-Report.pdf (accessed on 22 July 2012).

[43] Leahy TJ, Fiorini GL. Risk and Safety Working Group: Perspectives, accomplishments and activities. GIF Symposium, Paris, 9-10 September, 2009.

[44] GIF. Annual Report, 2011. Available from http://www.gen-4.org/GIF/About/index.htm (accessed on 22 July 2012).

[45] GIF - Economic Modeling Working Group. 2006. Software package including Users" Manual and Users" Guide posted on the GIF restricted web site.

[46] Delene JG, Hudson CR. Cost Estimate Guidelines for Advanced Nuclear Power Technologies, ORNL/TM-10071/R3, Oak Ridge National Laboratory, Oak Ridge, TN, US. 1993.

[47] ORNL. Nuclear Energy Cost Data Base: A Reference Data Base for Nuclear and Coal-fired Power plant Power Generation Cost Analysis (NECDB), DOE/NE-0095, prepared by Oak Ridge National Laboratory, Oak Ridge, TN, U.S. 1988.

[48] NEA. Economics of the Nuclear Fuel Cycle, OECD Nuclear Energy Agency (OECD/NEA), Paris, France. 1994. Available from www.nea.fr.

[49] NEA. Trends in the Nuclear Fuel Cycle, OECD Nuclear Energy Agency (OECD/NEA), Paris, France. 2000.

[50] Hejzlar P, Pope MJ, Williams WC, Driscoll MJ. Gas Cooled Fast Reactor for Generation IV Service, Progress in Nuclear Energy, 2005; 47 271-282.

[51] Carelli MD, Garrone P, Locatelli G, Mancini M, Mycoff C, Trucco P, Ricotti ME. Economic features of integral, modular, small-to-medium size reactors, Progress in Nuclear Energy, 2010; 52 403–414

[52] DOE (U.S. Department of Energy), Nuclear Energy Research Advisory Committee and the Generation IV International Forum. Generation IV Roadmap: Viability and Performance Evaluation Methodology Report. GIF002-13, DOE Nuclear Energy Research Advisory Committee and the Generation IV International Forum, Washington. 2002.

[53] IAEA (International Atomic Energy Agency). Proliferation Resistance Fundamentals for Future Nuclear Energy Systems. IAEA Department of Safeguards, IAEA, Vienna, Austria. 2002.

[54] Bari RA, PPeterson IU Therios, JJ Whitlock. "Proliferation Resistance and Physical Protection Evaluation Methodology Development and Applications." Proceedings of the Generation IV International Forum Symposium, Paris, France, September 9-10, 2009.

[55] Budlong Sylvester KW, Ferguson CD, Garcia E, Jarvinen GD, Pilat JF, Tape JW. Report of an Elicitation on an Example Sodium Fast Reactor (ESFR) System in Support of the Proliferation Resistance and Physical Protection (PR&PP) Working Group. Los Alamos National Laboratory White Paper, Los Alamos, New Mexico, USA. 2008.

[56] Charlton WS, LeBouf RF, Gariazzo C, Ford DG, Beard C, Landsberger S, Whitaker M. Proliferation Resistance Assessment Methodology for Nuclear Fuel Cycles. Nucl. Tech. 2007; 157 143-156.

[57] Greneche, D. 2008. A Practical Tool to Assess the Proliferation Resistance of Nuclear Systems: The SAPRA Methodology. ESARDA Bulletin, No. 39, p. 45. Available from http://esarda2.jrc.it/db_proceeding/mfile/B_2008-039-07.pdf.

[58] Slessarev I. On nuclear power intrinsically protected against long-lived actinide wastes, weapons proliferation and heavy accidents. Annals of Nuclear Energy 2006; 33 325-334.

[59] Penner SS, Seiser R, Schultz KR. Steps toward passively safe, proliferation-resistant nuclear power. Progress in Energy and Combustion Science, 2008; 34 275-287.

[60] Lennox TA, Millington DN, Sunderland DN. Plutonium management and Generation IV systems. Progress in Nuclear Energy, 2007; 49 589-596.

[61] Vanderhaegen M, Janssens-Maenhout G. Considerations on safety and proliferation-resistant aspects for the MSBR design, Nuclear Engineering and Design, 2010; 240 482–488.

[62] Uri U, Engel JR.Non-proliferation attributes of molten salt reactors. Nuclear Engineering and Design, 2000; 201 327–334.

[63] Acton J. The Myth of Proliferation Resistant Technology. Bulletin of the Atomic Scientists 2009; 65(6) 49-59.

Section 2

Nuclear Reactors Technology Research Across the World

Chapter 8

Thermal Hydraulics Prediction Study for an Ultra High Temperature Reactor with Packed Sphere Fuels

Motoo Fumizawa

Additional information is available at the end of the chapter

http://dx.doi.org/10.5772/53371

1. Introduction

Nowadays, very high temperature gas-cooled reactor (VHTR) design studies and utilization studies are investigated energetically in the some international conference, especially in the ICONE20 conference (Allelein, 2012, Iwatsuki, 2012, Sato, 2012). The VHTR project so called GIF is developing the design study to establish 1,000 °C as a coolant outlet temperature and to realize the hydrogen production (GIF-002-00,2002, Shiozawa,2005), where GIF is the Generation IV International Forum. For a long time, a fundamental design study has been carried out in the field of the high temperature gas-cooled reactor i.e. HTGR (Fumizawa, 1989b, Fumizawa, 2007), which showed that a coolant outlet temperature was around 900 °C. The interest of HTGR is increasing in many countries as a promising energy future option. There are currently two research reactors of HTGR type that are being operated in Japan and China. The inherent safety of HTGR is due to the large heat capacity and negative temperature reactivity coefficient. The high temperature heat supply can achieve more effective utilization of nuclear energy. For example, high temperature heat supply can provide for hydrogen production, which is expected as alternative energy source for oil. Also, outstanding thermal efficiency will be achieved at about 900 °C with a Brayton-cycle gas turbine plant.

However, the highest outlet coolant temperature of 1316 °C had been achieved by UHTREX, in Los Alamos Scientific Laboratory at the end of 1960's (El-Wakil, 1982, Hoglund,1966). It was a small scale Ultra High Temperature Nuclear Reactor (UHTR). The coolant outlet temperature would be higher than 1000 °C in the UHTR. The UHTREX adopted the hollow rod type fuel; the highest fuel temperature was 1,582 °C, which indicated that the value was over the current design limit. According to the handy calculation, it was derived that the pebble type fuel was superior to the hollow type in the field of fuel surface heat transfer condition (Fumizawa, 1989a).

In the present study, the fuels have changed to the pebble type so called the porous media. In order to compare the present pebble bed reactor and UHTREX, a calculation based on HTGR-GT300 was carried out in the similar conditions with UHTREX i.e. the inlet coolant temperature of 871°C, system pressure of 3.45 MPa and power density of 1.3 w/cm^3. The main advantage of pebble bed reactor (PBR) is that high outlet coolant temperature can be achieved due to its large cooling surface and high heat transfer coefficient making possible to get high thermal efficiency. Besides, the fuel loading and discharging procedures are simplified; the PBR system makes it possible that the frequent load and discharge are easier than the other reactor system loaded block type fuel without reactor shutdown. This report presents thermal-hydraulic calculated results for a concept design PBR system of 300MWth of the modular HTGR-GT300 with the pebble types of fuel element as shown in Figure 1. A calculation for comparison with UHTREX have been carried out and presented as well.

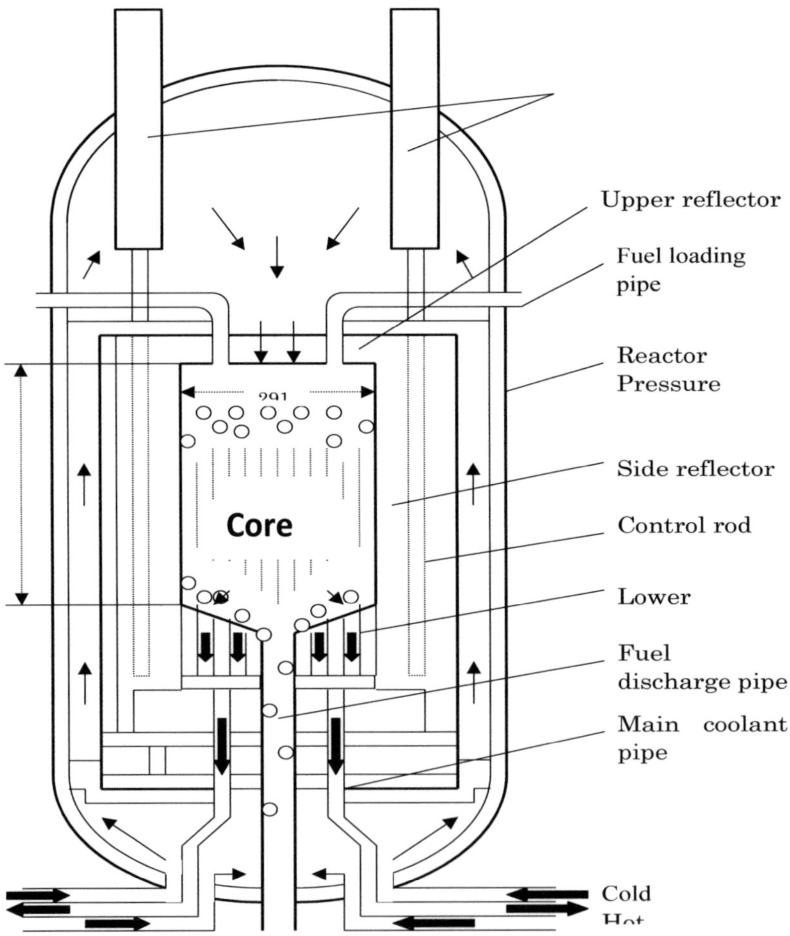

Figure 1. A concept of pebble bed reactor of HTGR –GT300

http://dx.doi.org/10.5772/53371

2. Porosity experiment

It is very important to measure the porosity of fuel in the reactor core. The porosity experiments carried out in both conditions of fully shaken down to the closest possibly packing and non-vibration, because the reactor is normally operated in the condition of non-vibration. The porosity was measured by packed iron balls in the graduated cylinder and the number of iron balls, which was counted with the accurate electronic balance of PM-6100 produced by Mettler Inst. Corp. As the result, the relation of porosity and diameter ratio was shown in Figure 2. Porosity ε changes from 0.4 to 0.7 in the present experiment. Porosity ε will vary from 0.4 to 0.55 at the large diameter ratio beyond 8. In the figure, it is clear that the value of porosity of non-vibration is larger than the case of fully vibrating condition (El-Wakil, 1982). The difference is very important to analyze the fuel temperature and thermal-hydraulic calculation

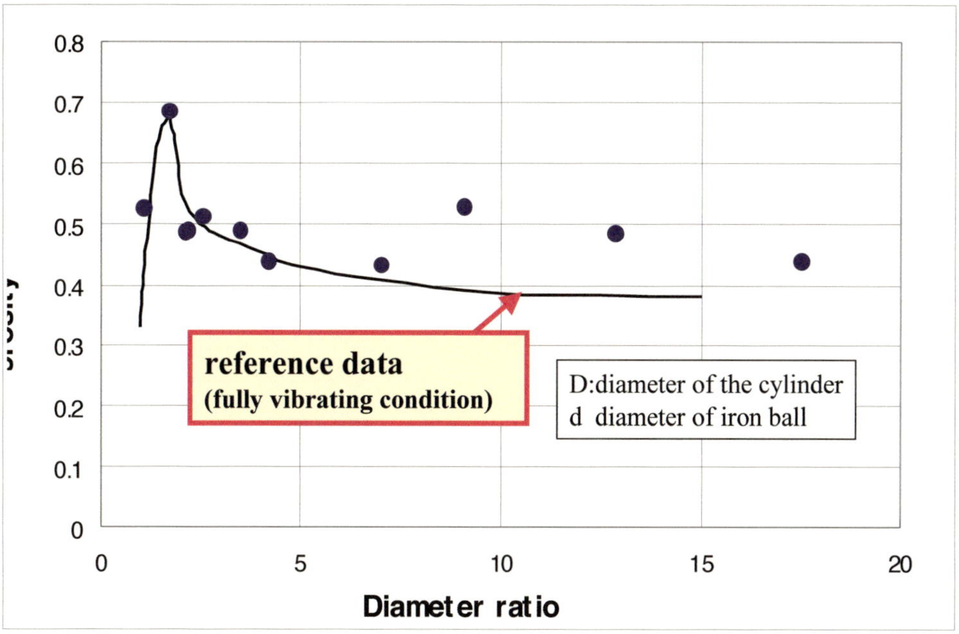

Figure 2. The relation of porosity and diameter ratio. Solid line was quoted from the reference (El-Wakil, 1982).

3. Reactor description

3.1. Concept of modular HTGR-GT300 and GT-600

A concept of pebble-bed type HTGR are shown in Figures 2 and 3 with the main nuclear and thermal-hydraulic specifications presented in Table 1. In the case that the thermal power is 300MW (GT-300), the average power density changes to 4.8 MW/m^3. The coolant gas enters from outer shell of primary coolant coaxial tube to the pressure vessel at temperature of 550°C and pressure of 6 MPa, follows the peripheral region of side reflectors up to the top and goes downward through the reactor active core. The outlet coolant goes out through the inner shell of primary coolant tube at temperature of 900°C. The cylindrical core is formed by the blocks of graphite reflector with the height of 9.4m and the diameter of 2.91m. There exist the holes in the reflector that some of them used for control rod channels and the others used for boron ball insertion in case of accident. In the case that the thermal power is 600MW (GT-600), the average power density changes to 9.6 MW/m^3.

3.2. Fuel element

The two types of pebble fuel elements, consisting of fuel and moderator, are shown in Figure 3. One is solid type where radius of inner graphite rco=0, and the other is shell type fuel element. The fuel compacts are a mixture of coated particles (Fumizawa, 1989a).

Thermal power (MW)	300 / 600
Coolant	Helium
Inlet coolant temperature (℃)	550
Outlet coolant temperature (℃)	900
Coolant Pressure (MPa)	6
Total coolant flow rate (kg/s)	172.1 / 344.2
Effective core coolant flow rate *Weff* (kg/s)	141.2 / 282.4
Core diameter (m)	2.91
Core height (m)	9.4
Uranium enrichment (wt %)	10
Average power density(MW/m^3)	4.8 / 9.6
Fuel type (for standard case)	6cm diameter pebble

Table 1. Major nuclear and thermal-hydraulic specification of GT-300 and GT-600

http://dx.doi.org/10.5772/53371

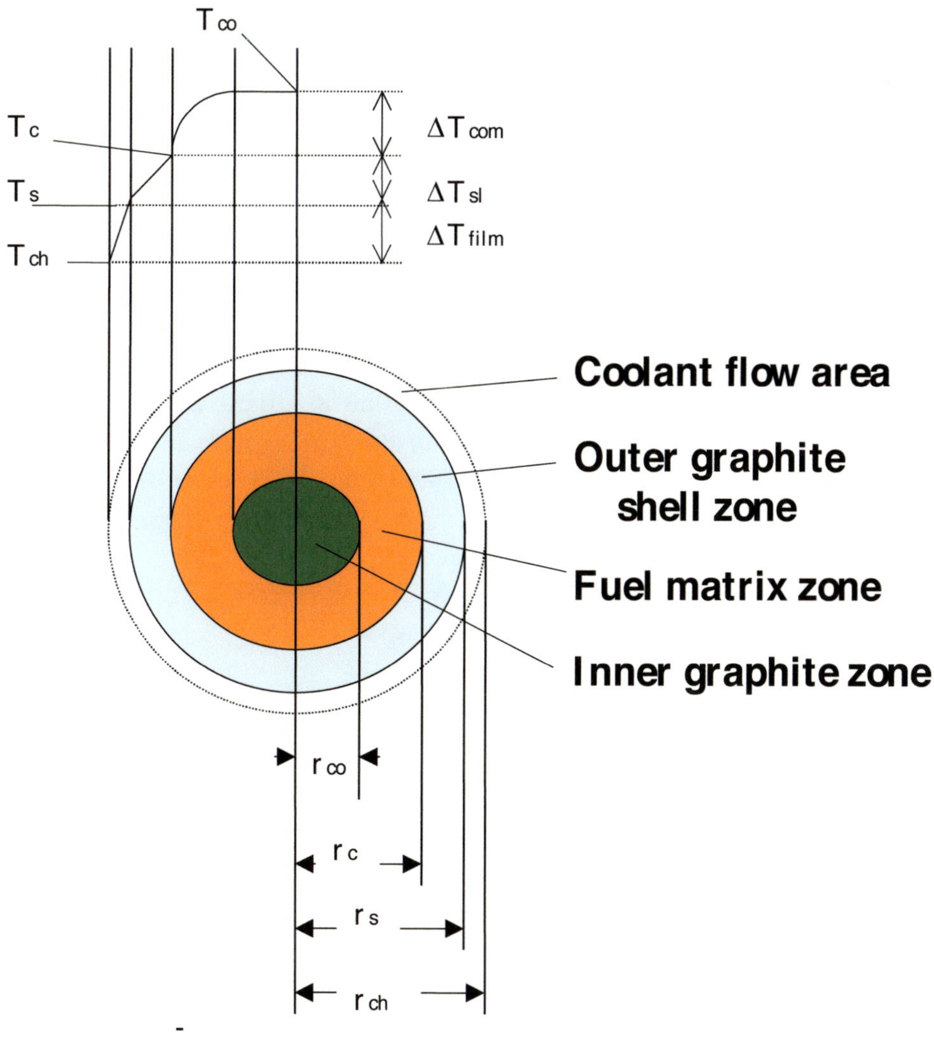

Figure 3. Relation of shell type fuel element and temperature difference, in the case that no inner graphite zone is called the solid type, i.e., rco=0

4. Thermlhydraulic analysis

4.1. PEBTEMP code

A one-dimensional thermal-hydraulic computer code was developed that was named PEBTEMP (Fumizawa, 2007) as shown in Figure 4. The code solves for the temperature of fuel element, coolant gas and core pressure drop using assumed power, power distribution, inlet and outlet temperature, the system pressure, fuel size and fuel type as input data.

The options for fuel type are of the pebble type; the multi holes blockk type and the pin-in-block type. The power distribution for cases of cosine and exponential is available. The users can calculate for the other distributions by preparing the input file.

The maximum fuel temperature will be calculated in PEBTEMP as follows:

$$T_f(z) = T_{in} + \Delta T_{cl}(z) + \Delta T_{film}(z) + \Delta T_{sl}(z) + \Delta T_{com} \tag{1}$$

Where $T_f(z)$: fuel temperature at the center of fuel element i.e. the maximum fuel temperature; ΔT_{cl}: gas temperature increment from inlet to height z; T_{in}: gas inlet temperature; $\Delta T_{film}(z)$: temperature difference between fuel element surface and coolant gas at z; $\Delta T_{sl}(z)$: temperature difference between fuel matrix surface and fuel element surface; $\Delta T_{com}(z)$: temperature difference between maximum fuel temperature and outer surface fuel temperature; q''' : power density; A_f: fuel element surface area; z: axial distance from the top of the core; g: coolant mass flow rate; C_p: coolant heat capacity.

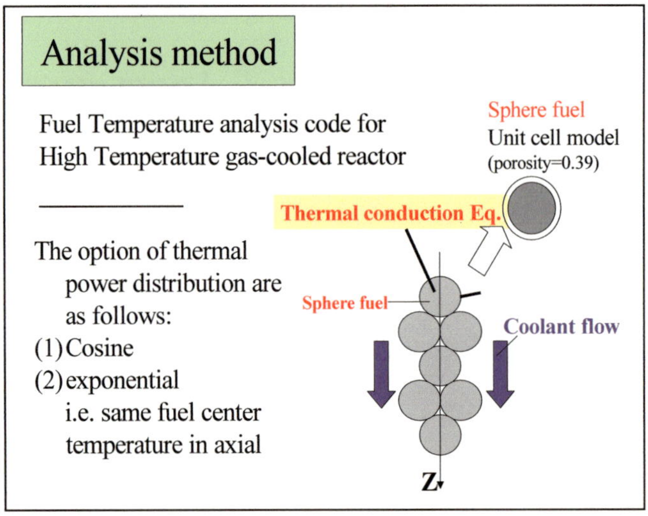

Figure 4. Analysis method of thermal-hydraulic computer code PEBTEMP

4.1.1. Temperature difference in the spherical fuel element

Figure 3 shows fuel configuration of the solid type and the shell type fuel element. In the solid type, ΔT_{com} is given as follows;

$$\begin{aligned} \Delta T_{com}(z) &= T_{co} - T_c \\ &= \frac{q'''(z) r_c^2}{6\lambda_c} \end{aligned} \tag{2}$$

http://dx.doi.org/10.5772/53371

In case of shell type fuel element, ΔT_{com} can be calculated by following expression;

$$\begin{aligned}\Delta T_{com}(z) &= T_{co} - T_c \\ &= \frac{q'''(z)}{6\lambda_c}\left(r_c^2 - 3r_{co}^2 + \frac{2r_{co}^3}{r_c} \right)\end{aligned} \tag{3}$$

4.1.2. Film temperature difference

The film temperature differences are calculated as follows;

$$\begin{aligned}\Delta T_{film} &= T_s - T_{ch} \\ &= \frac{q'''(z) r_c^3}{3 r_s^2 h}\end{aligned} \tag{4}$$

4.1.3. Heat transfer coefficient

Heat transfer coefficient h in Equation (4) is calculated using following correlation (Heil, 1969):

$$h = 0.68 \rho v_s C_p \,\mathrm{Re}^{-0.3}\,\mathrm{Pr}^{-0.66} \tag{5}$$

$$\mathrm{Re} = \frac{\rho v_s d}{(1-\varepsilon)\mu} \tag{6}$$

Where, ρ: coolant density; v_s: coolant velCity; Re: Reynolds number; Pr: Prandtl number; ε: porosity; d: fuel element diameter and μ: viscosity of fluid.

4.1.4. Pressure drop

Pressure drop through the core expresses by following correlation (Heil, 1969):

$$\Delta P = 6.986 \frac{(1-\varepsilon)^{n+1}}{\varepsilon^3} \mathrm{Re}_p^{-n} \rho v_s^2 \frac{H}{d} K + \Delta P_a \tag{7}$$

$$n = 0.22 \tag{8}$$

$$K = 1 - (1 + n + 3\frac{1-\varepsilon}{\varepsilon})0.26\frac{d}{R} \tag{9}$$

$$\mathrm{Re}_p = \frac{\rho v_s d}{\eta} \tag{10}$$

Where, H: core height; R: core radius and ΔP_a: acceleration pressure drop.

4.2. Effective flow rate calculation

As many blocks of graphite form the reflector, there exist gaps by which the coolant flow may pass through as shown in Figure 5. Actually, only one portion of coolant passes through the reactor core from the top to bottom. This portion is called effective flow rate and can be calculated iteratively in the code. It is well known that the pressure drop in the PBR core is higher than that of the hollow type or the multi-hole type fuel core. Considering the pressure drop, the empirical equation used in this code is as follows (Fumizawa, 1989a):

$$W_{eff} = 0.98 - 0.012\Delta P \tag{11}$$

Where, W_{eff}: effective coolant flow rate that has dimensionless value due to the normalization by the total coolant flow rate (kg/s).

ΔP: pressure drop through the core (kPa)

Figure 6 shows the flowchart of iterative calculation with effective coolant flow rate. The pressure drop effect for effective coolant flow rate in the PBR core is presented in the author's former paper (Fumizawa, 2007).

5. Calculation results

5.1. Calculation results for HTGR-GT300

Figure 7 shows the coolant pressure as a function of fuel diameter with different effective coolant flow rate from 0.5 – 1.0. This figure shows that the coolant pressure strongly depends on the effective coolant flow rate as well as fuel element diameter especially in case of the diameter smaller than 4cm. The pressure drop is inversely proportional with fuel element diameter.

Figure 8 shows the dependence of maximum fuel temperature on fuel diameter and leakage flow. The maximum fuel temperature of solid type and shell type fuel elements was calculated for different fuel diameter from 3.5 cm to 10 cm and the coolant flow leakage was taken into account. According to Eq. (9), the core adopted small diameter fuel element means high leakage coolant flows. In the viewpoint of the thermal-hydraulics, the diameter of solid type fuel around 5 cm has the lowest fuel temperature. The fuel temperature in case of shell type fuel is lower than that of solid type, and the thinner layer of fuel matrix the lower fuel temperature can be achieved.

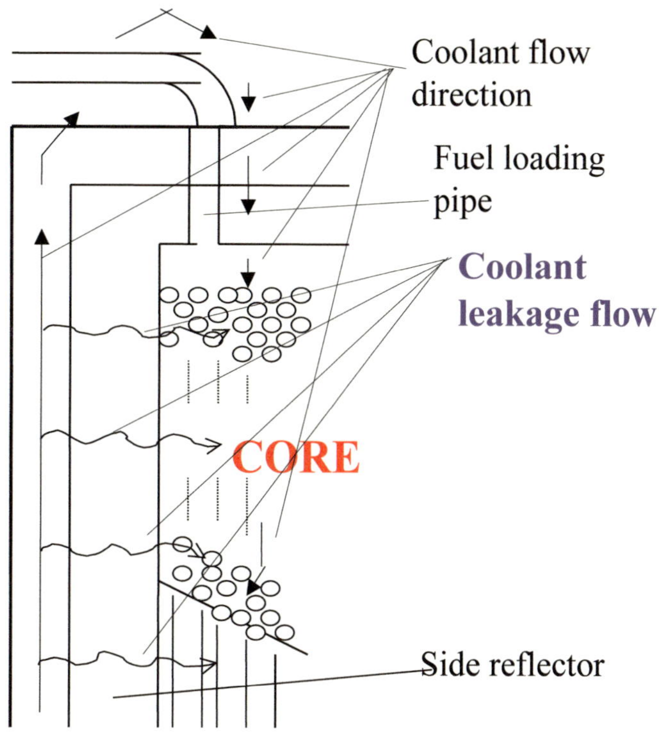

Figure 5. Coolant leakage flow concept

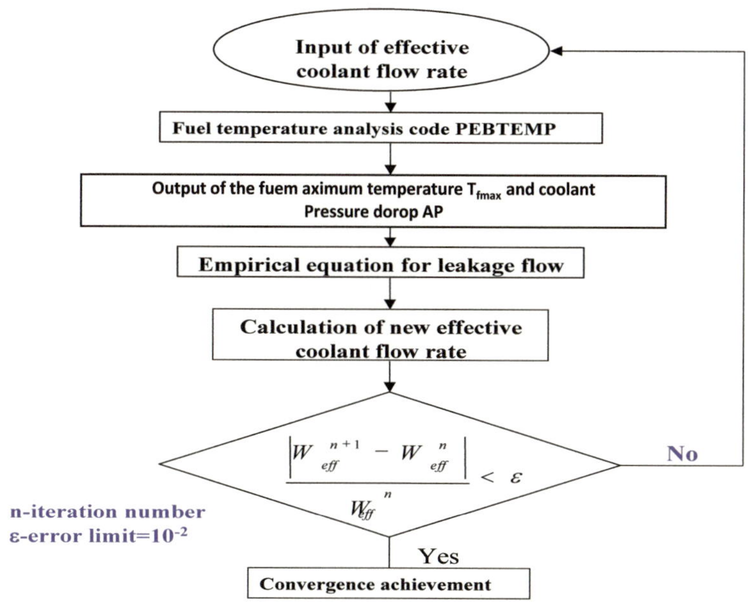

Figure 6. Flowchart of iterative calculation

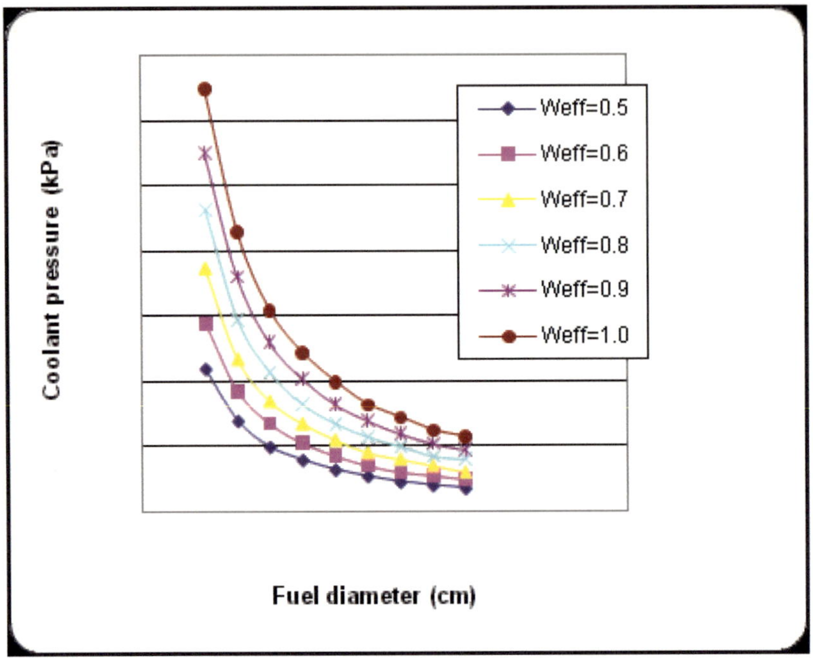

Figure 7. Coolant pressure as a function of fuel diameter for different effective coolant flow rate

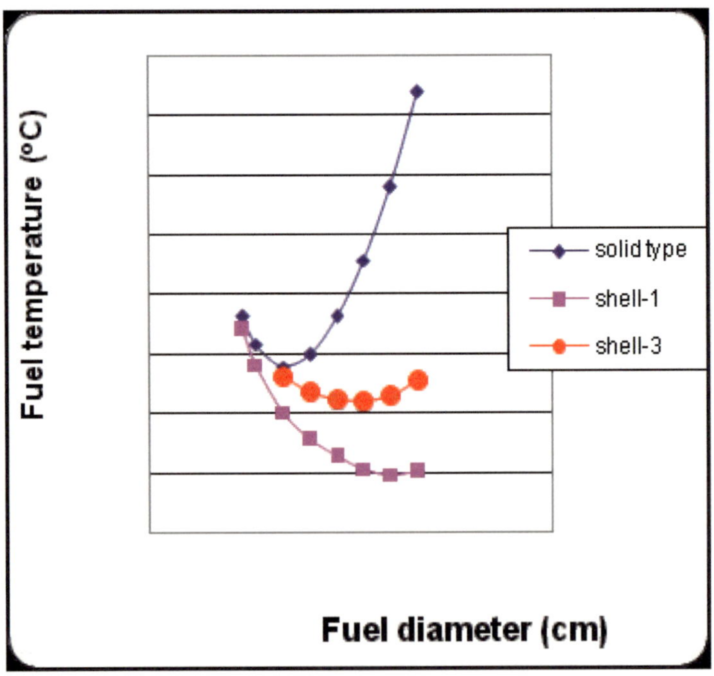

Figure 8. Maximum fuel temperature dependence on fuel diameter and leakage flow

http://dx.doi.org/10.5772/53371

5.2. Comparison between the fuel temperatures of both cores

The high power density effects are shown in Figures 9 and 10, which indicate the dependence of maximum fuel temperature on outlet coolant temperature for GT-300 and GT-600 with different porosity and W_{eff}=1. The maximum fuel temperature for GT-600 is 168 °C higher than that for GT-300 where the outlet coolant temperature is 900 °C and the porosity is 0.39. The maximum fuel temperature for GT-600 is 163 °C higher than that for GT-300 where the outlet coolant temperature is 1150 °C and the porosity is 0.39.

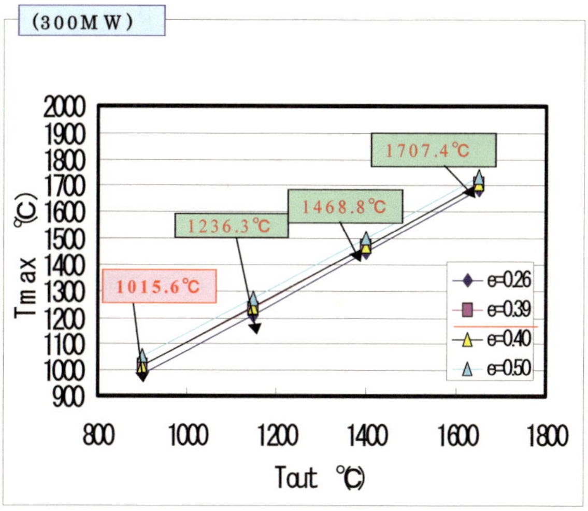

Figure 9. Dependence of maximum fuel temperature on outlet coolant temperature for GT-300 with different porosity and W_{eff}=1.

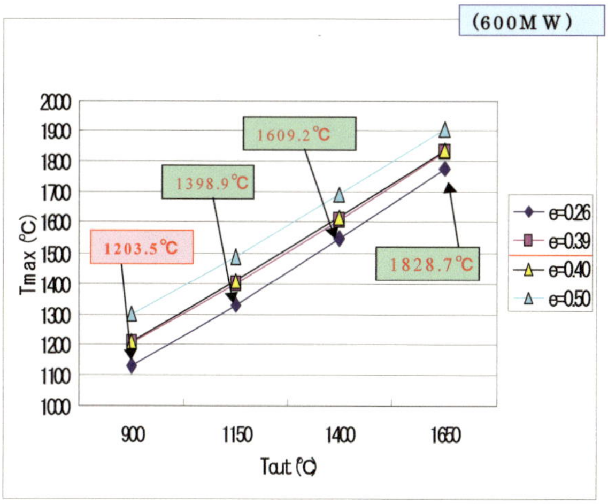

Figure 10. Dependence of maximum fuel temperature on outlet coolant temperature for GT-600 with different porosity and W_{eff}=1.

5.3. Calculation for a New Ultra High Temperature Reactor Experiment (NUHTREX)

The highest outlet coolant temperature has been achieved in case of UHTREX,i.e.,Ultra High Temperature Reactor Experiment (Fumizawa, 1989b) at Los Alamos Scientific Laboratory. UHTREX is the 3 MWth gas cooled reactor loaded hollow cylinder fuel element type in the hollow cylinder graphite bl Ck. The coolant (Helium) flows horizontally outward from inner hollow through fuel channel. The maximum outlet coolant temperature of 1316°C was recorded on June 24, 1969 (Fumizawa, 2000).

In order to compare the present PBR case with the UHTREX case, a calculation for a HTGR-GT300 was carried out using conditions similar to the UHTREX case, i.e. an inlet coolant temperature of 871°C, system pressure of 3.45 MPa and power density of 1.3 W/cm^3. The hot channel factor of 1.0, 1.1, 1.2, and 1.3 are chosen for the present calculation. The calculated results are presented in Figure 11. The results indicate that the fuel temperature of the present PBR case is much lower value compared to that of the UHTREX case, i.e. 1582 °C. Therefore, the present PBR system design will be named as NUHTREX i.e. New UHTREX, which will be one of the UHTR. Figure 12 shows the maximum fuel temperature in various core porosity and fuel diameter. High fuel temperature obtained from the high porosity and large fuel diameter, where W_{eff}=1.0, solid type fuel, standard case.

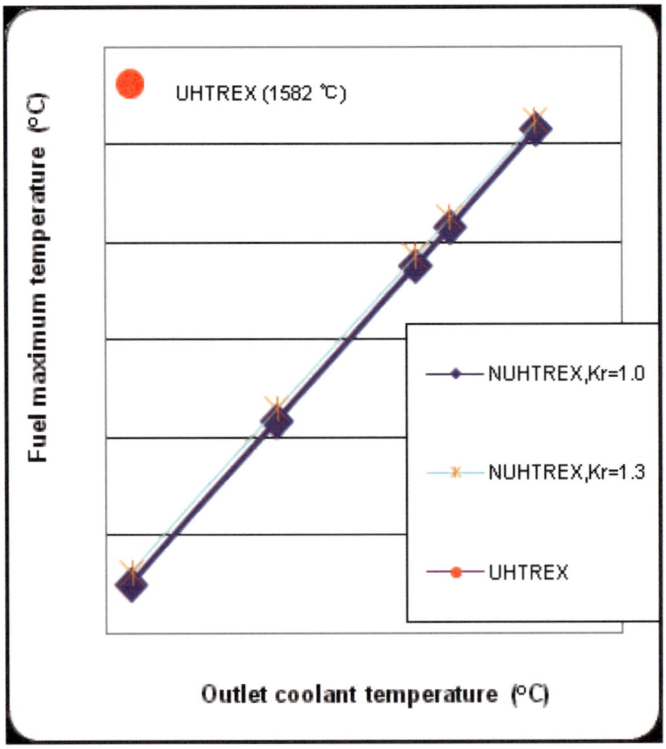

Figure 11. Dependence of maximum fuel temperature on outlet coolant temperature for NUHTREX with different hot channel factor (Kr) and UHTREX case.

http://dx.doi.org/10.5772/53371

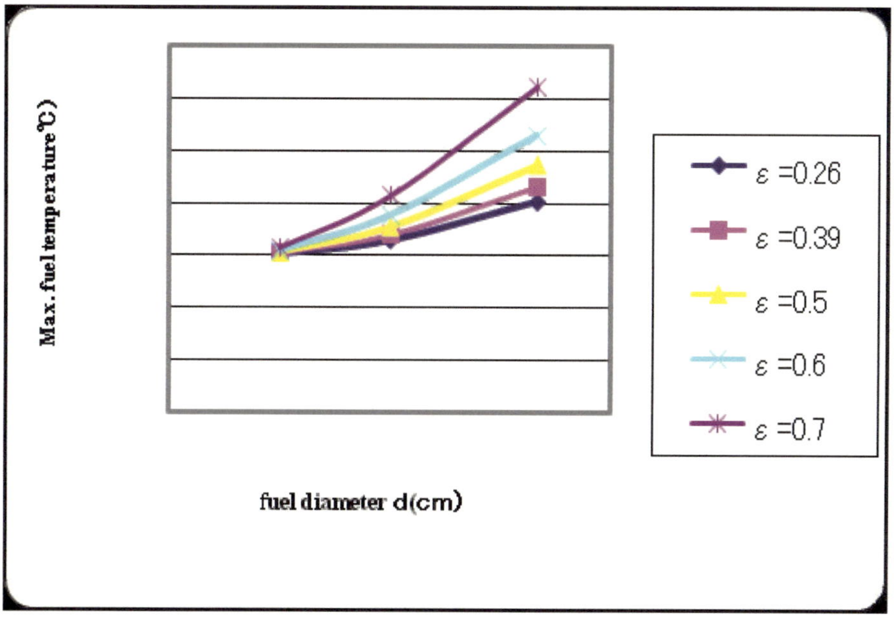

Figure 12. The maximum fuel temperature in various core porosity and fuel diameter

6. Conclusions

From present study, following items are summarized. It is clear that the value of porosity of non-vibration is larger than the case of fully vibrating condition. High fuel temperature obtained from the high porosity and large fuel diameter. In the viewpoint of the thermal-hydraulics, it is clarified that the diameter of solid type fuel around 5 cm has the lowest fuel temperature. The fuel temperature in case of shell type fuel is lower than that of solid type, because the thinner layer of fuel matrix the lower fuel temperature can be achieved. The maximum fuel temperature (Tmax) for GT-600 is higher than that for GT-300, for instance, Tmax for GT-600 is 168 °C higher than that for GT-300 where the outlet coolant temperature is 900 °C and the porosity is 0.39.

Nomenclature

A_f: fuel element surface area; (m^2)

C_p: coolant heat capacity; (J/kgK)

g: coolant mass flow rate; (kg/s)

H: core height; (m)

h : heat transfer coefficient; (W/m^{2}K)

q''': power density; (W/m^3)

R: core radius; (m)

Re: Reynolds number

$T_f(z)$: fuel temperature at the center of fuel element, i.e., the maximum fuel temperature; (°C)

T_{in}: gas inlet temperature; (°C)

T_{out}: gas outlet temperature; (°C)

W_{eff}: effective coolant flow rate, dimensionless value due to the normalization

z: axial distance from the top of the core; (m)

ΔP: pressure drop through the core (kPa)

ΔP_a: acceleration pressure drop; (kPa)

ΔT_{cl}: gas temperature increment from inlet to height z; (°C)

$\Delta T_{com}(z)$: temperature difference between maximum fuel temperature and outer surface fuel temperature; (°C)

$\Delta T_{film}(z)$: temperature difference between fuel element surface and coolant gas at z; (°C)

$\Delta T_{sl}(z)$: temperature difference between fuel matrix surface and fuel element surface; (°C)

Author details

Motoo Fumizawa

Shonan Institute of Technology, Japan

References

[1] Allelein, H. J, et al. Experimental investigation and analytical improvements for HTR pebble bed core",ICONE20-Power2012-54040, (2012). Allelein, 2012), 1-10.

[2] El-Wakil, M. M. Nuclear Energy Conversion, Thomas Y. Crowell Company Inc., USA ((1982). El-Wakil, 1982)

[3] Fumizawa, Motoo et alThe Conceptual Design of High Temperature Engineering Test Reactor Upgraded through Utilizing Pebble-in-block Fuel", JAERI-M 89-222 ((1989). Fumizawa, 1989a)

[4] Fumizawa, Motoo et alEffective Coolant Flow Rate of Flange Type Fuel Element for Very High Temperature Gas-Cooled Reactor ", J. of AESJ (1989). Fumizawa, 1989b), 31, 828-836.

[5] Fumizawa, Motoo et alPreliminary Study for Analysis of Gas-cooled Reactor with Sphere Fuel Element ", AESJ Spring MTG, I66 ((2000). Fumizawa, 2000)

[6] Fumizawa Motoo"Porosity effect in the core thermal hydraulics for an Ultra High Temperature Gas-cooled Nuclear Reactor", Memories of Shonan Institute of Technology, (2007). Fumizawa, 2007), 41(1), 1-8.

[7] GIF- A Technology Roadmap for Generation IV Nuclear Energy Systems, Generation IV International Forum ((2002). http://gif.inel.gov/roadmap/GIF-002-00,2002)

[8] Heil, J, et al. Zusammenstellung von Gleichung fuer den Druckverlust und den Warmeubergang fuer Stroemungen in Kugelschuttungen und prismatischen kanalen", IRE/I-20 ((1969). Heil, 1969)

[9] Hoglund, B. M. UHTREX Operation Near", Power Reactor Technology ((1966). Hoglund,1966), 9, 1.

[10] Iwatsuki, J, et al. ThermoChemical hydrogen production is process",ICONE20-Power2012-54095, (2012). (Iwatsuki, 2012), 1-3.

[11] Sato, H, et al. Control strategies for VHTR gas-turbine system with dry cooling",ICONE20-Power2012-54351, (2012). Sato, 2012), 1-8.

[12] Shiozawa,Shusaku et alThe HTTR Project as the World Leader of HTGR Research and Development", J. of AESJ (2005). Shiozawa,2005), 47, 342-349.

[13] UHTREX: "Alive and Running with Coolant at 2400 °F"Nuclear News ((1969). UHTREX,1969)

Chapter 9

Benefits in Using Lead-208 Coolant for Fast Reactors and Accelerator Driven Systems

Georgy L. Khorasanov and Anatoly I. Blokhin

Additional information is available at the end of the chapter

http://dx.doi.org/10.5772/52321

1. Introduction

The chapter is dedicated to the analysis of benefits in using lead enriched with the stable lead isotope, Pb-208, instead of natural mix of stable lead isotopes, Pb-204, Pb-206, Pb-207 and Pb-208, as heavy liquid metal coolant for core and top-lateral blankets of critical reactors on fast neutrons (FR) and for subcritical blanket of accelerator driven system (ADS). Pb-208 has very low cross sections, less than 0.3 mbarn, of neutron radiation capture below 50 keV of neutron energy and high threshold, around 2.6 MeV, for inelastic interaction of neutrons with Pb-208 nuclei. These features of Pb-208 as coolant allow reaching economy of neutrons in FR and ADS core and blanket, hardening neutron spectra, favorable conditions for performance of fuel breeding and minor actinides incineration.

2. Small neutron absorption in FR and ADS coolant from lead-208

In the range of neutron energies below 20 MeV microscopic cross sections of radiation neutron capture by the lead isotope Pb-208 are smaller than the cross sections of radiation neutron capture by natural lead, Pb-nat, which is proposed as heavy liquid metal coolants for future lead fast reactors (LFRs). This difference is especially large, by 3-4 orders of magnitude, for intermediate and low energy neutrons, E_n <50 keV. Share of neutrons with energies less than 50 keV is usually about 20-25% of all neutrons in FR or ADS cores and it increases in lateral and topical blankets of FRs. In Fig. 1 cross sections of radiation neutron capture by the lead isotopes Pb-204, Pb-206, Pb-207, Pb-208 and the natural mix of lead isotopes Pb-nat in the ABBN-93 system [1] of 28 neutron energy groups are resulted. The cross sections are received on the basis of files of the evaluated nuclear data for the ENDF/B-VII.0 version library.

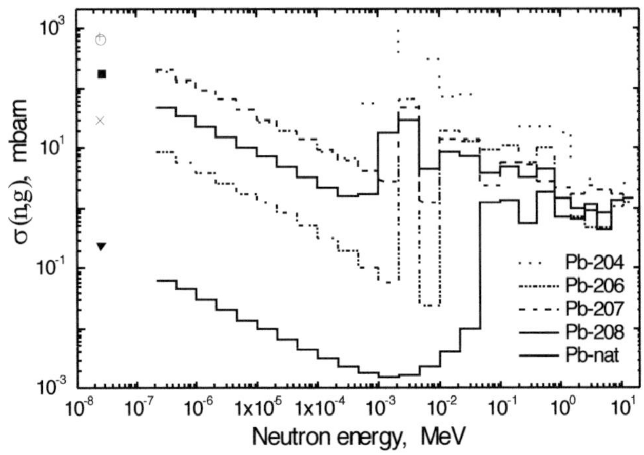

Figure 1. Microscopic cross sections of radiation neutron capture, σ(n,g), by stable lead isotopes and natural mix of lead isotopes taken from the ENDF/B-VII.0 library. Cross sections are represented in the ABBN-93 system of 28 neutron energy groups.

It is visible, that practically for all of the 28 neutron energy groups of the ABBN-93 system, the cross sections of radiation neutron capture by the lead isotope Pb-208 are less than the cross sections of radiation neutron capture by the mix of lead isotopes Pb-nat, and this difference is especially strong, of 3-4 orders of magnitudes, for intermediate and low energy neutrons,E_n<50 keV.

In this chapter section benefits of the use of molten Pb-208 instead of Pb-nat as coolant are demonstrated on the model of the ADS blanket offered for transmutation minor actinides (MA) in a number of nowadays developed scenarios of destruction nuclear plant waste products. Calculations of neutron and physical characteristics of the ADS subcritical core have been performed by means Monte Carlo technique with the MCNP/4C code [2] and NJOY program specially developed with the help of the library of cross sections on the basis of ENDF/B-VII.0 evaluated nuclear data files. As initial data the following ones were taken [3]:

- Annular core with a source of neutrons – a target on its axis,
- Calculated with the Monte-Carlo technique a spectrum of spallation neutrons in the target consisted of the modified lead-bismuth eutectic – Pb-208 (80%)-Bi (20%),
- Proton beam energy– E_p=600 MeV,
- The effective multiplication factor for the subcritical core cooled by Pb-208 –K_{ef} =0.970,
- Thermal capacity delivered in the subcritical core in the rated regime – N=80 MW,
- Core coolant – lead-208, Pb-208 (100%), or natural lead, Pb-nat (100%).

For reduction the core dimensions and minimization the quantity of the coolant, a mix of mononitrides of the depleted uranium, U-238, and plutonium from the PWR spent nuclear fuel and MA as the ADS core fuel was considered. Pu and MA contents in the uranium-plutonium mix were accepted equal to 15%.

The calculated basic technical parameters of the 80 $MW_{thermal}$ Pb-208/Pb-nat cooled ADS core, satisfying the initial data, are resulted in Table 1.

Parameters	Values
Subcritical core thermal power	80 MW
Annular core outer diameter	123.7 cm
Annular core inner diameter	56.0 cm
Annular core height	110.0 cm
Core fuel	$(U+Pu+MA)^{15}N$
Total fuel inventory	5410 kg
Total heavy metal inventory	5090 kg
Total Pu and Minor Actinides inventory	810 kg
Mean pin linear power	188 W/cm
Mean volume power density	118 W/cm^3
Effective multiplication factor for the core cooled by Pb-208/Pb-nat	K_{ef}=0.970 for Pb-208 K_{ef}=0.953 for Pb-nat
Proton beam energy	600 MeV
Proton beam current required to deliver 80 $MW_{thermal}$ core power	I_p=2.8 mA for Pb-208 I_p=4.3 mAfor Pb-nat
Proton beam power required to deliver 80 $MW_{thermal}$ core power	N_p=1.68 MW for Pb-208 N_p=2.58 MW for Pb-nat

Table 1. Parameters of the 80 $MW_{thermal}$Pb-208/Pb-nat cooled ADS core.

On the basis of the microscopic cross sections, σ(n, g), in 28 group approximation and neutron spectra calculated by means the MCNP/4C code for subcritical core cooled with molten Pb-208 or Pb-nat the one-group cross sections, $<\sigma(n, g)>=\sum\sigma_n\ \varphi_n/\sum\varphi_n$, of radiation neutron capture by the isotope Pb-208 and by the mix of isotopes Pb-nat averaged over corresponding neutron spectra of subzones 1-9 of the ADS subcritical core have been calculated. The ADS annular subcritical core was uniformly divided in 9 subzones. One-group cross sections $<\sigma(n, g)>$ of radiation neutron capture by Pb-208 and Pb-nat averaged over corresponding neutron spectra for subzones 1-9 of the ADS core are given in Table 2.

Target-source of neutrons	Subzone 1 **0.86**/5.08	Subzone 2 **0.85**/5.22	Subzone 3 **0.74**/5.86
	Subzone 4 **0.92**/4.50	Subzone 5 **0.90**/4.83	Subzone 6 **0.83**/5.50
	Subzone 7 **0.82**/5.62	Subzone 8 **0.83**/4.74	Subzone 9 **0.73**/5.24

Table 2. One-group cross sections of neutron radiation capture by coolants:Pb-208 (bold) /Pb-nat in the ADS subcritical core subzones. Cross sections in mbarns are given.

The mean value of $<\sigma(n, g)>$ for Pb-208 calculated from the data given in the Table 2 for subzones 1-9 is equal to 0.83 mbarn, that is by 6.2 times smaller than the mean value of $<\sigma(n, g)>$ for Pb-nat (5.18 mbarn). This is evidence of small neutron absorption in the coolant from lead-208. This allows using the neutron surplus for reducing initial fuel load, increasing plutonium breeding, etc.

A possibility of using a low neutron absorbing coolant from lead-208 in the project of critical FR RBEC-M having 900 MW thermal power and fueled with uranium–plutonium [4] was also analyzed. In this reactor project the eutectic of lead (45%) and bismuth (55%) is envisaged. Pb-Bi coolant is characterized by relatively low melting temperature but has no potential in its using in full scaled nuclear power due to low stocks of bismuth and its high post radiation radio toxicity via polonium-210. As is known next generation of LFRs and ADSs, for example the BREST and the EFIT, are planned to be cooled with natural lead.

Neutron and physical characteristics of two reactors were calculated. Reactor RBEC-M is cooled with lead-bismuth as it has been firstly designed at the Kurchatov Institute, the Russian Federation. The second version of the reactor, RBEC-M, is cooled with lead-208 [5, 6]. These reactors are distinguished only by materials of coolant and coolant's temperatures are closely spaced. All other characteristics of these two reactors are similar. In these conditions the masses of uranium-plutonium fuel to reach criticality were calculated. The economy of the fuel load in the critical reactor RBEC-M cooled with lead-208 has been recognized. Then neutron fluxes in the core, lateral and low topical blankets of these two critical reactors were calculated. Calculations have been performed with the Monte Carlo code MCNP/4C, the special program NJOY and in using the neutron cross section data from the library ENDF/B-VII.0.

In Fig. 2 the microscopic cross sections of radiation capture of neutrons by lead-208 and Pb-Bi are given. The cross sections are taken from the ENDF/B-VII.0 library and they are represented in the ABBN-93 28 neutron energy group's approximation.

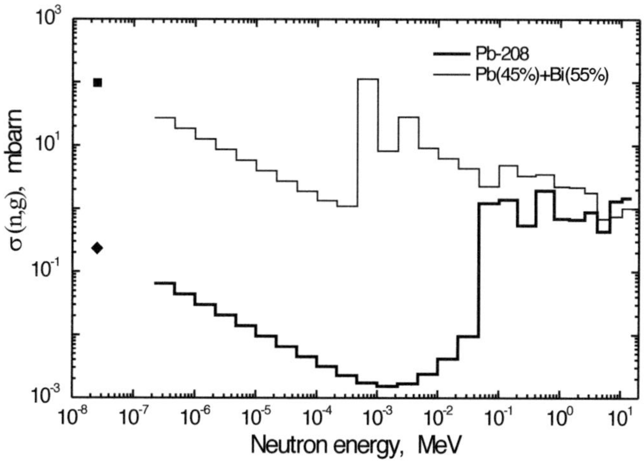

Figure 2. Microscopic cross sections of radiation capture of neutrons by lead-208 and Pb-Bi. Data are taken from the ENDF/B-VII.0 library.

From Fig. 2 follows that for all of neutron energies represented, the microscopic cross sections of neutron capture by Pb-Bi are larger in comparison with Pb-208. For energies below E_n<50 keV the cross sections of neutron radiation capture by lead-208 are by 3-5 orders of magnitude smaller than the cross section of neutron capture by Pb-Bi.

In Table 3 the one-group cross sections of neutron radiation capture by two various coolants, Pb-Bi and Pb-208, in RBEC-M neutron spectra are presented.

Name of reactor	Small fuel enrich-ment zone	Middle fuel enrich-ment zone	Big fuel enrich-ment zone	Lateral blanket zone	Topical blanket below small enrich-ment zone	Topical blanket below middle enrich-ment zone	Topical blanket below big enrich-ment zone
RBEC-M (lead 45%-bismuth 55%)	**3.7119**	**3.6239**	**3.6640**	**4.8288**	**5.3238**	**5.2248**	**5.4097**
RBEC-M (lead-208 100%)	**0.9330**	**0.9419**	**0.9393**	**0.8660**	**0.8087**	**0.8122**	**0.7901**

Table 3. One-group cross sections of neutron radiation capture by coolants (Pb-Bi or Pb-208) of FR RBEC-M. Cross sections in millibarns are given.

From Table 3 follows that reactor RBEC-M with coolant from lead-208 is characterized with minimum value of mean one-group cross sections of neutron capture by the coolant in the core, $<\sigma>=0.94$ millibarns. The corresponding mean cross section for the RBEC-M reactor core cooled by Pb-Bi is equal to 3.67 millibarns. As concerns the lateral and topical blankets, the one-group cross section of neutron capture by Pb-Bi exceeds the value of corresponding cross section for Pb-208 by 6-7 times. The small value of cross sections and corresponding excess of neutrons in the zones of the reactor RBEC-M cooled with Pb-208 can be used for reducing the fuel load in the reactor core, conversion of depleted uranium into plutonium and transmutation of minor actinides and long-lived fission products immersed in the lateral and topical zones.

3. Gain in core fuel loading due to excess of neutrons

As it follows from Table 1, in the ADS with subcritical blanket of 80 MW thermal power the calculated effective neutron multiplication factor, K_{ef}, increases from $K_{ef}=0.953$ for coolant from Pb-nat to $K_{ef}=0.970$ for Pb-208 as coolant. In this replacement of the coolant proton beam power required to deliver 80 MW thermal blanket power will be reduced from 2.58 MW for Pb-nat to 1.68 MW for Pb-208. The gain in proton beam power equal to 0.9 MW arising in using low neutron absorbing coolant from Pb-208 is very valuable taking into account the high cost of 600 MeV energy proton beam power, approximately $100 millions/MW according Ref. 7.

In LFR conditions the behavior of K_{ef} is very similar. In LFR RBEC-M of 900 MW thermal power in replacement its standard lead-bismuth coolant with lead-208 K_{ef} increases from its standard value, $K_{ef}=1.00957$, to the value $K_{ef}=1.0246$. It was calculated that to leave this coefficient at the standard level, i.e. to ensure the criticality, $K_{ef}=1.00957$, the quantity of power grade plutonium in uranium-plutonium nitride fuel must be decreased from 3,595 kg to 3,380 kg, i.e. Pu enrichment in the fuel will be decreased from its initial value of 13.7% down to 13.0%. The economy in initial core fuel loading equal to 215 kg power grade plutonium per 340 MW of electrical power is very valuable taking into account that by this time the power grade plutonium stockpile reprocessed from light water reactors (LWRs) is expected to be insufficient for possible introducing in future a plenty of FRs in the countries quickly increasing their nuclear power. Now in France, the most advanced country on nuclear fuel reprocessing; only 8.5 tons/year of power grade Pu are available after reprocessing 1150 tons of spent fuel from LWRs.

4. Increasing the fuel breeding gain in FRs and ADSs cooled with lead-208

The excess of neutrons due to their small absorption in lead-208 can be used for fuel breeding and transmutation of long-lived radiotoxic fission products. As an example, we assume the radiation capture of neutrons by uranium-238 leading to creation of plutonium-239 [9].

http://dx.doi.org/10.5772/52321

The affectivity of this process will be as large as the value of one-group cross section of radiation neutron capture by uranium-238 nucleus is great. At neutron energies near to E_n=5-10 eV microscopic cross sections of neutron capture by uranium-238 have maximum equal to 170 barns as it can be seen from Fig. 3.

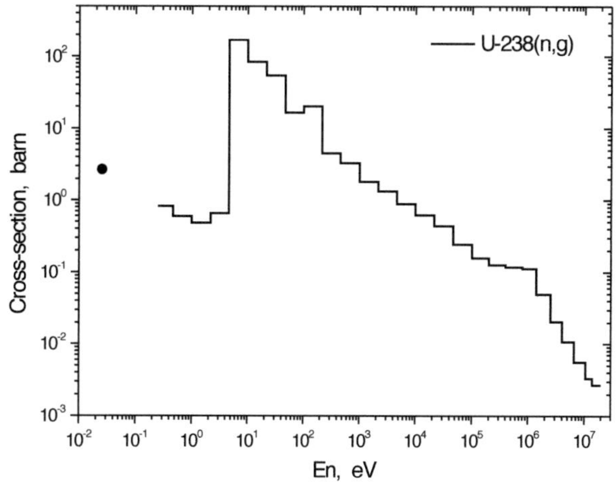

Figure 3. Microscopic cross sections of neutron radiation capture by U-238; taken from ENDF/B-VII.0 library.

That is why if the neutron spectra contains an increased share of neutrons of small and intermediate energies the corresponding one-group will be larger. In Table 4 the one-group cross sections of neutron radiation capture by U-238 in the ADS subcritical blanket cooled with Pb-208 or Pb-nat are given.

Target-source of neutrons	Subzone 1 **0.34**/0.32	Subzone 2 **0.45**/0.43	Subzone 3 **1.23**/0.58
	Subzone 4 **0.27**/0.23	Subzone 5 **0.22**/0.28	Subzone 6 **0.67**/0.35
	Subzone 7 **0.55**/0.46	Subzone 8 **0.46**/0.33	Subzone 9 **1.18**/0.61

Table 4. One-group cross sections of neutron radiation capture, <σ(n,g)>, by U-238 in the ADS subcritical core subzones cooled with Pb-208 (bold) /Pb-nat. Cross sections in barns are given.

The mean value of U-238 one-group radiation neutron capture cross section averaged over subzones 1-9 is equal to 0.6 barns for coolant from Pb-208 and to 0.4 barns for coolant from Pb-nat. The difference is due to excess of neutrons in the subzones 3, 6, 9 with smaller neutron energies at blanket's periphery and corresponding greater share of low energy neutrons.

In Fig. 4and Fig.5 the results of estimation of Pu-239 accumulation and U-238 incineration in the subzone 3 of the 80 MW thermal ADS are given. These calculations were performed using ACDAM code developed in the IPPE [8].

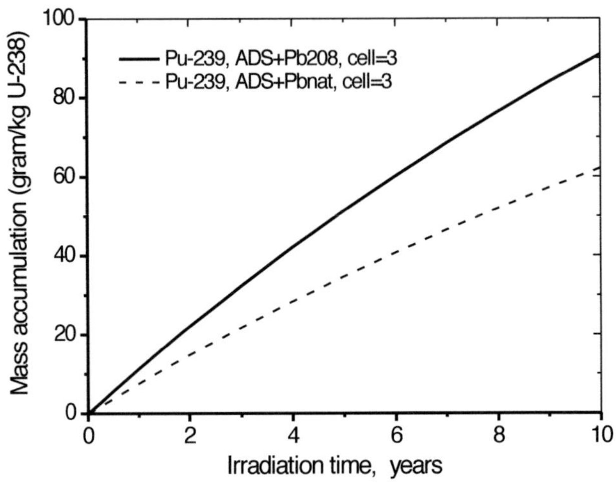

Figure 4. Accumulation of Pu-239 in grams in inserting 1 kg of U-238 in the neutron spectrum of the 80 MW ADS subzone 3. The bold line corresponds to ADS cooled with Pb-208 and the dash line - to ADS cooled with Pb-nat.

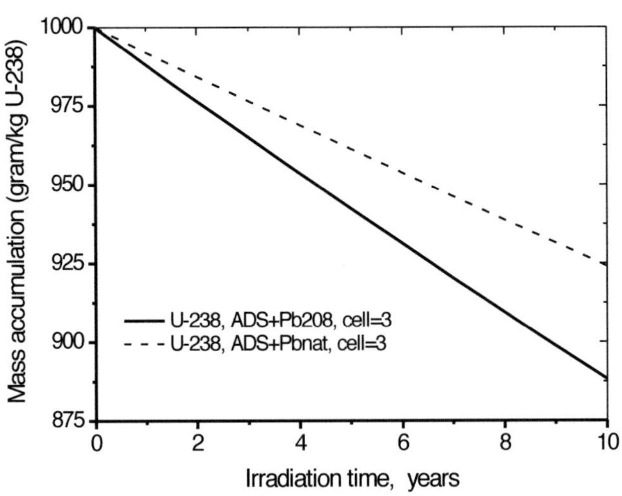

Figure 5. Mass decreasing of U-238 from its initial mass of 1 kg inserted in the neutron spectrum of the 80 MW ADS subzone 3. The bold line corresponds to ADS cooled with Pb-208 and the dash line - to ADS cooled with Pb-nat.

Similar behavior of U-238 one-group neutron capture cross section can be seen for LFR RBEC-M. The RBEC-M core is divided into 4 cells, its positions are shown in Table 5.

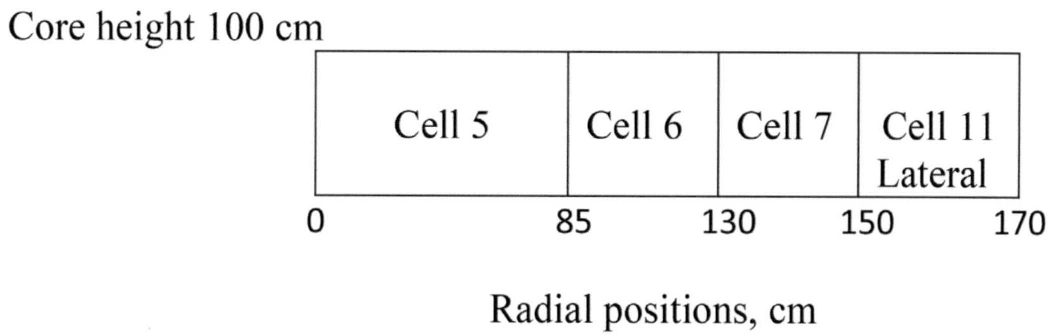

Table 5. Cell positions in the RBEC-M core.

In Fig.6 and Fig.7 neutron fluxes ratio acting in the cell number 5 and lateral cell number 11 cooled with Pb-208/Pb-Bi are given.

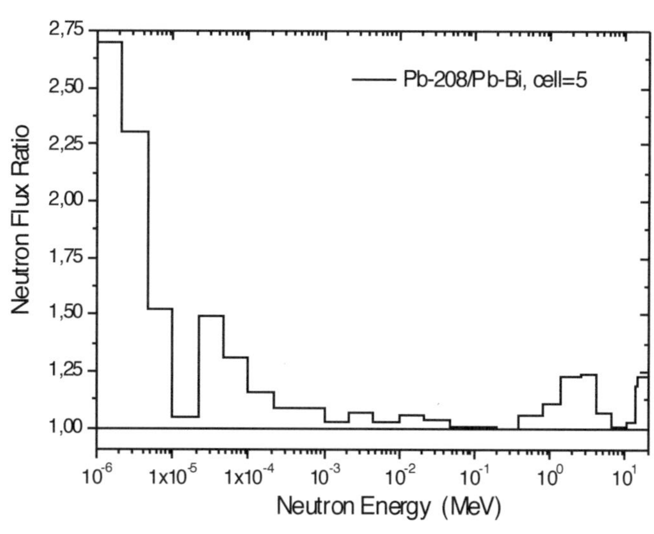

Figure 6. Neutron fluxes ratio in linear scale for RBEC-M cell 5 cooled with Pb-208/Pb-Bi.

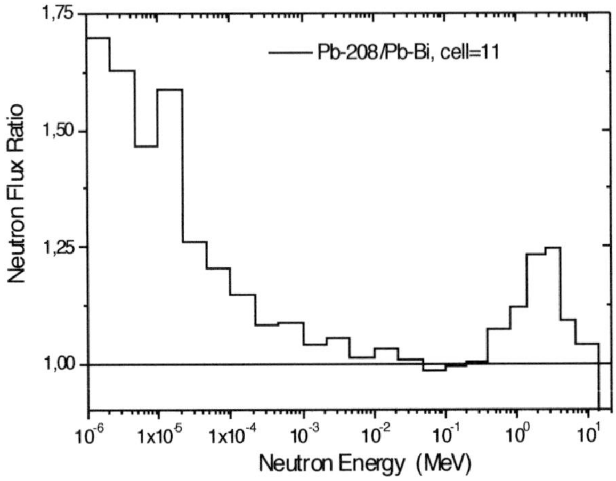

Figure 7. Neutron fluxes ratio in linear scaleforRBEC-M cell 11 (lateral blanket) cooled with Pb-208/Pb-Bi.

It can be seen that in using Pb-208 as coolant the share of main part of neutrons is increasing as well as the share of relatively small part of neutrons of low energy is also increasing. The last circumstance is due to small absorption of neutrons below their energy of 50 keV.

In Table 6 the one-group cross sections of neutron radiation capture by U-238 are given.

	Cell 5	Cell 6	Cell 7	Cell 11 - Lateral
<σ(n,g)/>, barns	**0.273**/0.273	**0.262** /0.261	**0. 271**/0.269	**0.575**/0.532

Table 6. One-group cross sections of neutron radiation capture, by U-238 in the RBEC-M core cells cooled with Pb-208 (bold) /Pb-Bi.

It is visible that one-group cross sections of neutron radiation capture, <σ(n,g)>, by U-238in the RBEC-M lateral blanket is equal to 0.575 barns which is very close to the value of <σ(n,g)>=0.6 barns for ADS blanket.

The one-group cross sections for this nuclide averaged over neutron spectra of LFRs and ADSs cooled with lead-208 are approximately of 0.6 barns which are comparable with the one-group cross sections for typical breeders.

5. Hardening of ADS and FR neutron spectra in using lead-208 instead of natural lead or lead-bismuth

In nuclear power installations with fast neutrons, ADSs and FRs, the mean energy of core neutrons does not exceed 0.5 MeV, while the mean energy of fission neutrons emitted by uranium-235, for example, is equal to 1.98 MeV. In Fig. 8 typical spectrum of neutrons in the core of lead fast reactor and spectrum of fission neutrons emitted by uranium-235 are given.

It is visible that LFR neutron spectrum is strongly moderated as compared with the spectrum of fission neutrons. Neutron moderation is due to interaction of neutrons with fuel, structural materials and coolant.

Meanwhile hard spectrum of neutrons in ADS and FR core is preferable for incineration of minor actinides (MA). Incineration of long-lived radio toxic MA – neptunium, americium and curium – is one of the key problem of the nuclear power engineering. The world fleet of light water reactors (LWR) produces about 3.2 tons of MA per year as wastes. It is expected to incinerate MA in future ADSs which will be able consuming MA in quantities of 40% of fuel heavy atoms (h. a.). But ADS installation creation is a very expensive way and it needs a long time. It exists an opportunity to incinerate MA in FR core but from reactor's safety point of view the FR core can be loaded with MA in quantities not more than 2.5% of h. a. As alternative it is possible to load a radial blanket with the fuel having MA of 10% of h. a. But to avoid curium-242 accumulation during americium transmutation it is desirable to incinerate MA via their fission. In Fig. 9 microscopic fission cross section for one of MA, namely americium-241, is given.

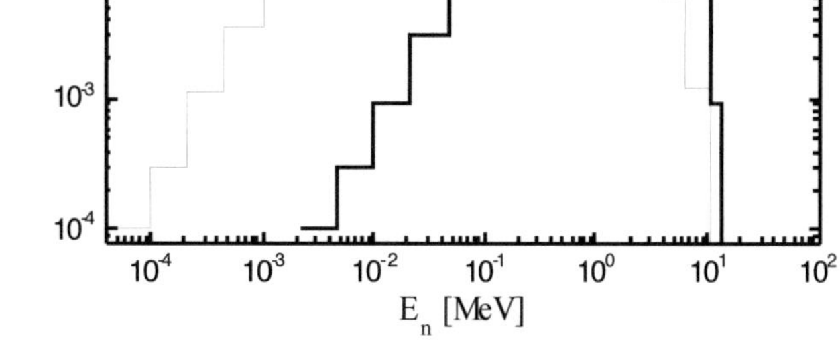

Figure 8. Neutron spectrum in the fuel zone of the 700 $MW_{thermal}$LFR (2) and spectrum of U-235 fission neutrons (1) in the ABBN-93 neutron energy group system.

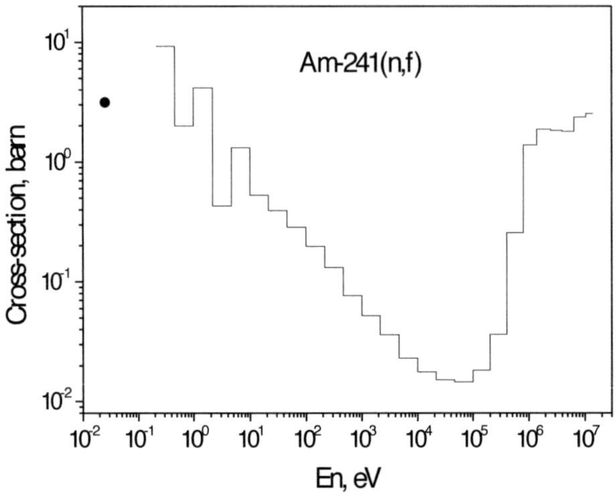

Figure 9. Microscopic fission cross sections for americium-241 taken from ENDF/B-VII.0 library.

It is visible that for fast neutron region there is a threshold equal to 0.1 MeV above which Am-241 fission cross sections are growing up to 2-3 barns at E_n=1-10 MeV. In the range of thermal and intermediate neutron energies, E_n<10 keV, Am-241 fission cross sections are also large enough but at these energies neutron capture cross sections are too large as it is shown in Fig. 10.

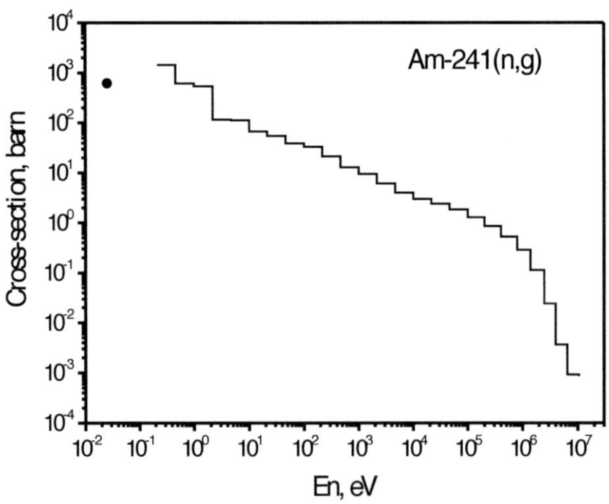

Figure 10. Microscopic neutron radiation capture cross sections for Am-241 taken from ENDF/B-VII.0 library.

For estimation of Am-241 incineration capability, the neutron spectra in 80 MW ADS subcritical blanket were calculated using MCNP-4C code and nuclear data library ENDF/B-VII.0. Calculations were performed in the ABBN-energy group structure system with 28 neutron energy groups.On the base of calculated neutron spectra for above mentioned 9 subzones of the ADS blanket the mean energies of neutrons and one-group fission cross sections for Am-241 were calculated. In Table 7 one-group fission cross sections calculated for Am-241 are given.

	Subzone 1 **0.2854/**0.3310	Subzone 2 **0.2886**/0.1984	Subzone 3 **0.2197**/0.1168
Target- source of neutrons	Subzone 4 **0.3335**/0.3725	Subzone 5 **0.3681/**0.3334	Subzone 6 **0.2429**/0.2630
	Subzone 7 **0.2245/0.**1711	Subzone 8 **0.2483**/0.2594	Subzone 9 **0.2194**/0.1932

Table 7. One-group Am-241 fission cross sections in barns for subzones 1-9 of the 80 MW subcritical blanket cooled with Pb-208 (bold) and Pb-nat.

As follows from these calculations, the replacement of coolant from Pb-nat with Pb-208 in ADS blanket leads to increasing the mean neutron energy averaged over 1-9 subzones from its value 0.3785 to 0.4026 MeV, i.e. on 6.4%. In this case the one-group Am-241 fission cross section averaged over 1-9 subzones increases from 0.2488 to 0.2700 barns, i.e. on 8.5%. In Fig. 11 the calculated dependence of one-group fission cross sections for Am-241 upon mean neutron energy is given. It is visible that these cross sections are growing as the neutron energy increases. From this Figure it can be concluded that there is a relatively high reserve to reach the maximum available mean neutron energy equal to 1.98 MeV which corresponds to mean neutron energy of neutrons emitted by uranium-235.

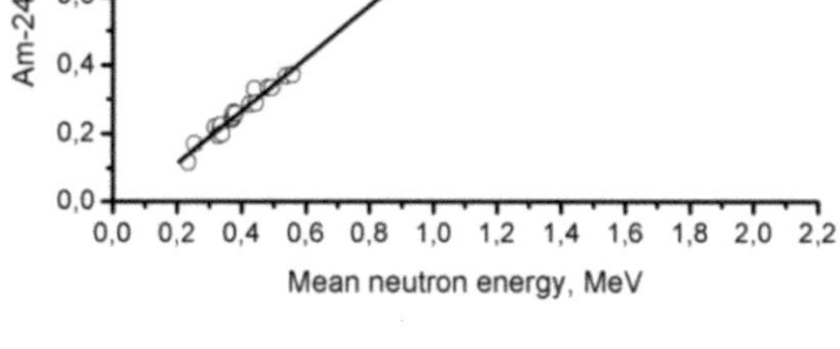

Figure 11. The calculated dependence of one-group fission cross sections for Am-241 upon mean neutron energy.

Similar calculations for LFR RBEC-M have been performed. For RBEC-M core and its lateral blanket the neutron spectra were calculated using MCNP5 code and nuclear data library ENDF/B-VII.0. Calculations were performed in the ABBN-energy group structure system with 28 neutron energy groups. On the base of calculated neutron spectra for several zones of the LFR core and lateral blanket the mean energies of neutrons were calculated. As follows from these calculations, the replacement of coolant from Pb-Bi with Pb-208 in RBEC-M leads to increasing the core mean neutron energy from its standard value 0.3992 MeV to 0.4246 MeV, i.e. on 6.4%. The one-group fission cross section for Am-241 included in core fuel in small quantities is increasing under these conditions from 0.2716 to 0.2981 barns, i.e. on 9.8%.

As concerns the RBEC-M lateral blanket, in switching coolant from Pb-Bi to Pb-208, its mean neutron energy increases from 0.2509 to 0.2662 MeV, i.e. on 6.1%.Under these conditions the one-group fission cross section for Am-241 is increasing up to 10%.

Thus, it can be concluded in nuclear installations with fast neutrons the replacement of natural lead or lead-bismuth with lead-208 as coolant leads to increasing the mean energy of neutrons approximately on 6% and corresponding increasing the one-group fission cross section of Am-241 on 8-10%. It must be mentioned once more that incineration of Am-241 via fission is more preferable than its transmutation because it allows avoiding creation of Cm-242 which has relatively high thermal emission that creates some difficulties in handling spent fuel containing MA.

6. Conclusion

It is shown that the one-group cross sections of neutron radiation capture, $\langle\sigma(n,g)\rangle$, for Pb-208 used as ADS and FR core coolant are equal to 0.8-0.9 mbarns, which are by 4-7 times smaller than the mean value of $\langle\sigma(n,g)\rangle$ for Pb-nat or Pb-Bi used as ADS and FR core coolant and by 2-3 times are smaller than for sodium coolant.

The mean value of $\langle\sigma(n, g)\rangle$ for U-238 leading to conversion into Pu-239 for ADS and FR core cooled with Pb-208 is equal to 0.6 barns which is comparable with the value of the same nuclide one-group cross section $\langle\sigma(n, g)\rangle$ for neutron spectrum of the FR core cooled with sodium.

In the ADS with subcritical blanket of 80 MW power the calculated effective neutron multiplication factor, K_{ef}, increases on 1.7% in replacement of coolant from Pb-nat to Pb-208. In this replacement the proton beam power required to deliver 80 MW blanket power will be reduced to 0.9 MW.

In FR conditions the behavior of K_{ef} is very similar. In LFR RBEC-M of 900 MW thermal power in replacement its standard lead-bismuth coolant with lead-208 K_{ef} increases on 1.5%. This replacement leads to decreasing power grade Pu enrichment in the fuel from its initial value of 13.7% to 13.0% that means economy in fuel loading equal to 630 kg of power grade plutonium per 1 GW electrical.

The replacement of coolant from Pb-nat with Pb-208 in ADS blanket leads also to increasing the mean neutron energy averaged over the blanket from 0.3785 to 0.4026 MeV, i.e. on 6.4%. In the FR RBEC-M the replacement of coolant from Pb-Bi with Pb-208 leads to increasing the core mean neutron energy from its standard value 0.3992 MeV to 0.4246 MeV, i.e. also on 6.4%.

The harder spectrum of neutrons in ADS and FR core is for example preferable for incineration of Am-241 via its fission to avoid Cm-242 accumulation. It is shown that in neutron spectrum hardening on 6.4% the one-group Am-241 fission cross section averaged over ADS blanket or FR core increases on 8-10% from its initial value of 0.24 barns. In further increasing FR core neutron energy, if possible, the Am-241 fission cross section might be increased up to the value of 1.4 barns.

The advantages of the neutron and physical characteristics of molten Pb-208 allow considering it as a perspective material as coolant for next generation fast reactors.

The possibility of using Pb-208 as coolant in commercial fast critical or subcritical reactors requires a special considering but relatively high content of this isotope in natural lead, 52.3%, and perspectives of using high performance photochemical technique of lead isotope separation [10, 11] allow expecting to obtain in future such a material in large quantities and under economically acceptable price. Besides, the principal possibility of acquisition of radiogenic lead containing high enriched lead-208, up to 93%, exists [12].

Acknowledgements

The study was supported by the Russian Fundamental Research Fund grants, the last of which is RFBR # 08-08-92201-GFEN_a.

Author details

Georgy L. Khorasanov* and Anatoly I. Blokhin

*Address all correspondence to: khorasan@ippe.ru

State Scientific Centre of the Russian Federation – Institute for Physics and Power Engineering named after A.I. Leypunsky (IPPE), Russia

References

[1] Manturov, G. N., Nikolaev, M. N., & Tsiboulia, A. M. (1996). Group Constant System ABBN-93. Part 1: Nuclear Constants for Calculation of Neutron and Photon Emission

Fields.Issues of Atomic Science and Technology, Series: Nuclear Constants, (in Russian).(1), 59.

[2] Briesmeister, J. F. (1987). MCNP-4 General Monte Carlo Code for Neutron and Photon Transport. LA-M-Rev.2., 7396.

[3] Khorasanov, G. L., Korobeynikov, V. V., Ivanov, A. P., & Blokhin, A. I. (2009). Minimization of an initial fast reactor uranium-plutonium load by using enriched lead-208 as coolant. *Nuclear Engineering and Design*, 239(9), 1703-1707, 0029-5493.

[4] Alekseev, P. N., Mikityuk, K. O., Vasiljev, A. V., Fomichenko, P. A., Shchepetina, T. D., & Subbotin, S. A. Optimization of Conceptual Decisions for the Lead-Bismuth Cooled Fast Reactor RBEC-M. Atomnaya Energiya, (2004), 97(2), 115-125. http://www.iaea.org/Nuclear Power /SMR/ crpi25001 /html/.

[5] 2010 Macroscopic cross sections of neutron radiation capture by Pb-208, U-238 and Tc-99 nuclides in the accelerator driven subcritical core cooled with molten Pb-208. In CD-ROM Proceedings of the International Conference PHYSOR 2010 Advances in Reactor Physics to Power the Nuclear Renaissance, Pittsburgh, Pennsylvania, USA, May 9-14, Paper #286 at the Session 5C "Advanced Reactors Design".

[6] (2012). One-Group Fission Cross Sections for Plutonium and Minor Actinides Inserted in Calculated Neutron Spectra of Fast Reactor Cooled with Lead-208 or Lead-Bismuth Eutectic In CD-ROM Proceedings of the International Conference PHYSOR 2012 Advances in Reactor Physics- Linking Research, Industry, and Education,Knoxville, Tennessee, USA, April 15-20, , Paper #106 at the Session "Reactor concepts and designs".

[7] Biarrotte, J.L. The LINAC accelerator for MYRRHA. In: Proc.of the Topical Day "From MYRRHA towards XT-ADS", (2004). SCK-CEN, Mol, Belgium.

[8] Blokhin, A.I. et al. Code ACDAM for investigation the nuclear properties of materials under long time neutron irradiation. Perspective materials, (2010), (2), 46-55 (In Russian), (0102-8978X)

[9] Khorasanov, G. L., Blokhin, A. I., & Low, A. Neutron Absorbing Coolant for Fast Reactors & Accelerator Driven Systems In the book: "Cooling Systems: Energy, Engineering and Applications", Series: Mechanical Engineering Theory and Applications. Editors: Aaron I. Shanley, Nova Science Publishers, Inc., New York, (2011). Chapter 4, 978-1-61209-379-6, 89-98.

[10] ISTC #2573 project: "Investigation of Processes of High- Performance Laser Separation of Lead Isotopes by Selective Photoreactions for Development of Environmentally Clean Perspective Power Reactor Facilities", Project Manager: A.M. Yudin (Saint-Petersburg, Efremov Institute, NIIEFA), Project Submanagers: G.L. Khorasanov (Obninsk, Leypunsky Institute, IPPE) and P.A. Bokhan (Novosibirsk, Institute for Semiconductor Physics, ISP), , 2004-2005.

[11] Bortnyansky, A.L., V.L.Demidov, S.A. Motovilov, F.P. Podtikan, Yu.I. Savchenko, V.A. Usanov, A.M. Yudin, B.P. Yatsenko. Experimental Laser Complex for Lead Isotope Separation by means Selective Photochemical Reactions. In: Proceeding of the X International Conference "Physical and Chemical Processes on Selection of Atoms and Molecules", Moscow, Russia, 3-7 October 2005, 76-82, 2005, (in Russian).

[12] Khorasanov, Georgy, L., Anatoly, I. Blokhin and Anton A. Valter (2012). New Coolant from Lead Enriched with the Isotope Lead-208 and Possibility of Its Acquisition from Thorium Ores and Minerals for Nuclear Energy Needs, Nuclear Reactors, Amir ZacariasMesquita (Ed.), 978-9-53510-018-8InTech, Available from: http://www.intechopen.com/articles/show/title/a-new-coolant-from-lead-enriched-with-the-isotope-lead- -and-possibility-of-its-acquisition-from-thorium-ores-and-minerals.

Chapter 10

Nuclear Power as a Basis for Future Electricity Production in the World: Generation III and IV Reactors

Igor Pioro

Additional information is available at the end of the chapter

http://dx.doi.org/10.5772/51916

1. Introduction

It is well known that the electrical-power generation is the key factor for advances in any other industries, agriculture and level of living (see Table 1). In general, electrical energy can be produced by: 1) non-renewable sources such as coal, natural gas, oil, and nuclear; and 2) renewable sources such as hydro, wind, solar, biomass, geothermal and marine. However, the

No.	Country	Watts per person	Year	HDI* (2010)
1	Norway	2812	2005	1
2	Finland	1918	2005	16
3	Canada	1910	2005	8
4	USA	1460	2011	4
5	Japan	868	2005	11
6	France	851	2005	14
7	Germany	822	2009	10
8	Russia	785	2010	65
9	European Union	700	2005	
10	Ukraine	446	2005	69
11	China	364	2009	89
12	India	51	2005	119

* HDI – Human Development Index by United Nations; The HDI is a comparative measure of life expectancy, literacy, education and standards of living for countries worldwide. It is used to distinguish whether the country is a developed, a developing or an under-developed country, and also to measure the impact of economic policies on quality of life. Countries fall into four broad human-development categories, each of which comprises ~42 countries: 1) Very high – 42 countries; 2) high – 43; 3) medium – 42; and 4) low – 42.

Table 1. Electrical-energy consumption per capita in selected countries (Wikipedia, 2012).

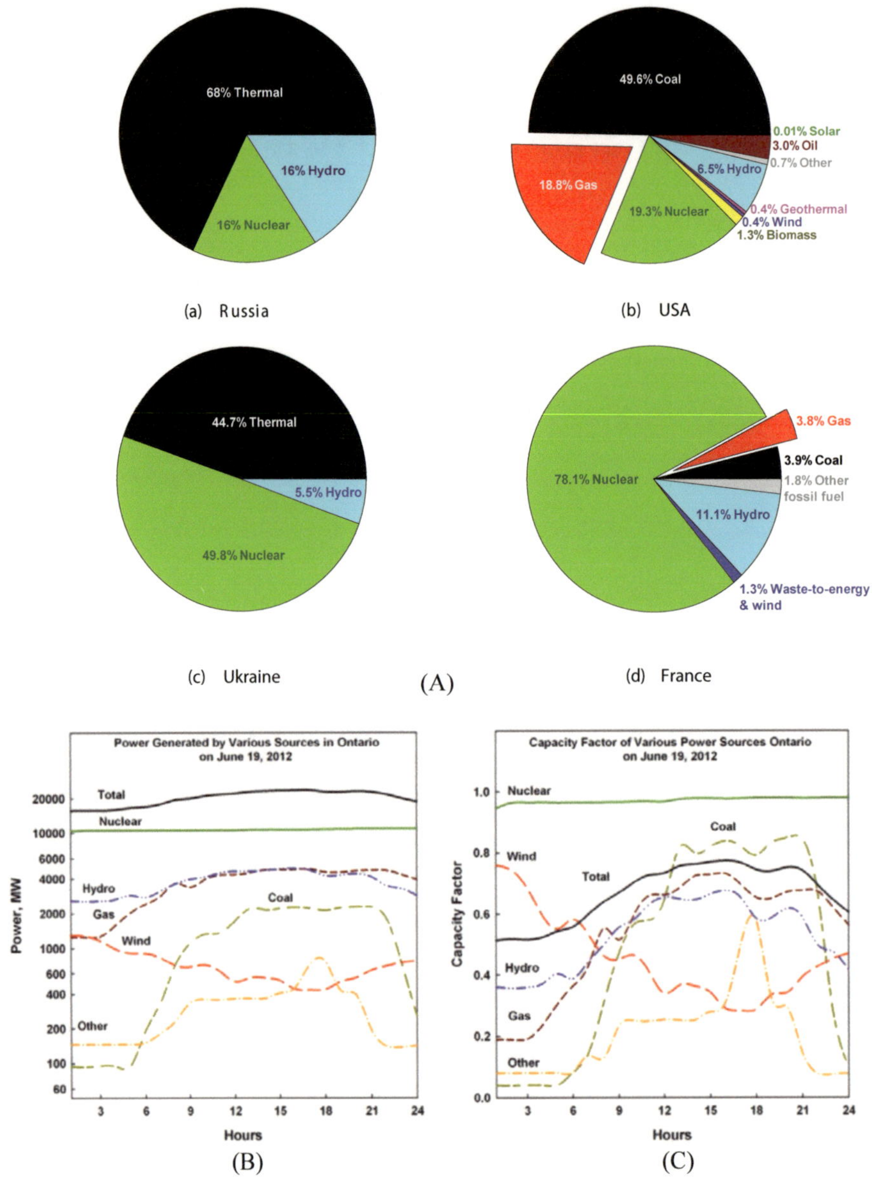

*The net capacity factor of a power plant (Wikipedia, 2012) is the ratio of the actual output of a power plant over a period of time and its potential output if it had operated at full nameplate capacity the entire time. To calculate the capacity factor, take the total amount of energy the plant produced during a period of time and divide by the amount of energy the plant would have produced at full capacity. Capacity factors vary significantly depending on the type of fuel that is used and the design of the plant. Typical capacity factors for modern NPPs can be within 90%, thermal and hydro-electric power plants can be on average within 45% (can vary within 10 – 99% depending on local conditions), wind power plants – 20 – 40% and photovoltaic solar power plants – 15 – 20%.

Figure 1. (A). Electricity production by source in selected countries (data from 2005 – 2010 presented here just for reference purposes) (Wikipedia, 2012). (B). Power generated by various sources in the Province of Ontario (Canada) on June 19, 2012 (based on data from http://ieso.ca/imoweb/marketdata/genEnergy.asp). (C). Capacity factors* of various power sources in the Province of Ontario (Canada) on June 19, 2012 (based on data from http://ieso.ca/imoweb/marketdata/genEnergy.asp).

http://dx.doi.org/10.5772/51916

main sources for electrical-energy production are: 1) thermal - primary coal and secondary natural gas; 2) nuclear and 3) hydro. The rest of the sources might have visible impact just in some countries (see Figure 1). In addition, the renewable sources such as wind (see Figure 1b,c) and solar are not really reliable sources for industrial power generation, because they depend on Mother nature and relative costs of electrical energy generated by these and some other renewable sources with exception of large hydro-electric power plants can be significantly higher than those generated by non-renewable sources. Therefore, thermal and nuclear electrical-energy production will be considered further.

2. Thermal power plants

In general, the major driving force for all advances in thermal and Nuclear Power Plants (NPPs) is thermal efficiency. Ranges of thermal efficiencies of modern power plants are listed in Table 2 for references purposes.

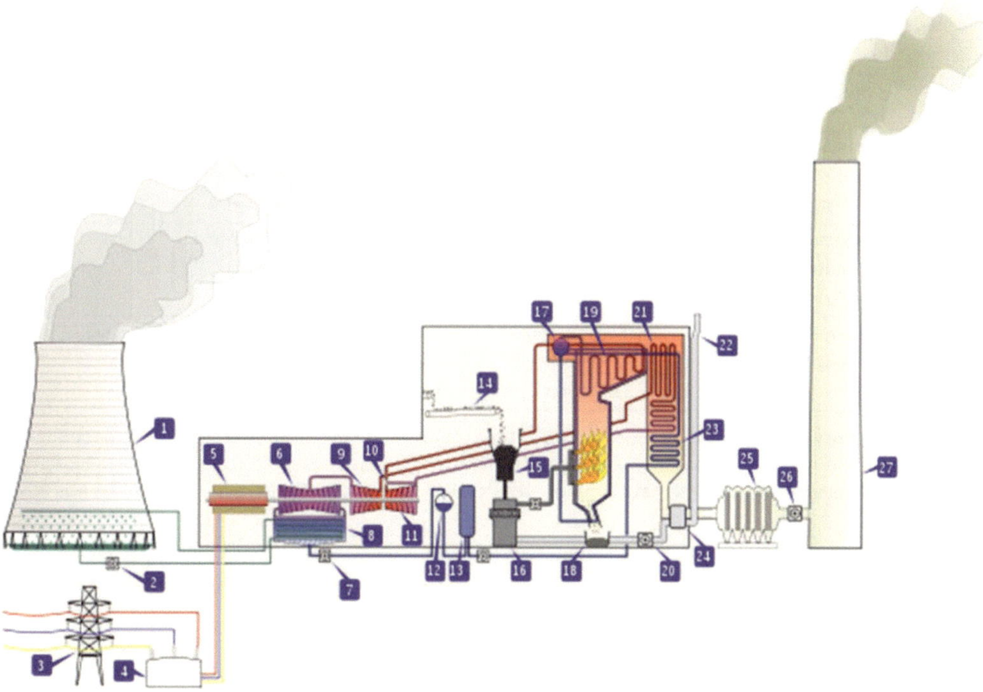

1) Cooling tower; 2) Cooling-water pump; 3) Transmission line (3-phase); 4) Step-up transformer (3-phase); 5) Electrical generator (3-phase); 6) Low-pressure steam turbine; 7) Condensate pump; 8) Surface condenser; 9) Intermediate-pressure steam turbine; 10) Steam control valve; 11) High-pressure steam turbine; 12) Deaerator; 13) Feedwater heater; 14) Coal conveyor; 15) Coal hopper; 16) Coal pulverizer; 17) Boiler steam drum; 18) Bottom-ash hopper; 19) Superheater; 20) Forced-draught (draft) fan; 21) Reheater; 22) Combustion-air intake; 23) Economiser; 24) Air preheater; 25) Precipitator; 26) Induced-draught fan; and 27) Flue-gas stack.

Figure 2. Typical scheme of coal-fired thermal power plant (Wikipedia, 2012):

No	Power Plant	Gross Efficiency %
1	Combined-cycle power plant (combination of Brayton gas-turbine cycle (fuel natural or Liquefied Natural Gas (LNG); combustion-products parameters at the gas-turbine inlet: $T_{in} \approx 1650°C$) and Rankine steam-turbine cycle (steam parameters at the turbine inlet: $T_{in} \approx 620°C$ ($T_{cr} = 374°C$)) (see Figure 8).	Up to 62
2	Supercritical-pressure coal-fired thermal power plant (new plants) (Rankine-cycle steam inlet turbine parameters: $P_{in} \approx$ 25-38 MPa ($P_{cr} = 22.064$ MPa), $T_{in} \approx$ 540-625°C ($T_{cr} = 374°C$) and $T_{reheat} \approx$ 540-625°C) (see Figures 2 and 3).	Up to 55
3	Subcritical-pressure coal-fired thermal power plant (older plants) (Rankine-cycle steam: $P_{in} \approx 17$ MPa, $T_{in} \approx 540°C$ ($T_{cr} = 374°C$) and $T_{reheat} \approx 540°C$) (see Figure 2).	Up to 40
4	Carbon-dioxide-cooled reactor (Advanced Gas-cooled Reactor (AGR) (see Figure 12)) NPP (Generation III, current fleet) (reactor coolant – carbon dioxide: $P \approx 4$ MPa and $T_{in} / T_{out} \approx 290 / 650°C$; secondary Rankine-cycle steam: $P_{in} \approx 17$ MPa ($T_{sat} \approx 352°C$) and $T_{in} \approx 560°C$ ($T_{cr} = 374°C$))	Up to 42
5	Sodium-cooled Fast Reactor (SFR) NPP (see Figure 15) (Generation III and IV, currently just one reactor – BN-600 operates in Russia) (reactor coolant – liquid sodium: $P \approx 0.1$ MPa and $T_{max} \approx$ 500-550°C; secondary Rankine-cycle steam: $P_{in} \approx 14$ MPa ($T_{sat} \approx 337°C$) and $T_{in} \approx 505°C$ ($T_{cr} = 374°C$)).	Up to 40
6	Pressurized Water Reactor (PWR) NPP (Generation III+, to be implemented within next 1–10 years) (reactor coolant – light water: $P \approx 16$ MPa ($T_{sat} = 347°C$) and $T_{out} \approx 327°C$; secondary Rankine-cycle steam: $P_{in} \approx 7.8$ MPa and $T_{in} = T_{sat} \approx 293°C$).	Up to 36-38
7	PWR NPP (see Figure 9) (Generation III, current fleet) (reactor coolant – light water: $P \approx 16$ MPa ($T_{sat} = 347°C$) and $T_{in} / T_{out} \approx 290 / 325°C$; secondary Rankine-cycle steam: $P_{in} \approx 7.2$ MPa and $T_{in} = T_{sat} \approx 288°C$).	32-36
8	Boiling Water Reactor (BWR) NPP (see Figure 10) (Generation III, current fleet) (reactor coolant light water; direct cycle; steam parameters at the turbine inlet: $P_{in} \approx 7.2$ MPa and $T_{in} = T_{sat} \approx 288°C$). Advanced BWR (ABWR) NPP (Generation III+) has approximately the same thermal efficiency.	~34
9	RBMK reactor (boiling reactor, pressure-channel design) NPP (see Figure 14) (Generation II and III, current fleet) (reactor coolant light water; direct cycle; steam parameters at the turbine inlet: $P_{in} \approx 6.6$ MPa and $T_{in} = T_{sat} \approx 282°C$).	~32
10	Pressurized Heavy Water Reactor (PHWR) NPP (see Figure 11) (Generation III, current fleet) (reactor coolant – heavy water: $P_{in} \approx 11$ MPa; $P_{out} \approx 10$ MPa ($T_{sat} = 311°C$) and $T_{in} / T_{out} \approx 265 / 310°C$; secondary Rankine-cycle steam (light water): $P_{in} \approx 4.6$ MPa and $T_{in} = T_{sat} \approx 259°C$).	~32

[1]Gross thermal efficiency of a unit during a given period of time is the ratio of the gross electrical energy generated by a unit to the thermal energy of a fuel consumed during the same period by the same unit. The difference between gross and net thermal efficiencies includes internal needs for electrical energy of a power plant, which might be not so small (5% or even more).

Table 2. Typical ranges of thermal efficiencies (gross[1]) of modern thermal and nuclear power plants (shown just for reference purposes).

3. Coal-fired thermal power plants

For thousands years, mankind used and still is using wood and coal for heating purposes. For about 100 years, coal is used for generating electrical energy at coal-fired thermal power plants worldwide. All coal-fired power plants (see Figure 2) operate based on, so-called, steam Rankine cycle, which can be organized at two different levels of pressures: 1) older or smaller capacity power plants operate at steam pressures no higher than 16 – 17 MPa and 2) modern

http://dx.doi.org/10.5772/51916

large capacity power plants operate at supercritical pressures from 23.5 MPa and up to 38 MPa (see Figure 3). Supercritical pressures[1] mean pressures above the critical pressure of water, which is 22.064 MPa (see Figure 4). From thermodynamics it is well known that higher thermal efficiencies correspond to higher temperatures and pressures (see Table 2). Therefore, usually subcritical-pressure plants have thermal efficiencies of about 34 – 40% and modern supercritical-pressure plants – 45 – 55%. Steam-generators outlet temperatures or steam-turbine inlet temperatures have reached level of about 625°C (and even higher) at pressures of 25 – 30 (35 – 38) MPa. However, a common level is about 535 – 585°C at pressures of 23.5 – 25 MPa (see Figure 3).

In spite of advances in coal-fired power-plants design and operation worldwide they are still considered as not environmental friendly due to producing a lot of carbon-dioxide emissions as a result of combustion process plus ash, slag and even acid rains (Pioro et al., 2010). However, it should be admitted that known resources of coal worldwide are the largest compared to that of other fossil fuels (natural gas and oil).

For better understanding specifics of supercritical water compared to water at subcritical pressures it is important to define special terms and expressions used at these conditions. For better understanding of these terms and expressions Figures 4 – 7 are shown below.

4. Definitions of selected terms and expressions related to critical and supercritical regions (Pioro and Mokry, 2011a)

Compressed fluid is a fluid at a pressure above the critical pressure, but at a temperature below the critical temperature.

Critical point (also called a *critical state*) is a point in which the distinction between the liquid and gas (or vapour) phases disappears, i.e., both phases have the same temperature, pressure and specific volume or density. The *critical point* is characterized by the phase-state parameters T_{cr}, P_{cr} and V_{cr} (or ρ_{cr}), which have unique values for each pure substance.

Near-critical point is actually a narrow region around the critical point, where all thermophysical properties of a pure fluid exhibit rapid variations.

Pseudocritical line is a line, which consists of pseudocritical points.

Pseudocritical point (characterized with P_{pc} and T_{pc}) is a point at a pressure above the critical pressure and at a temperature ($T_{pc} > T_{cr}$) corresponding to the maximum value of the specific heat at this particular pressure.

Supercritical fluid is a fluid at pressures and temperatures that are higher than the critical pressure and critical temperature. However, in the present chapter, a term *supercritical fluid* includes both terms – a *supercritical fluid* and *compressed fluid*.

1 See some explanations on supercritical-pressures specifics at the end of this section.

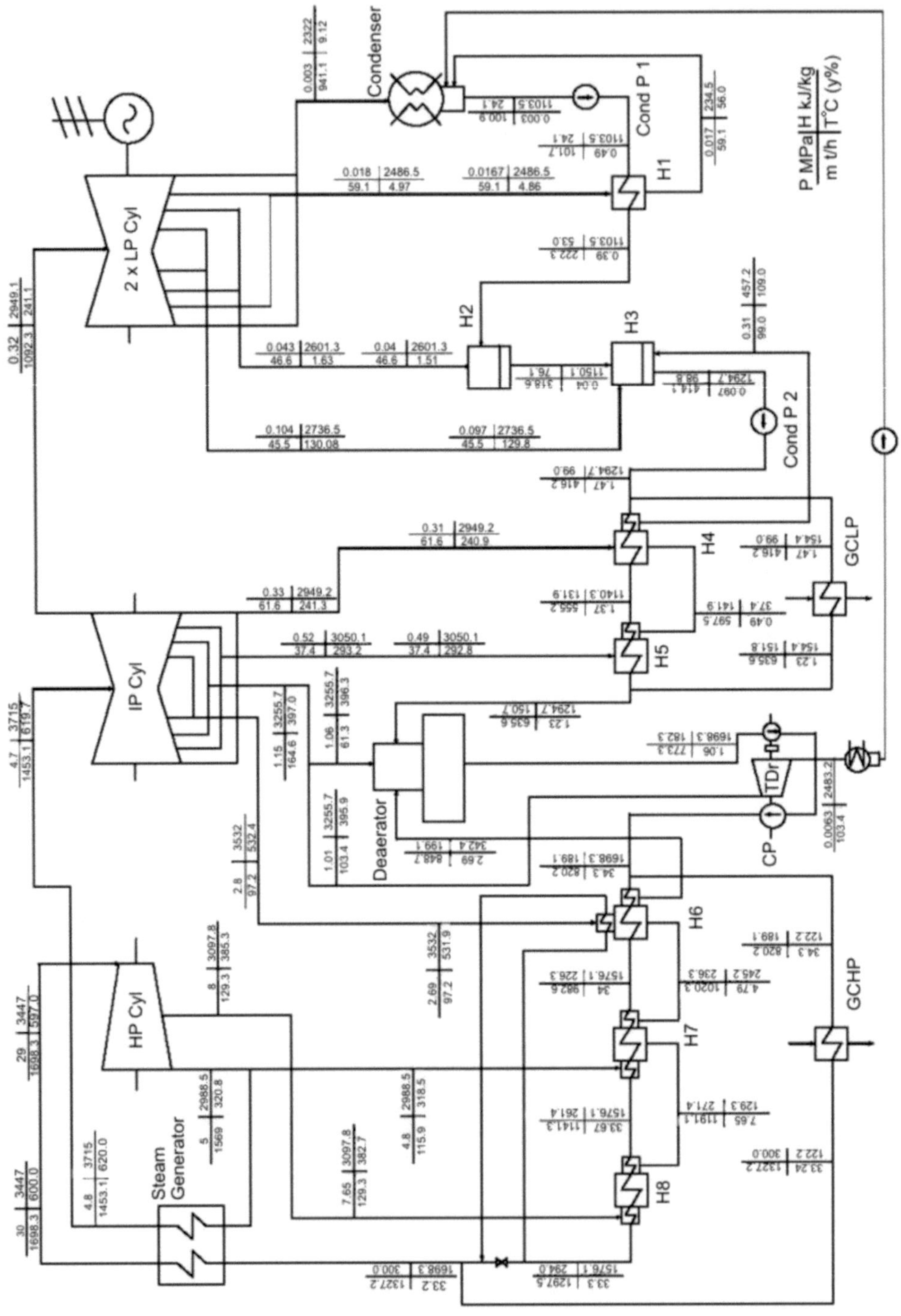

Figure 3. Supercritical-pressure single-reheat regenerative cycle 600-MW$_{el}$ Tom'-Usinsk thermal power plant (Russia) layout (Kruglikov et al., 2009): Cond P – Condensate Pump; CP – Circulation Pump; Cyl – Cylinder; GCHP – Gas Cooler of High Pressure; GCLP – Gas Cooler of Low Pressure; H – Heat exchanger (feedwater heater); HP – High Pressure; IP – Intermediate Pressure; LP – Low Pressure; and TDr – Turbine Drive.

http://dx.doi.org/10.5772/51916

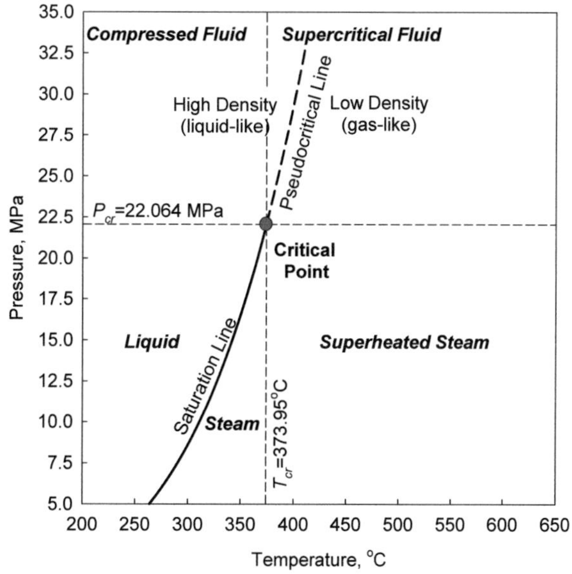

Figure 4. Pressure-Temperature diagram for water.

Supercritical "steam" is actually supercritical water, because at supercritical pressures fluid is considered as a single-phase substance. However, this term is widely (and incorrectly) used in the literature in relation to supercritical "steam" generators and turbines.

Superheated steam is a steam at pressures below the critical pressure, but at temperatures above the critical temperature.

General trends of various properties near the critical and pseudocritical points (Pioro et al., 2011; Pioro and Mokry, 2011a; Pioro and Duffey, 2007) can be illustrated on a basis of those of water. Figure 5 shows variations in basic thermophysical properties of water at a supercritical pressure of 25 MPa (also, in addition, see Figure 6). Thermophysical properties of 105 pure fluids including water, carbon dioxide, helium, refrigerants, etc., 5 pseudo-pure fluids (such as air) and mixtures with up to 20 components at different pressures and temperatures, including critical and supercritical regions, can be calculated using the NIST REFPROP software (2010).

At critical and supercritical pressures a fluid is considered as a single-phase substance in spite of the fact that all thermophysical properties undergo significant changes within critical and pseudocritical regions (see Figure 5). Near the critical point, these changes are dramatic. In the vicinity of pseudocritical points, with an increase in pressure, these changes become less pronounced (see Figure 6).

At supercritical pressures properties such as density (see Figure 5) and dynamic viscosity undergo a significant drop (near the critical point this drop is almost vertical) within a very narrow temperature range, while the kinematic viscosity and specific enthalpy (see Figure 5)

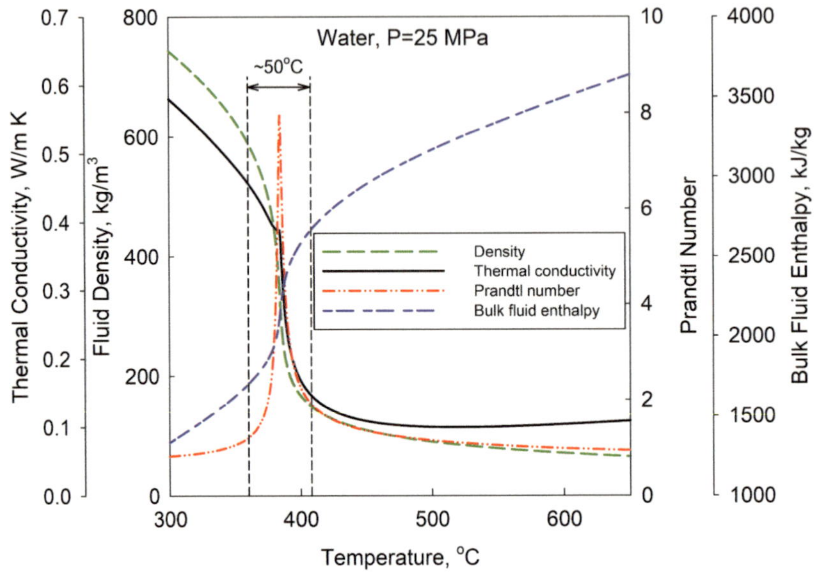

Figure 5. Variations of selected thermophysical properties of water near pseudocritical point: Pseudocritical region at 25 MPa is about ~50°C.

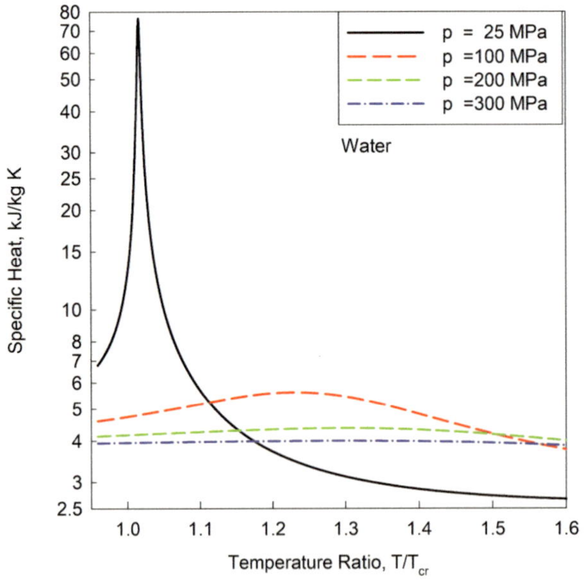

Figure 6. Specific heat variations at various supercritical pressures: Water.

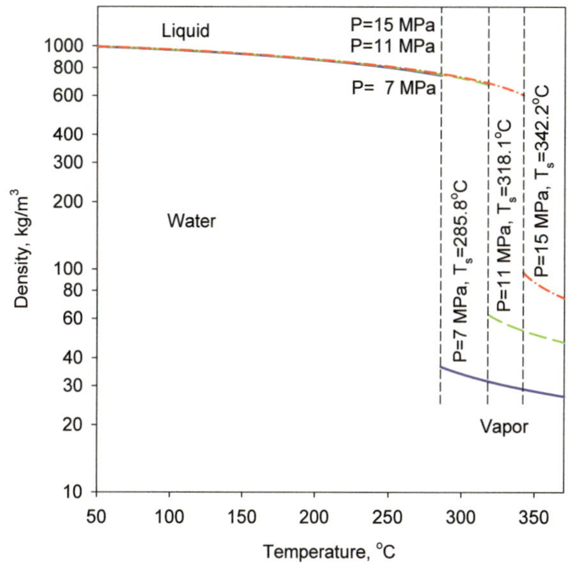

Figure 7. Density variations at various subcritical pressures for water: Liquid and vapour.

undergo a sharp increase. The volume expansivity, specific heat, thermal conductivity and Prandtl number have peaks near the critical and pseudocritical points (see Figures 5 and 6). Magnitudes of these peaks decrease very quickly with an increase in pressure (see Figure 6). Also, "peaks" transform into "humps" profiles at pressures beyond the critical pressure. It should be noted that the dynamic viscosity, kinematic viscosity and thermal conductivity (see Figure 5) undergo through the minimum right after critical and pseudocritical points.

The specific heat of water (as well as of other fluids) has a maximum value in the critical point. The exact temperature that corresponds to the specific-heat peak above the critical pressure is known as a pseudocritical temperature (see Figure 4). At pressures approximately above 300 MPa (see Figure 6) a peak (here it is better to say "a hump") in specific heat almost disappears, therefore, such term as a *pseudocritical point* does not exist anymore. The same applies to the *pseudocritical line*. It should be noted that peaks in the thermal conductivity and volume expansivity may not correspond to the pseudocritical temperature (Pioro et al., 2011; Pioro and Mokry, 2011a; Pioro and Duffey, 2007).

In general, crossing the pseudocritical line from left to right (see Figure 4) is quite similar as crossing the saturation line from liquid into vapour. The major difference in crossing these two lines is that all changes (even drastic variations) in thermophysical properties at supercritical pressures are gradual and continuous, which take place within a certain temperature range (see Figure 5). On the contrary, at subcritical pressures there is properties discontinuation on the saturation line: one value for liquid and another for vapour (see Figure 7). Therefore, supercritical fluids behave as single-phase substances (Gupta et al., 2012). Also, when dealing with supercritical fluids we usually apply the term *"pseudo"* in front of a *critical point*, *boiling*, *film boiling*, etc. Specifics of heat transfer at supercritical pressures can be found in Pioro et al. (2011), Mokry et al. (2011), Pioro and Mokry (2011b), and Pioro and Duffey (2007).

5. Combined-cycle thermal power plants

Natural gas is considered as a relatively "clean" fossil fuel compared to coal and oil, but still emits a lot of carbon dioxide due to combustion process when it used for electrical generation. The most efficient modern thermal power plants with thermal efficiencies within a range of 50 – 62% are, so-called, combined-cycle power plants, which use natural gas as a fuel (see Figure 8).

In spite of advances in thermal power plants design and operation, they still emit carbon dioxide into atmosphere, which is currently considered as one of the major reasons for a climate change. In addition, all fossil-fuel resources are depleting quite fast. Therefore, a new reliable and environmental friendly source for the electrical-energy generation should be considered.

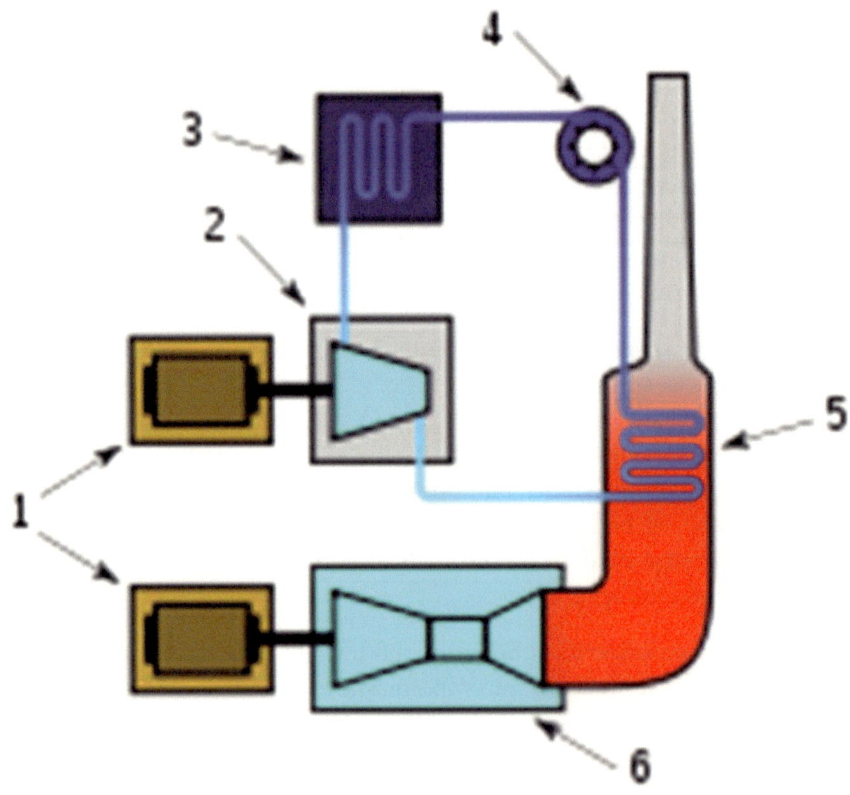

Figure 8. Working principle of combined-cycle thermal power plant (gas turbine (Brayton cycle) and steam turbine (Rankine cycle) plant) (Wikipedia, 2012): 1 electrical generators; 2 steam turbine; 3 condenser; 4 circulation pump; 5 steam generator / exhaust-gases heat exchanger; and 6 gas turbine.

6. Nuclear power plants

6.1. Modern nuclear reactors

Nuclear power is also a non-renewable source as the fossil fuels, but nuclear resources can be used for significantly longer time than some fossil fuels plus nuclear power does not emit carbon dioxide into atmosphere. Currently, this source of energy is considered as the most viable one for electrical generation for the next 50 – 100 years.

For better understanding specifics of current and future nuclear-power reactors it is important to define their various classifications.

6.2. Classifications of nuclear-power reactors

1. By neutron spectrum: (a) thermal (the vast majority of current nuclear-power reactors), (b) fast (currently, only one nuclear-power reactor is in operation in Russia: SFR – BN-600), and (c) interim or mixed spectrum.
2. By reactor-core design:

i. Neutron-core design: (a) homogeneous, i.e., the fuel and reactor coolant are mixed together (one of the Generation IV nuclear-reactors concepts) and (b) heterogeneous, i.e., the fuel and reactor coolant are separated through a sheath or cladding (currently, all nuclear-power reactors);

ii. General core design: (a) Pressure-Vessel (PV) (the majority of current nuclear-power reactors including PWRs, BWRs, etc.) and (b) Pressure-Channel (PCh) or Pressure-Tube (PT) reactors (CANDU ((CANada Deuterium-Uranium) reactors, RBMKs), EGPs (Power Heterogeneous Loop reactor (in Russian abbreviations)), etc.).

3. By coolant:

i. Water-cooled reactors: (a) Light-Water (H_2O) Reactors (LWRs) - PWRs, BWRs, RBMKs, EGPs, and (b) heavy-water (D_2O) reactors – mainly CANDU-type reactors.

ii. Gas-cooled reactors: Carbon-dioxide-cooled reactors (Magnox[2] reactors (Gas-Cooled Reactors (GCRs)) and AGRs) and helium-cooled reactors (two Generation IV nuclear-reactor concepts); (c) liquid-metal-cooled reactors: SFR, lead-cooled and lead-bismuth-cooled reactors (Generation IV nuclear-reactor concepts); (d) molten-salt-cooled reactors (one of Generation IV nuclear-reactor concepts); and (e) organic-fluids-cooled reactors (existed only as experimental reactors some time ago).

4. By type of a moderator (Kirillov et al., 2007): (a) liquid moderator (H_2O and D_2O are currently used in nuclear-power reactors as moderators) and (b) solid moderator (graphite[3] (RBMKs, EGPs, Magnox reactors (GCRs), AGRs), zirconium hydride (ZrH_2), beryllium (Be) and beryllium oxide (BeO)).

---

2 In this reactor the fuel-rod sheath is made of magnesium alloy known by the trade name as "Magnox", which was used as the name of the reactor (Hewitt and Collier, 2000).

5. By application: (a) power reactors (PWRs, BWRs, CANDU reactors, GCRs, AGRs, RBMKs, EGPs, SFR from current fleet) (b) research reactors (for example, NRU (National Research Universal) (AECL, Canada, http://www.aecl.ca/Programs/NRU.htm), etc.), (c) transport or mobile reactors (submarines and ships (icebreakers, air-carriers, etc.), (d) industrial reactors for isotope production (for example, NRU), etc., and (e) multipurpose reactors (for example, NRU, etc.).

6. By number of flow circuits: (a) single-flow circuit (once-through or direct-cycle reactors) (BWRs, RBMKs, EGPs); (b) double-flow circuit (PWRs, PHWRs, GCRs, AGRs) and (c) triple-flow circuit (usually SFRs).

7. By fuel enrichment: (a) Natural-Uranium fuel (NU) (99.3$\%_{wt}$ of non-fissile isotope uranium-238 (^{238}U) and 0.7% of fissile isotope uranium-235 (^{235}U)) (CANDU-type reactors, Magnox reactors), (b) Slightly-Enriched Uranium (SEU) (0.8 – 2$\%_{wt}$ of ^{235}U), (c) Low-Enriched Uranium (LEU) (2 – 20% of ^{235}U) (the vast majority of current nuclear-power reactors: PWRs, BWRs, AGRs, RBMKs, EGPs), and (d) Highly-Enriched Uranium (HEU) (>20$\%_{wt}$ of ^{235}U) (can be SFR).

8. By used fuel (Peiman et al., 2012): (a) Conventional nuclear fuels (low thermal conductivity): Uranium dioxide (UO_2, used in the vast majority of nuclear-power reactors), Mixed OXides (MOX) (($U_{0.8}Pu_{0.2})O_2$, where 0.8 and 0.2 are the molar parts of UO_2 and PuO_2, used in some reactors) and thoria (ThO_2) (considered for a possible use instead of UO_2 in some countries, usually, with large resources of this type of fuel, for example, in India); and (b) alternative nuclear fuels (high thermal conductivity): Uranium dioxide plus silicon carbide (UO_2–SiC), uranium dioxide composed of graphite fibre (UO_2–C), uranium dioxide plus beryllium oxide (UO_2–BeO), uranium dicarbide (UC_2), uranium monocarbide (UC) and uranium mononitride (UN); the last three fuels are mainly intended for use in high-temperature Generation IV reactors.

First success of using nuclear power for electrical generation was achieved in several countries within 50-s, and currently, Generations II and III nuclear-power reactors are operating around the world (see Tables 3 and 4 and Figures 9-15). In general, definitions of nuclear-reactors generations are as the following: 1) Generation I (1950 – 1965) – early prototypes of nuclear reactors; 2) Generation II (1965 – 1995) – commercial power reactors; 3) Generation III (1995 – 2010) – modern reactors (water-cooled NPPs with thermal efficiencies within 30 – 36%; carbon-dioxide-cooled NPPs with the thermal efficiency up to 42% and liquid sodium-cooled NPPs with the thermal efficiency up to 40%) and Generation III+ (2010 – 2025) – reactors with improved parameters (evolutionary design improvements) (water-cooled NPPs with the thermal efficiency up to 38%) (see Table 5); and 4) Generation IV (2025 - …) – reactors in principle with new parameters (NPPs with the thermal efficiency of 43 – 50% and even higher for all types of reactors).

---

3 After the Chernobyl NPP severe nuclear accident in Ukraine in 1986 with the RBMK reactor, graphite is no longer considered as a possible moderator in any water-cooled reactors.

http://dx.doi.org/10.5772/51916

1. PWRs (see Figure 9 and Tables 6 and 7) – 267 (*268*) (248 (*247*) GW_{el}); forthcoming – 89 (93 GW_{el}).
2. BWRs or ABWRs (see Figure 10 and Table 8) – 84 (*92*) (85 (*78*) GW_{el}); forthcoming – 6 (8 GW_{el}).
3. GCRs (see Figures 12 and 13) – 17 (*18*) (9 GW_{el}), UK (AGRs (see Figure 12) – 14 and Magnox (see Figure 13) – 3); forthcoming – 1 (0.2 GW_{el}).
4. PHWRs (see Figure 11) – 51 (*50*) (26 (*25*) GW_{el}), Argentina 2, Canada 22, China 2, India 18, Pakistan 1, Romania 2, S. Korea 4; forthcoming – 9 (5 GW_{el}).
5. Light-water, Graphite-moderated Reactors (LGRs) (see Figure 14 and Table 6) – 15 (10 GW_{el}), Russia, 11 RBMKs and 4 EGPs[1] (earlier prototype of RBMK).
6. Liquid-Metal Fast-Breeder Reactors (LMFBRs) (see Figure 15 and Table 6) – 1 (0.6 GW_{el}), SFR, Russia; forthcoming – 4 (1.5 GW_{el}).

[1]EGP –channel-type, graphite moderated, light water, boiling reactor with natural circulation.

Table 3. Operating and forthcoming nuclear-power reactors (in total - 435 (*444*) (net 370 (*378*) GW_{el}) (Nuclear News, 2012); (*in Italic mode*) - number of power reactors before the Japan earthquake and tsunami disaster in spring of 2011) (Nuclear News, 2011).

No.	**Nation**	**# Units**	**Net GW_{el}**
1.	USA	104	103
2.	France	58	63
3.	Japan[1]	50 (*54*)	44 (*47*)
4.	Russia	33	24
5.	S. Korea	21 (*20*)	19 (*18*)
6.	Canada[2]	22	15
7.	Ukraine	15	13
8.	Germany	9 (*17*)	12 (*20*)
9.	UK	18 (*19*)	10
10.	China 14 (*13*)	11 (*10*)	

[1]Currently, i.e., in October of 2012, only 2 reactors in operation. However, more reactors are planned to put into operation.

[2]Currently, i.e., October of 2012, 18 reactors in operation and 4 already shut-down.

Table 4. Current nuclear-power reactors by nation (10 first nations) (Nuclear News, 2012); (*in Italic mode*) - number of power reactors before the Japan earthquake and tsunami disaster in spring of 2011) (Nuclear News, 2011).

ABWR – Toshiba, Mitsubishi Heavy Industries and Hitachi-GE (Japan-USA) (the only one Generation III+ reactor design already implemented in the power industry).

Advanced CANDU Reactor (ACR-1000) AECL, Canada.

Advanced Plant (AP-1000) – Toshiba-Westinghouse (Japan-USA) (6 under construction in China and 6 planned to be built in China and 6 – in USA).

Advanced PWR (APR-1400) – South Korea (4 under construction in S. Korea and 4 planned to be built in United Arad Emirates).

European Pressurized-water Reactor (EPR) AREVA, France (1 should be put into operation in Finland, 1 under construction in France and 2 in China and 2 planned to be built in USA).

VVER[1] (design AES[2]-2006 or VVER-1200 with ~1200 MW_{el}) – GIDROPRESS, Russia (2 under construction in Russia and several more planned to be built in various countries). Reference parameters of Generation III+ VVER (Ryzhov et al., 2010) are listed below:

Parameter	Value
Thermal power, MW_{th}	3200
Electric power, MW_{el}	1160
NPP thermal efficiency, %	36
Primary coolant pressure, MPa	16.2
Steam-generator pressure, MPa	7.0
Coolant temperature at reactor inlet, °C	298
Coolant temperature at reactor outlet, °C	329
NPP service life, years	50
Main equipment service life, years	60
Replaced equipment service life, years, not less than	30
Capacity factor, %	up to 90
Load factor, %	up to 92
Equipment availability factor	99
Length of fuel cycle, years	4-5
Frequency of refuellings, months	12-18
Fuel assembly maximum burn-up, MW day/kgU	up to 60-70
Inter-repair period length, years	4-8
Annual average length of scheduled shut-downs (for refuellings, scheduled maintenance work), days per year	16-40
Refueling length, days per year	≤16

http://dx.doi.org/10.5772/51916

Number of not scheduled reactor shutdowns per year	≤1
Frequency of severe core damage, 1/year	$<10^{6}$
Frequency of limiting emergency release, 1/year	$<10^{7}$
Efficient time of passive safety and emergency control system operation without operator's action and power supply, hour	≥24
OBE/SSE, magnitude of MSK-64 scale	6 and 7*
Compliance with EUR requirements, yes/no	Yes
*RP main stationary equipment is designed for SSE of magnitude 8.	

[1]VVER or WWER - Water Water Power Reactor (in Russian abbreviations).

[2]AES – Atomic Electrical Station (Nuclear Power Plant) (in Russian abbreviations).

Table 5. Selected Generation III+ reactors (deployment in 5–10 years).

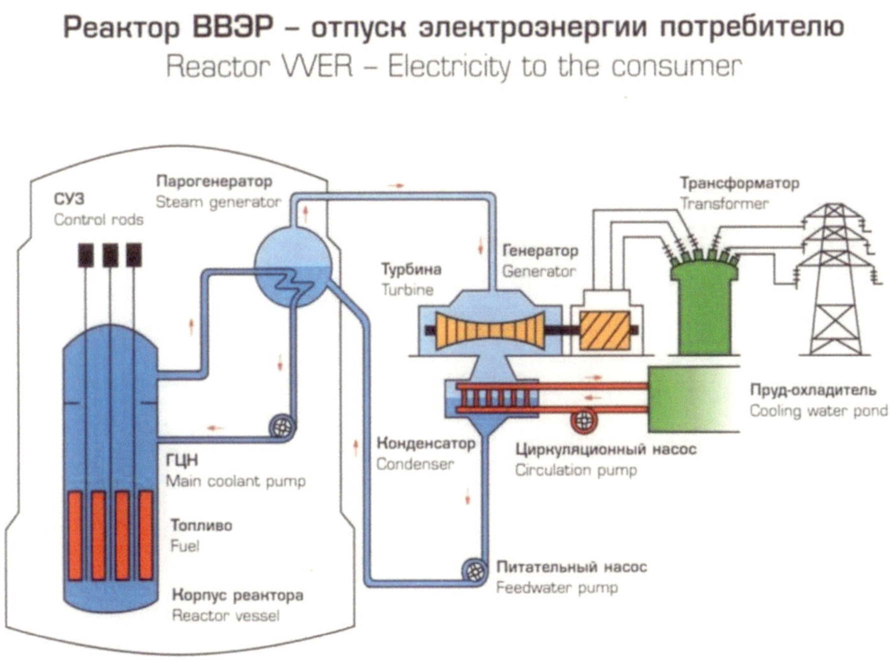

Figure 9. Scheme of typical Pressurized Water Reactor (PWR) (Russian VVER) NPP (ROSENERGOATOM, 2004) (courtesy of ROSENERGOATOM): General basic features – 1) thermal neutron spectrum; 2) uranium-dioxide (UO_2) fuel; 3) fuel enrichment about 4%; 4) indirect cycle with steam generator (also, a pressurizer required (not shown)), i.e., double flow circuit (double loop); 5) Reactor Pressure Vessel (RPV) with vertical fuel rods (elements) assembled in bundle strings cooled with upward flow of light water; 6) reactor coolant and moderator are the same fluid; 7) reactor coolant outlet parameters: Pressure 15 – 16 MPa (T_{sat} = 342 – 347°C) and temperatures inlet / outlet 290 – 325°C; and 8) power cycle - subcritical-pressure regenerative Rankine steam-turbine cycle with steam reheat[4] (working fluid - light water, turbine steam inlet parameters: Saturation pressure of 6 – 7 MPa and saturation temperature of 276 – 286°C).

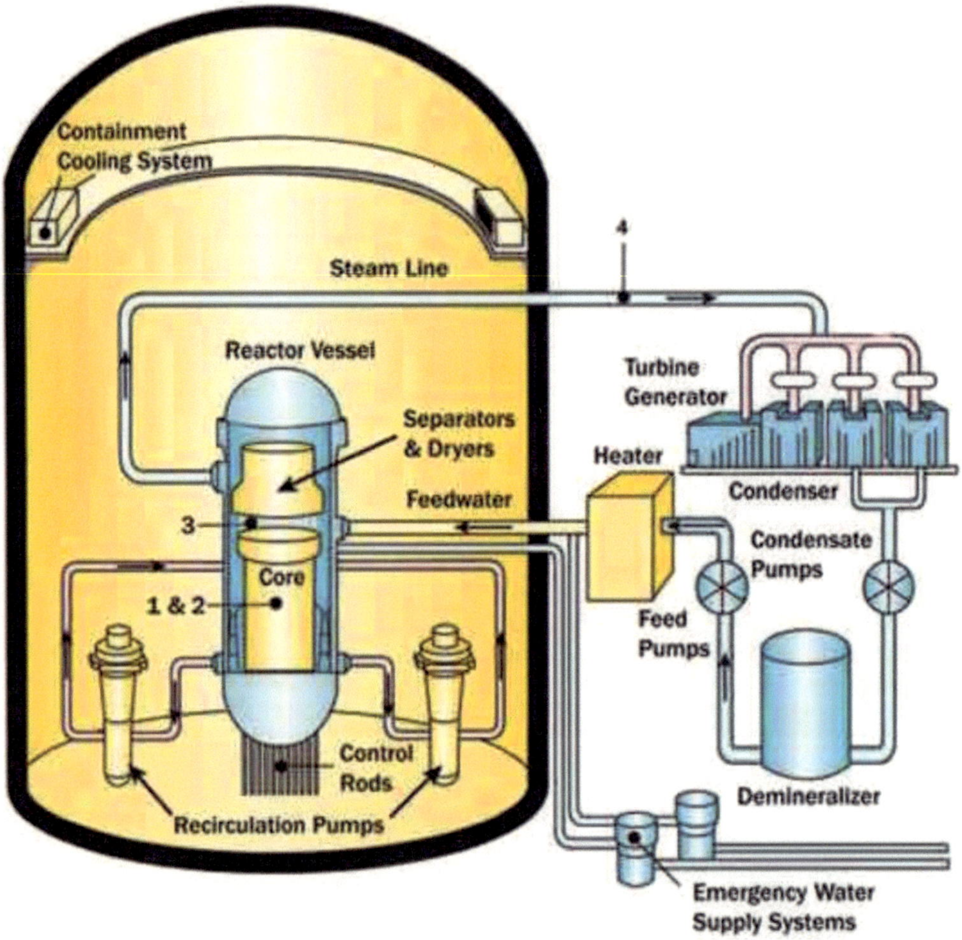

Figure 10. Scheme of typical Boiling Water Reactor (BWR) NPP (courtesy of NRC USA): General basic features – 1) thermal neutron spectrum; 2) uranium-dioxide (UO_2) fuel; 3) fuel enrichment about 3%; 4) direct cycle with steam separator (steam generator and pressurizer are eliminated), i.e., single-flow circuit (single loop); 5) RPV with vertical fuel rods (elements) assembled in bundle strings cooled with upward flow of light water (water and water-steam mixture); 6) reactor coolant, moderator and power-cycle working fluid are the same fluid; 7) reactor coolant outlet parameters: Pressure about 7 MPa and saturation temperature at this pressure is about 286°C; and 8) power cycle - subcritical-pressure regenerative Rankine steam-turbine cycle with steam reheat.

---

4 For the reheat the primary steam is used. Therefore, the reheat temperature is lower than the primary steam temperature. In general, the reheat parameters at NPPs are significantly lower than those at thermal power plants.

http://dx.doi.org/10.5772/51916

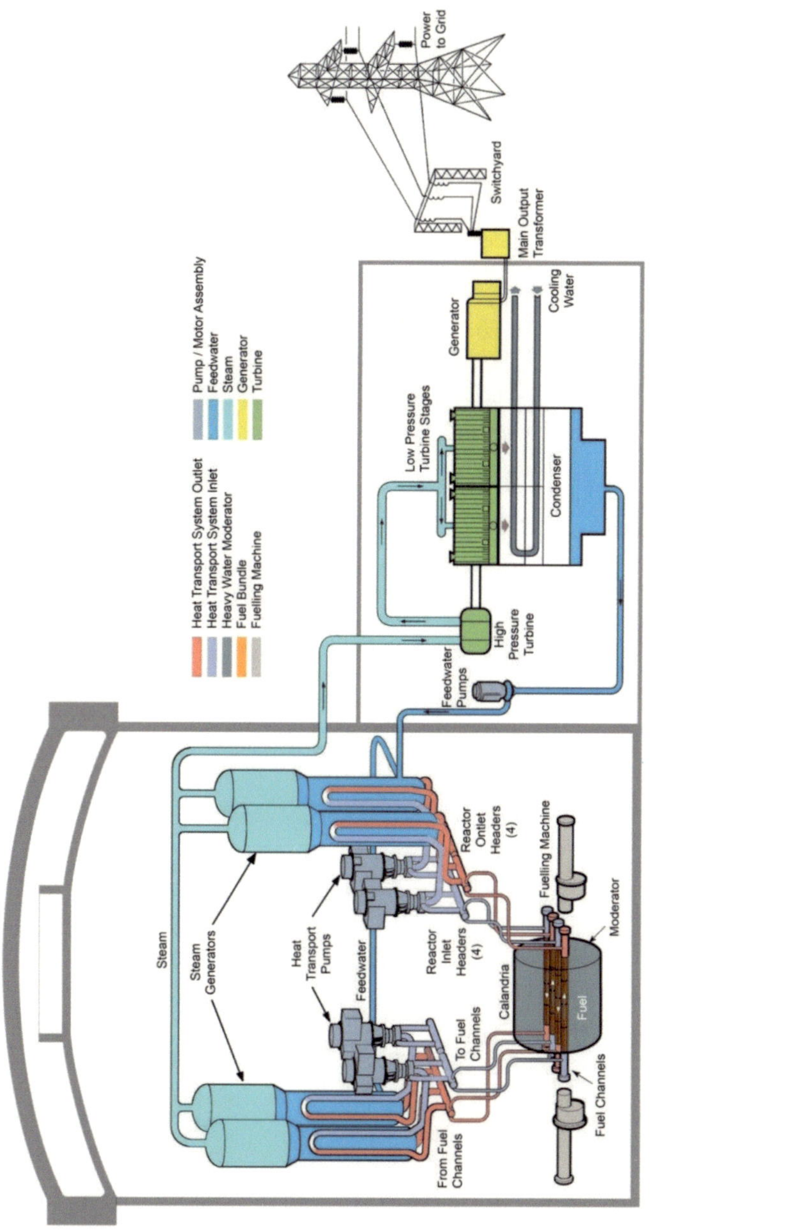

Figure 11. Scheme of CANDU-6 reactor (PHWR) NPP (courtesy of AECL): General basic features – 1) thermal-neutron spectrum; 2) natural uranium-dioxide (UO_2) fuel; 3) fuel enrichment about 0.7%; 4) indirect cycle with steam generator (also, a pressurizer required (not shown)), i.e., double-flow circuit (double loop); 5) pressure-channel design: Calandria vessel with horizontal fuel channels (see Figure 16c); 6) reactor coolant and moderator separated, but both are heavy water; 7) reactor coolant outlet parameters: Pressure about 9.9 MPa and temperature close to saturation (310°C); 8) on-line refuelling; and 9) power cycle - subcritical-pressure regenerative Rankine steam-turbine cycle with steam reheat (working fluid light water, turbine steam inlet parameters: Saturation pressure of ~4.6 MPa and saturation temperature of 259°C).

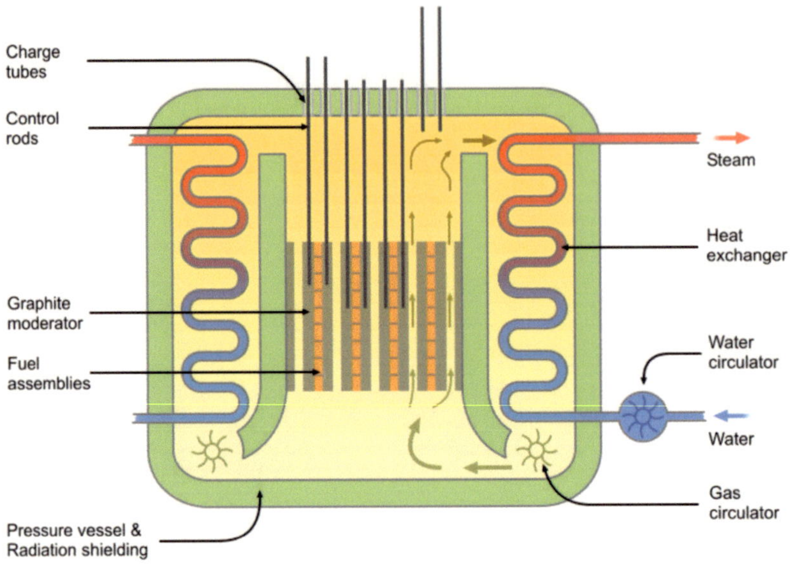

Figure 12. Scheme of Advanced Gas-cooled Reactor (AGR) (Wikimedia, 2012). Note that the heat exchanger is contained within the steel-reinforced concrete combined pressure vessel and radiation shield.

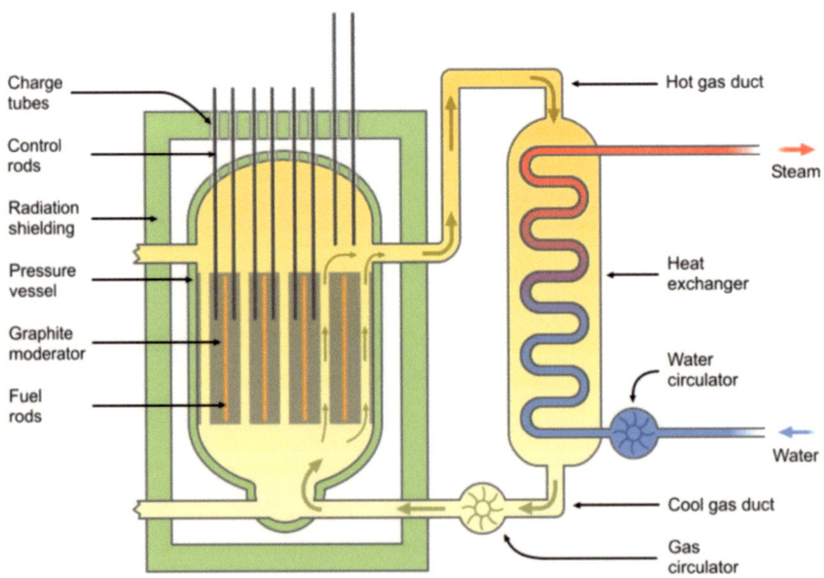

Figure 13. Scheme of Magnox nuclear reactor (GCR) showing gas flow (Wikipidea, 2012). Note that the heat exchanger is outside the concrete radiation shielding. This represents an early Magnox design with a cylindrical, steel, pressure vessel.

http://dx.doi.org/10.5772/51916

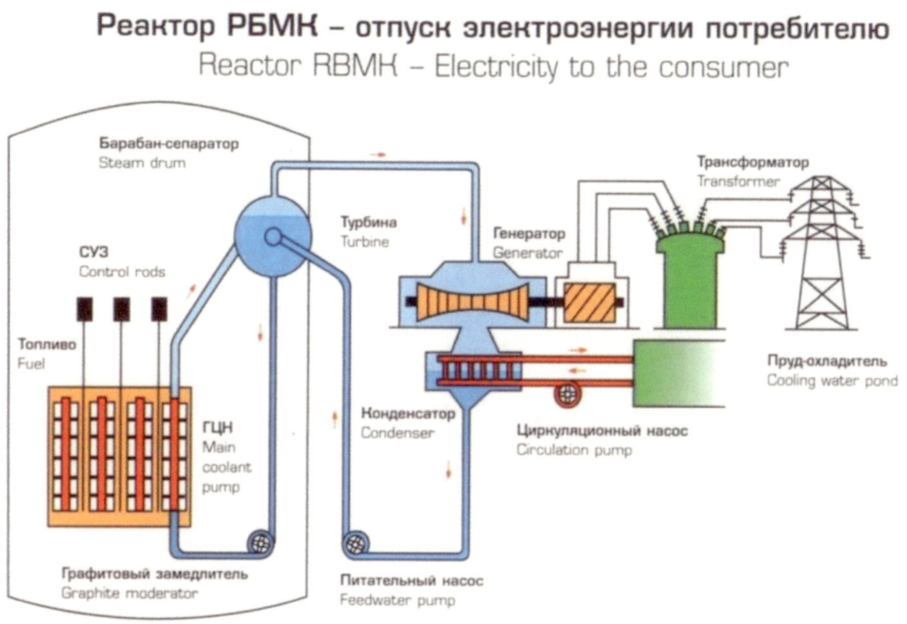

Figure 14. Scheme of Light-water Graphite-moderated Reactor (LGR) (Russian RBMK) NPP (ROSENERGOATOM, 2004) (courtesy of ROSENERGOATOM).

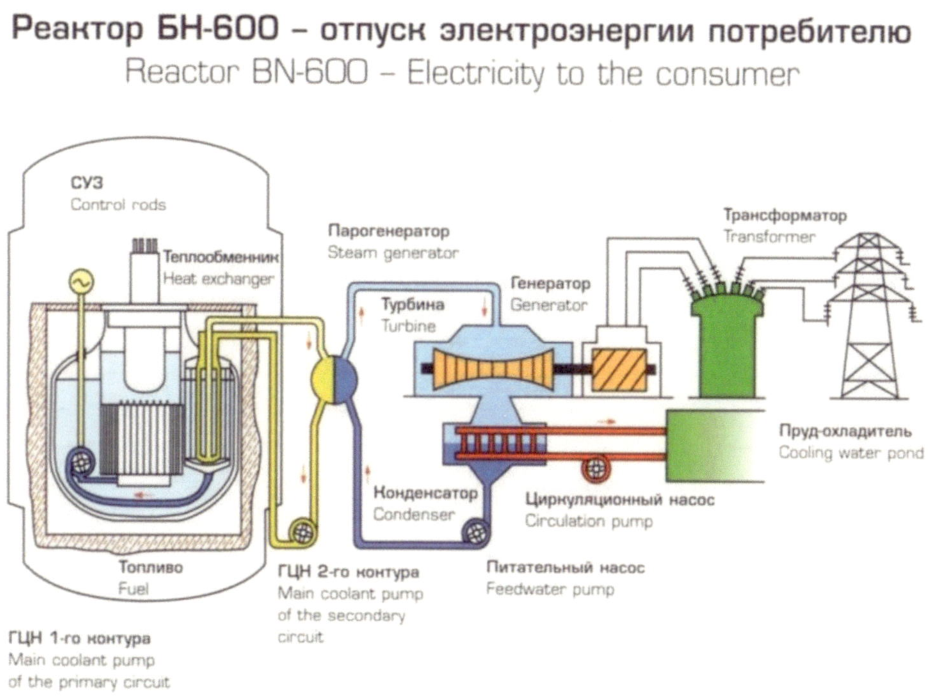

Figure 15. Scheme of Liquid-Metal Fast-Breeder Reactor (LMFBR) or SFR (Russian BN-600) NPP (ROSENERGOATOM, 2004) (courtesy of ROSENERGOATOM).

Parameter	VVER-440	VVER-1000 (Figure 9)	EGP-6	RBMK-1000 (Figure 14)	BN-600 (Figure 15)
Thermal power, MW_{th}	1375	3000	62	3200	1500
Electrical power, MW_{el}	440	1000	12	1000	600
Thermal efficiency, %	32.0	33.3	19.3	31.3	40.0
Coolant pressure, MPa	12.3	15.7	6.2	6.9	~0.1
Coolant massflow rate, t/s	11.3	23.6	0.17	13.3	6.9
Coolant inlet/outlet temperatures, °C	270/298	290/322	265	284	380/550
Steam massflow rate, t/s	0.75	1.6	0.026	1.56	0.18
Steam pressure, MPa	4.3	5.9	6.5	6.6	15.3
Steam temperature, °C	256	276	280	280	505
Reactor core: Diameter/Height m/m	3.8/11.8	4.5/10.9	4.2/3.0	11.8/7	2.1/0.75
Fuel enrichment, %	3.6	4.3	3.0;3.6	2.0-2.4	21;29.4
No. of fuel bundles	349	163	273	1580	369

Table 6. Major Parameters of Russian Power Reactors (Grigor'ev and Zorin, 1988).

Pressure Vessel (PV) ID, m	3.91
PV wall thickness, m	0.19
PV height without cover, m	10.8
Core equivalent diameter, m	2.88
Core height, m	2.5
Volume heat flux, MW/m^3	83
No. of fuel assemblies	349
No of rods per assembly	127
Fuel mass, ton	42
Part of fuel reloaded during year	1/3
Fuel	UO_2

Table 7. Additional parameters of VVER-1000.

Power	
Thermal output, MW_{th}	3830
Electrical output, MW_e	1330
Thermal efficiency, %	34
Specific power, kW/kg(U)	26
Power density, kW/L	56

http://dx.doi.org/10.5772/51916

Power	
Average linear heat flux, kW/m	20.7
Fuel-rod heat flux average/max, MW/m^2	0.51/1.12
Core	
Length, m	3.76
OD, m	4.8
Reactor-coolant system	
Pressure, MPa	7.17
Core massflow rate, kg/s	14,167
Core void fraction average/max	0.37 / 0.75
Feedwater inlet temperature, °C	216
Steam outlet temperature, °C	290
Steam outlet massflow rate, kg/s	2083
Reactor Pressure Vessel	
Inside Diameter, m	6.4
Height, m	22.1
Wall thickness, m	0.15
Fuel	
Fuel pellets	UO^2
Pellet OD, mm	10.6
Fuel rod OD, mm	12.5
Zircaloy sheath (cladding) thickness, mm	0.86

Table 8. Typical parameters of US BWR (Shultis and Faw, 2008).

Analysis of data listed in Table 3 shows that the vast majority nuclear reactors are water-cooled units. Only reactors built in UK are the gas-cooled type, and one reactor in Russia uses liquid sodium for its cooling.

UK carbon-dioxide-cooled reactors consist of two designs (Hewitt and Collier, 2000): 1) older design – Magnox reactor (GCR) (see Figure 13) and 2) newer design – AGR (see Figure 12). The Magnox design is a natural-uranium graphite-moderated reactor with the following parameters: Coolant – carbon dioxide; pressure - 2 MPa; outlet/inlet temperatures – 414/250°C; core diameter – about 14 m; height – about 8 m; magnesium-alloy sheath with fins; and thermal efficiency – about 32%. AGRs have the following parameters: Coolant – carbon dioxide; pressure - 4 MPa; outlet/inlet temperatures – 650/292°C; secondary-loop steam – 17 MPa and 560°C; stainless-steel sheath with ribs and hollow fuel pellets (see Figure 16b); enriched fuel 2.3%; and thermal efficiency – about 42% (the highest in nuclear-power industry so far). However, both these reactor designs will not be constructed anymore. They will just operate to the end of their life term and will be shut down. The same is applied to Russian RBMKs and EGPs.

Just for reference purposes, typical fuel elements (rods) / bundles of various reactors are shown in Figure 16, and typical sheath temperatures, heat transfer coefficients and heat fluxes are listed below.

Typical maximum sheath temperatures for steady operation (Hewitt and Collier, 2000)

BWR	300°C
PWR	320°C
Magnesium alloy (Magnox reactor)	450°C
AGR stainless steel	750°C
SFR	750°C

Typical heat-transfer-coefficient values for reactor coolants (Hewitt and Collier, 2000)

High-pressure carbon dioxide (forced convection)	1 kW/m^2K
Boiling water in a kettle (nucleate pool boiling)	10 kW/m^2K
Water (single-phase forced convection)	30 kW/m^2K
PHWR	46 kW/m^2K
Liquid sodium (forced convection)	55 kW/m^2K
Boiling water (flow boiling)	60 kW/m^2K

Typical Heat Fluxes (HFs) for steady operation (Hewitt and Collier, 2000)

	HF, MW/m^2	T_{sheath}-T_{fluid}, °C
Boiling water in a kettle	0.15	15
CANDU reactor	0.625	14
BWR	1.0	15
PWR	1.5	50
SFR	2.0	35

Scheme 1. Typical maximum sheath temperatures for steady operation (Hewitt and Collier, 2000)

All current NPPs and oncoming Generation III+ NPPs are not very competitive with modern thermal power plants in terms of their thermal efficiency, a difference in values of thermal efficiencies between thermal and NPPs can be up to 20 – 30% (see Table 2). Therefore, new generation NPPs should be designed and built in the nearest future.

7. Next generation nuclear reactors

The demand for clean, non-fossil based electricity is growing; therefore, the world needs to develop new nuclear reactors with higher thermal efficiencies in order to increase electricity generation and decrease detrimental effects on the environment. The current fleet of NPPs is classified as Generation II and III (just a limited number of Generation III+ reactors (mainly, ABWRs) operates in some countries). However, all these designs (here we are talking about

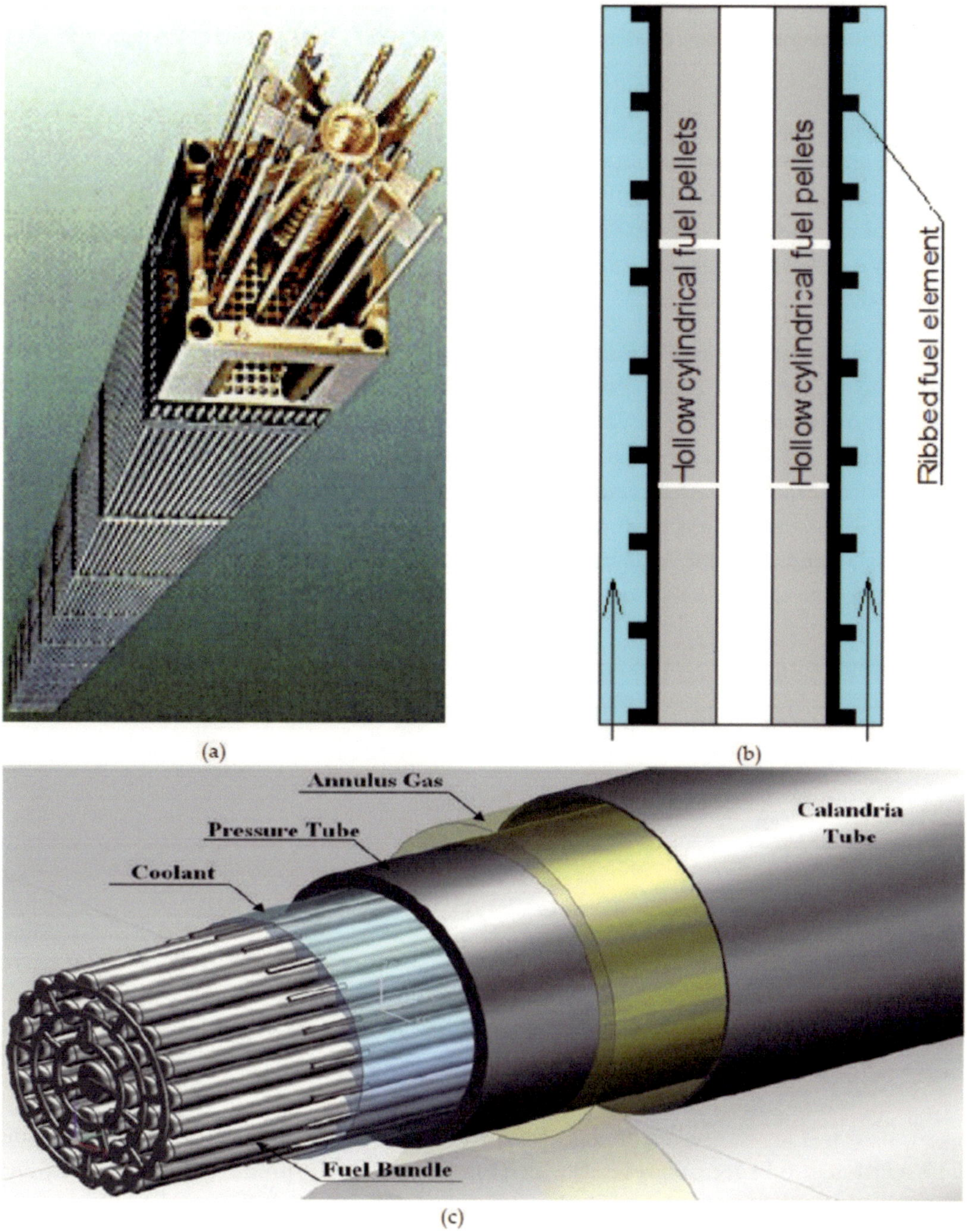

Figure 16. Typical PWR bundle string (courtesy of KAERI, http://www.nucleartourist.com/systems/pwrfuel1.htm) (a); AGR ribbed fuel element with hollow fuel pellet (Hewitt and Collier, 2000) (b); and CANDU reactor fuel channel (based on AECL design) (c).

only water-cooled power reactors) are not as energy efficient as they should be, because their operating temperatures are relatively low, i.e., below 350°C for a reactor coolant and even lower for steam.

Currently, a group of countries, including Canada, EU, Japan, Russia, USA and others have initiated an international collaboration to develop the next generation nuclear reactors (Generation IV reactors). The ultimate goal of developing such reactors is an increase in thermal efficiencies of NPPs from 30 – 36% to 45 - 50% and even higher. This increase in thermal efficiency would result in a higher production of electricity compared to current LWR technologies per 1 kg of uranium.

The Generation IV International Forum (GIF) Program has narrowed design options of nuclear reactors to six concepts. These concepts are: 1) Gas-cooled Fast Reactor (GFR) or just High Temperature Reactor (HTR), 2) Very High Temperature Reactor (VHTR), 3) Sodium-cooled Fast Reactor (SFR), 4) Lead-cooled Fast Reactor (LFR), 5) Molten Salt Reactor (MSR), and 6) SuperCritical Water-cooled Reactor (SCWR). Figures 17 – 24 show schematics of these concepts. These nuclear-reactor concepts differ one from each other in terms of their design, neutron spectrum, coolant, moderator, operating temperatures and pressures. A brief description of each Generation IV nuclear-reactor concept has been provided below.

Reactor Parameter	Unit	Reference Value
Reactor power	MW_{th}	600
Coolant inlet/outlet temperatures	°C	490/850
Pressure	MPa	9
Coolant massflow rate	kg/s	320
Average power density	MW_{th}/m^3	100
Reference fuel compound		UPuC/SiC (70/30%) with about 20% Pu
Net-plant efficiency	%	48

Table 9. Key-design parameters of Gas-cooled Fast Reactor (GFR) concept.

Gas-cooled Fast Reactor (GFR) or High Temperature Reactor (HTR) (see Figure 17 and Table 9.) is a fast-neutron-spectrum reactor, which can be used for the production of electricity and co-generation of hydrogen through thermochemical cycles or high-temperature electrolysis. The coolant is helium with inlet and outlet temperatures of 490 and 850°C, respectively. The net plant efficiency is about 48% with the direct Brayton helium-gas-turbine cycle. Table 9 lists a summary of design parameters for GFR (US DOE, 2002). However, due to some problems with implementation of the direct Brayton helium-gas-turbine cycle, the indirect Rankine steam cycle or even indirect supercritical carbon-dioxide Brayton gas-turbine cycle are also considered. The indirect cycles will be linked to the GFR through heat exchangers.

http://dx.doi.org/10.5772/51916

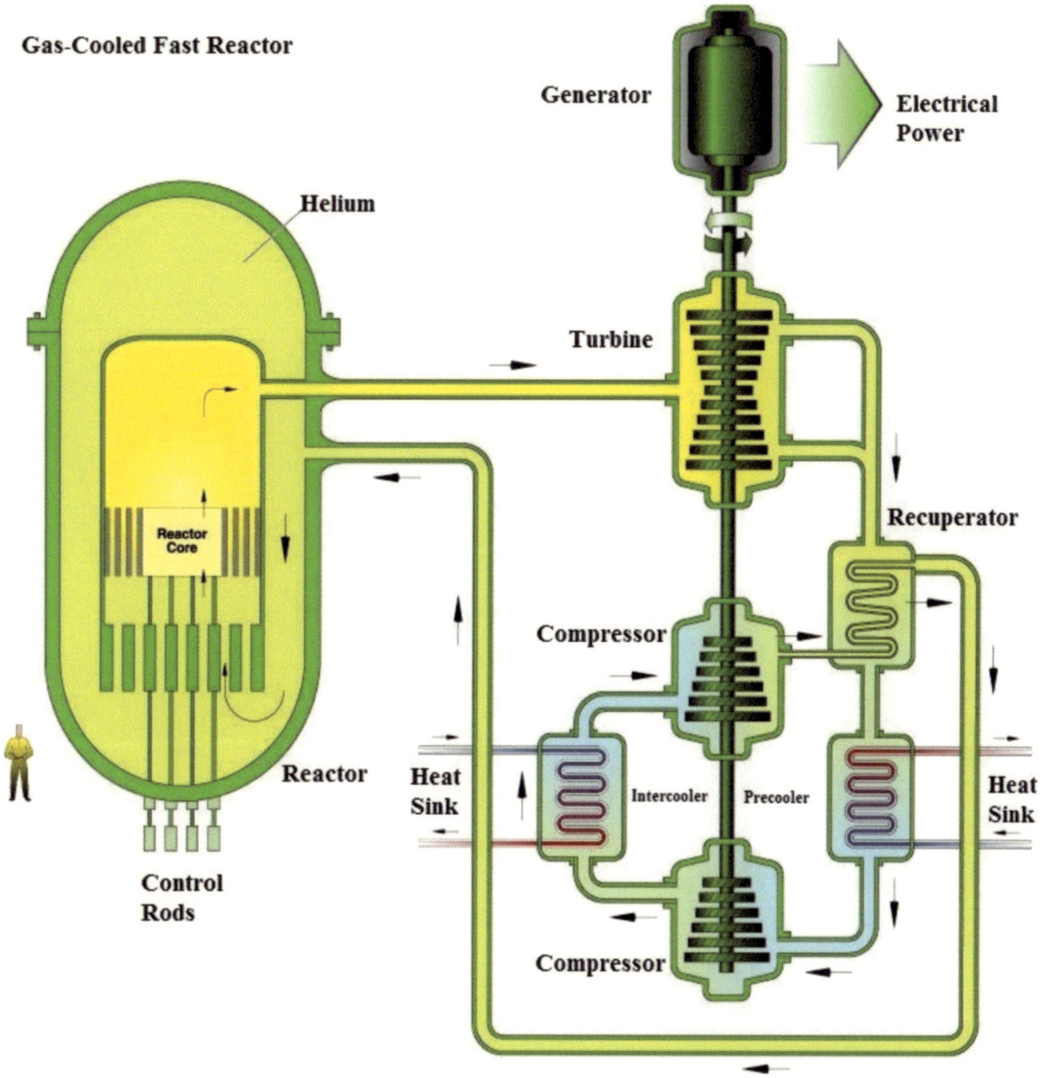

Figure 17. Scheme of Gas-cooled Fast Reactor (GFR) NPP concept (US DOE).

Very High Temperature Reactor (VHTR) (see Figure 18) is a thermal-neutron-spectrum reactor. The ultimate purpose of this nuclear-reactor design is the co-generation of hydrogen through high-temperature electrolysis. In a VHTR, graphite and helium have been chosen as the moderator and the coolant, respectively. The inlet and outlet temperatures of the coolant are 640 and 1000°C, respectively, at a pressure of 7 MPa (US DOE, 2002). Due to such high outlet temperatures, the thermal efficiency of VHTR will be above 50%. A summary of design parameters of VHTR are listed in Table 10 (US DOE, 2002).

In general, the US DOE supports research on several Generation IV reactor concepts (http://nuclear.energy.gov/genIV/neGenIV4.html). However, the priority is being given to the VHTR, as a system compatible with advanced electricity production, hydrogen co-generation and high-temperature process-heat applications.

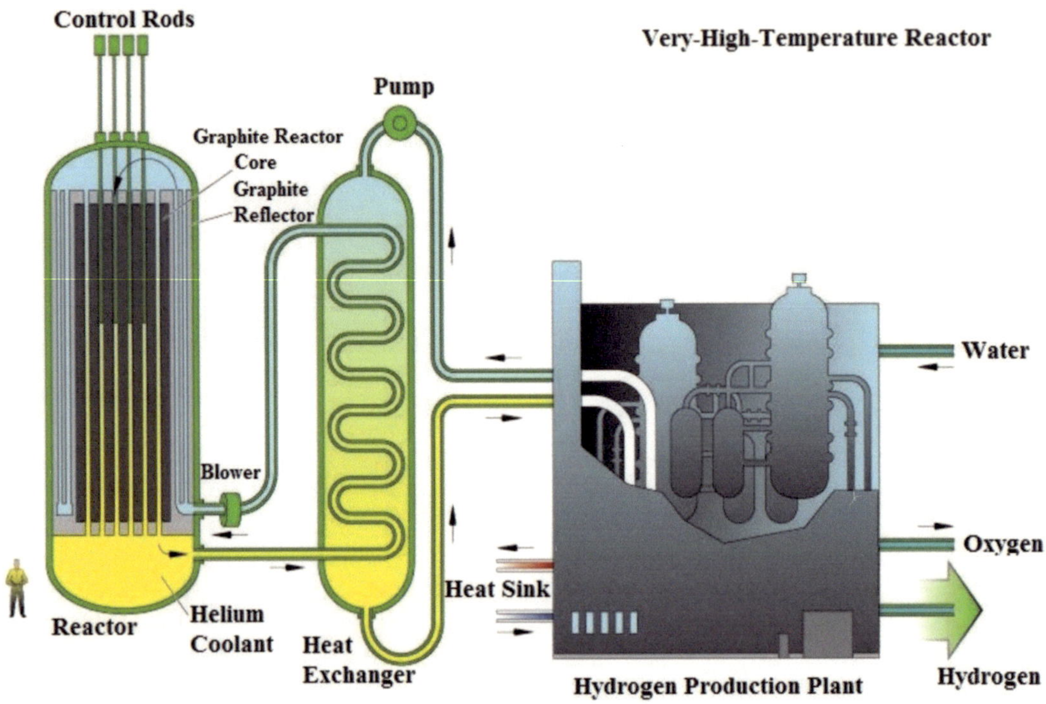

Figure 18. Scheme of Very High Temperature Reactor (VHTR) plant concept with co-generation of hydrogen (US DOE).

Reactor Parameter	Unit	Reference Value
Reactor power	MW_{th}	600
Average power density	MW_{th}/m^3	610
Coolant inlet/outlet temperatures	°C	640/1000
Coolant/Massflow rate	kg/s	Helium/320
Reference fuel compound		ZrC-coated particles in pins or pebbles
Net-plant efficiency	%	>50

Table 10. Key-design parameters of Very High Temperature Reactor (VHTR) concept.

Similar to GFR, SFR (see Figure 19) is a fast-neutron-spectrum reactor. The main objectives of SFR are the management of high-level radioactive wastes and production of electricity. SFR uses liquid sodium as a reactor coolant with an outlet temperature between 530 and 550°C at the atmospheric pressure. The primary choices of fuel for SFR are oxide and metallic fuels.

http://dx.doi.org/10.5772/51916

Table 11 lists a summary of design parameters of SFR (US DOE, 2002). The SFR concept is also on the priority list for the US DOE (http://nuclear.energy.gov/genIV/neGenIV4.html).

Currently, SFR is the only one Generation IV concept implemented in the power industry. Russia and Japan are leaders within this area. In particularly, Russia operates SFR at the Beloyarsk NPP (for details, see BN-600 in Table 6) and constructs even more powerful SFR – BN-850. Japan has operated SFR at the Monju NPP some time ago (http://en.wikipedia.org/wiki/Monju_Nuclear_Power_Plant). In Russia and Japan the SFRs are connected to the subcritical-pressure Rankine steam cycle through heat exchangers (see Figure 19). However, in the US and some other countries a supercritical carbon-dioxide Brayton gas-turbine cycle is considered as the power cycle for future SFRs, because carbon dioxide and sodium are considered to be more compatible than water and sodium. In general, sodium is highly reactive metal. It reacts with water evolving hydrogen gas and releasing heat. Due to that sodium can ignite spontaneously with water. Also, it can ignite and burn in air at high temperatures. Therefore, special precautions should be taken for safe operation of this type reactor. One of them is the triple-flow circuit with the intermediate sodium loop between the reactor coolant (primary sodium) and water as the working fluid in the power cycle.

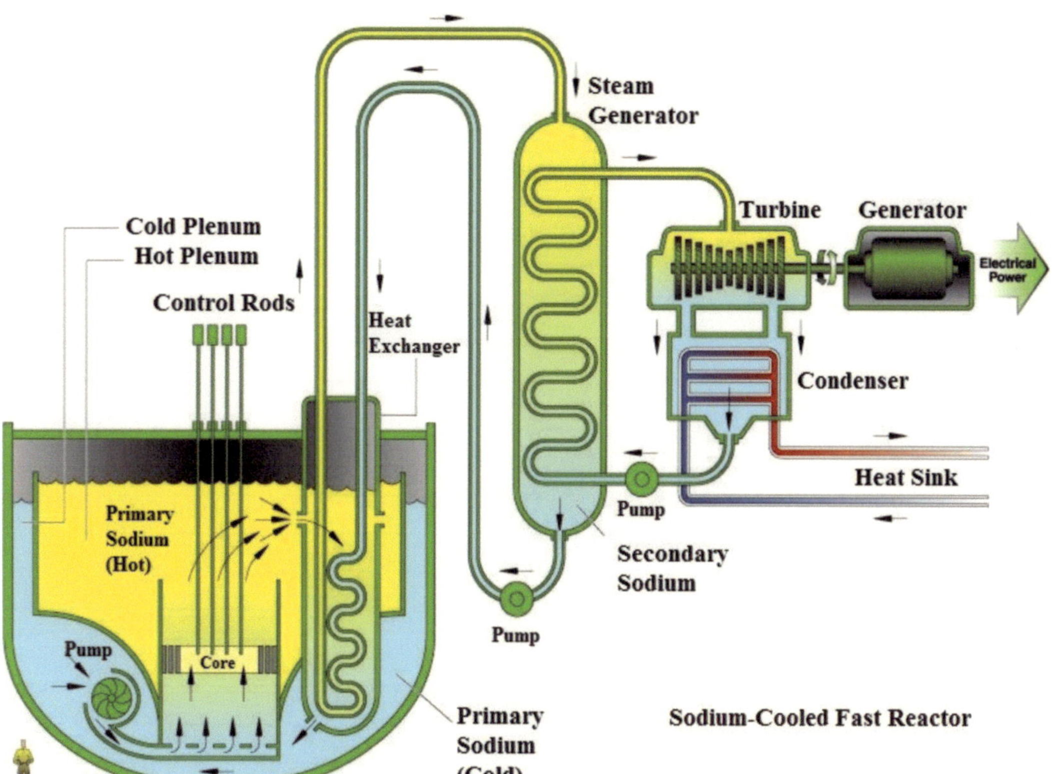

Figure 19. Scheme of Sodium Fast Reactor (SFR) NPP concept (US DOE, 2002).

Reactor Parameter	Unit	Reference Value
Reactor power	MW_{th}	1000–5000
Thermal efficiency	%	40–42%
Coolant		Sodium
Coolant melting/boiling temperatures	°C	98/883
Coolant density at 450°C	kg/m^3	844
Pressure inside reactor	MPa	~0.1
Coolant maximum outlet temperature	°C	530–550
Average power density	MW_{th}/m^3	350
Reference fuel compound		Oxide or metal alloy
Cladding		Ferritic or ODS ferritic
Average burnup	GWD/MTHM	~150–200

Table 11. Key-design parameters of SFR concept (also, see Table 6 for parameters of currently operating SFR BN-600).

Reactor Parameter	Unit	Brest-300	Brest-1200
Reactor power (thermal/electrical)	MW	700/300	2800/1200
Thermal efficiency	%	43	
Primary coolant		Lead	
Coolant melting/boiling temperatures	°C	328/1743	
Coolant density at 450°C	kg/m^3	10,520	
Pressure inside reactor	MPa	~0.1	
Coolant inlet/outlet temperatures	°C	420/540	
Coolant massflow rate	t/s	40	158
Maximum coolant velocity	m/s	1.8	1.7
Fuel		UN+PuN	
Fuel loading	t	16	64
Term of fuel inside reactor	years	5	5–6
Fuel reloading per year		1	
Core diameter/height	m / m	2.3/1.1	4.8/1.1
Number of fuel bundles		185	332
Fuel-rod diameter	mm	9.1; 9.6; 10.4	
Fuel-rod pitch	mm	13.6	
Maximum cladding temperature	°C	650	
Steam-generator pressure	MPa	24.5	
Steam-generator inlet/outlet temperatures	°C	340/520	
Steam-generator capacity	t/s	0.43	1.72
Term of reactor	years	30	60

Table 12. Key-design parameters of LFRs planned to be built in Russia (based on NIKIET data).

LFR (see Figure 20) is a fast-neutron-spectrum reactor, which uses lead or lead-bismuth as the reactor coolant. The outlet temperature of the coolant is about 550°C (but can be as high as 800°C) at an atmospheric pressure. The primary choice of fuel is a nitride fuel. The supercritical carbon-dioxide Brayton gas-turbine cycle has been chosen as a primary choice for the power cycle in US and some other countries, while the supercritical-steam Rankine cycle is considered as the primary choice in Russia (see Table 12).

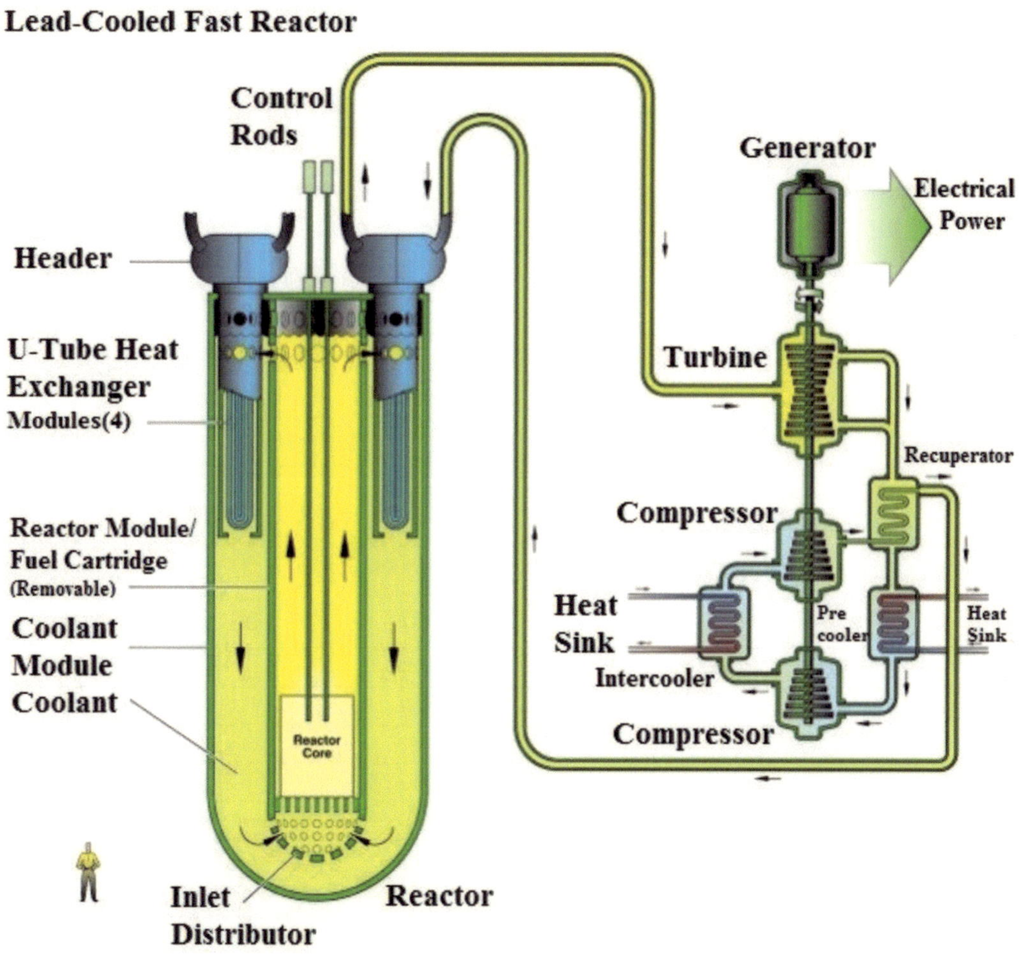

Figure 20. Scheme of Lead Fast Reactor (LFR) NPP concept (US DOE).

MSR (see Figure 21) is a thermal-neutron-spectrum reactor, which uses a molten fluoride salt with dissolved uranium while the moderator is made of graphite. The inlet temperature of the coolant (e.g., fuel-salt mixture) is 565°C while the outlet temperature reaches 700°C. However, the outlet temperature of the fuel-salt mixture can even increase to 850°C when co-generation of hydrogen is considered as an option. The thermal efficiency of the plant is between 45 and 50%. Table 13 lists the design parameters of MSR (US DOE, 2002).

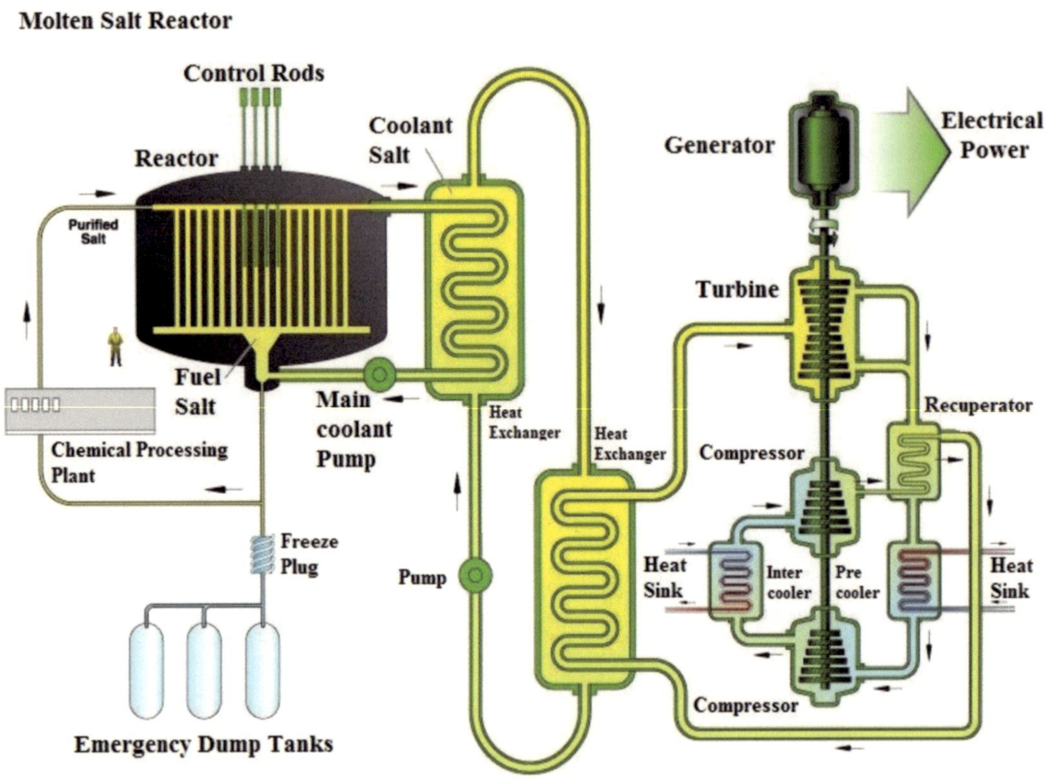

Figure 21. Scheme of Molten Salt Reactor (MSR) NPP concept (US DOE, 2002).

Reactor Parameter	Unit	Reference Value
Reactor power	MW_{el}	1000
Net thermal efficiency	%	4450
Average power density	MW_{th}/m^3	22
Fuel-salt inlet/outlet temperatures	°C	565/700 (800)
Moderator		Graphite
Neutron-spectrum burner		Thermal-Actinide

Table 13. Key-design parameters of MSR concept.

The design of SCWRs is seen as the natural and ultimate evolution of today's conventional water-cooled nuclear reactors (Schulenberg and Starflinger, 2012; Pioro, 2011; Oka et al., 2010; Pioro and Duffey, 2007):

1. Modern PWRs operate at pressures of 15 – 16 MPa.
2. BWRs are the once-through or direct-cycle design, i.e., steam from a nuclear reactor is forwarded directly into a turbine.

3. Some experimental reactors used nuclear steam reheat with outlet steam temperatures well beyond the critical temperature, but at pressures below the critical pressure (Saltanov and Pioro, 2011). And

4. Modern supercritical-pressure turbines, at pressures of about 25 MPa and inlet temperatures of about 600°C, operate successfully at coal-fired thermal power plants for more than 50 years.

Parameters	Unit	PV SCWR Concepts		
Country	**–**	**Russia**		**USA**
Spectrum	–	Thermal	Fast	Thermal
Power electrical	MW	1500	1700	1600
Thermal efficiency	%	34	44	45
Pressure	MPa	25	25	25
Coolant inlet/outlet temperatures	ºC	280/550	280/530	280/500
Massflow rate	kg/s	1600	1860	1840
Core height/diameter	m/m	3.5/2.9	4.1/3.4	4.9/3.9
Fuel	–	UO_2	MOX	UO_2
Enrichment	$\%_{wt}$	–	–	5
Maximum cladding temperature	ºC	630	630	–
Moderator	–	H_2O	–	H_2O

Table 14. Modern concepts of Pressure-Vessel Super Critical Water-cooled Reactors (PV SCWRs) (Pioro and Duffey, 2007).

In general, SCWRs can be classified based on a pressure boundary, neutron spectrum and/or moderator (Pioro and Duffey, 2007). In terms of the pressure boundary, SCWRs are classified into two categories, a) Pressure Vessel (PV) SCWRs (see Figure 22), and b) Pressure Tube (PT) or Pressure Channel (PCh) SCWRs (see Figures 23 and 24). The PV SCWR requires a pressure vessel with a wall thickness of about 50 cm in order to withstand supercritical pressures. Figure 22 shows a scheme of a PV SCWR NPP. Table 14 lists general operating parameters of modern PV-SCWR concepts. On the other hand, the core of a PT SCWR consists of distributed pressure channels, with a thickness of about 10 mm, which might be oriented vertically or horizontally, analogous to CANDU and RBMK reactors, respectively. For instance, SCW CANDU reactor (Figure 23) consists of 300 horizontal fuel channels with coolant inlet and outlet temperatures of 350 and 625°C at a pressure of 25 MPa (Pioro and Duffey, 2007). It should be noted that a vertical-core option (Figure 24) has not been ruled out; both horizontal and vertical cores are being studied by the Atomic Energy of Canada Limited (AECL). Table 15 provides information about modern concepts of PT SCWR.

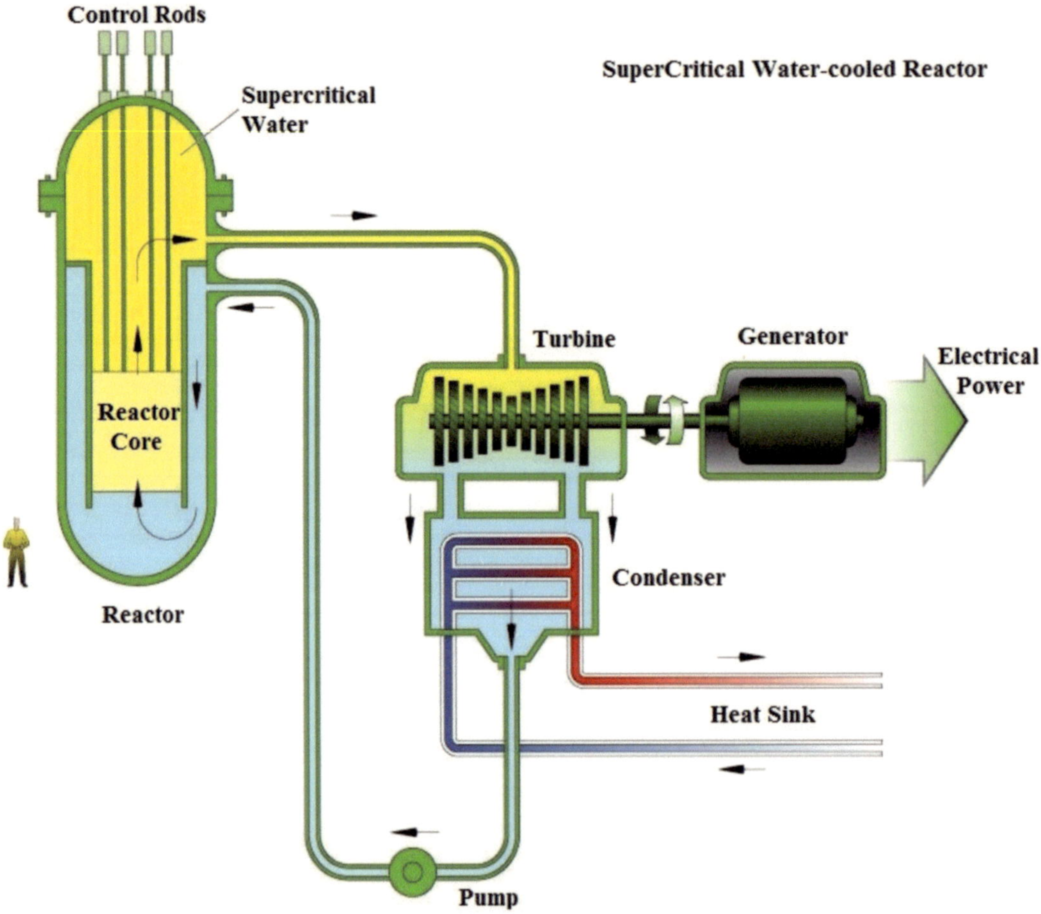

Figure 22. Schematic of Pressure-Vessel Super Critical Water-cooled Reactor (PV SCWR) NPP concept (US DOE, 2002).

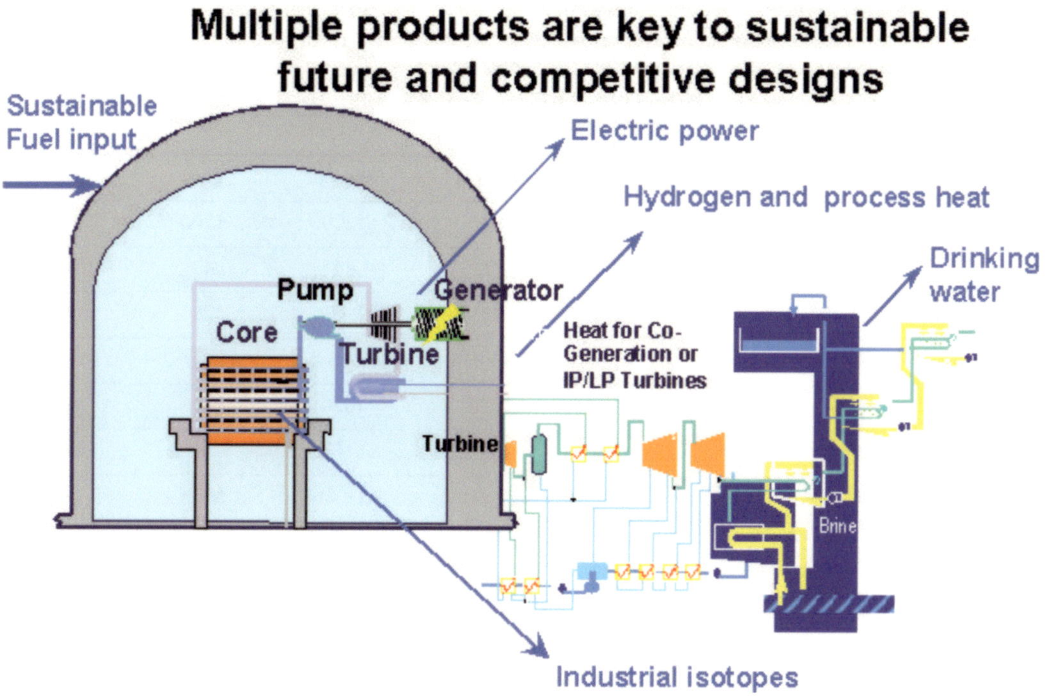

Figure 23. General scheme of Pressure-Tube (PT) SCW-CANDU-reactor NPP concept (courtesy of Dr. R. Duffey, AECL).

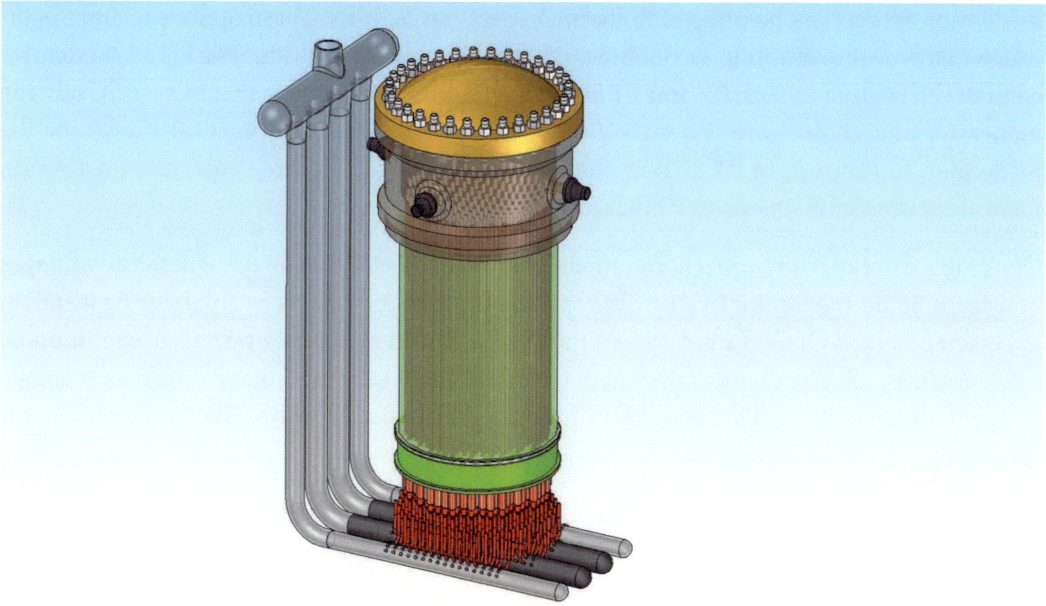

Figure 24. Vertical core-configuration option of PT SCW-CANDU-reactor concept (courtesy of AECL).

Parameters	Unit	PT-SCWR concepts			
Country	–	**Canada (Figures 23 and 24)**	**Russia (NIKIET)**		
Spectrum	–	Thermal	Thermal	Fast	Thermal
Power electrical	MW_{el}	1220	1200	1200	850
Thermal efficiency	%	48	44	43	42
Coolant pressure	MPa	25	24,5	25	25
Coolant temperature	ºC	350–625	270–545	400–550	270–545
Mas flowrate	kg/s	1320	1020	–	922
Core height/diameter	m/m	/7	6/12	3.5/11	5/6.5
Fuel	–	UO_2/Th	UCG	MOX	UO_2
Enrichment	$\%_{wt}$	4	4,4	–	6
Maximum cladding temperature	ºC	850	630	650	700
Moderator	–	D_2O	Graphite	–	D_2O

Table 15. Modern concepts of PT SCWRs (Pioro and Duffey, 2007).

In terms of the neutron spectrum, most SCWR designs are a thermal spectrum; however, fast-spectrum SCWR designs are possible (Oka et al., 2010). In general, various liquid or solid moderator options can be utilized in thermal-spectrum SCWRs. These options include light-water, heavy-water, graphite, beryllium oxide, and zirconium hydride. The liquid-moderator concept can be used in both PV and PT SCWRs. The only difference is that in a PV SCWR, the moderator and coolant are the same fluid. Thus, light-water is a practical choice for the moderator. In contrast, in PT SCWRs the moderator and coolant are separated. As a result, there are a variety of options in PT SCWRs.

One of these options is to use a liquid moderator such as heavy-water. One of the advantages of using a liquid moderator in PT SCWRs is that the moderator acts as a passive heat sink in the event of a Loss Of Coolant Accident (LOCA). A liquid moderator provides an additional safety feature[5], which enhances the safety of operation. On the other hand, one disadvantage of liquid moderators is an increased heat loss from the fuel channels to the liquid moderator, especially, at SCWR conditions.

The second option is to use a solid moderator. Currently, in RBMK reactors and some other types of reactors such as Magnox, AGR, and HTR, graphite is used as a moderator. However, graphite may catch fire at high temperatures at some conditions. Therefore, other materials such as beryllium, beryllium oxide and zirconium hydride may be used as solid moderators.

5 Currently, such option is used in CANDU-6 reactors.

In this case, heat losses can be reduced significantly. On the contrary, the solid moderators do not act as a passive-safety feature.

High operating temperatures in SCWRs lead to high fuel centreline temperatures. Currently, UO_2 has been used in LWRs, PHWRs, etc. However, the uranium-dioxide fuel has a lower thermal conductivity, which results in high fuel centerline temperatures. Therefore, alternative fuels with high thermal-conductivities such as UO_2-BeO, UO_2-SiC, UO_2 with graphite fibre, UC, UC_2, and UN might be used (Peiman et al., 2012).

However, the major problem for SCWRs development is reliability of materials at high pressures and temperatures, high neutron flux and aggressive medium such as supercritical water. Unfortunately, up till now nobody has tested candidate materials at such severe conditions.

8. Conclusions

1. Major sources for electrical-energy production in the world are: 1) thermal - primary coal and secondary natural gas; 2) nuclear and 3) hydro.
2. In general, the major driving force for all advances in thermal and nuclear power plants is thermal efficiency. Ranges of gross thermal efficiencies of modern power plants are as the following: 1) Combined-cycle thermal power plants – up to 62%; 2) Supercritical-pressure coal-fired thermal power plants – up to 55%; 3) Carbon-dioxide-cooled reactor NPPs – up to 42%; 4) Sodium-cooled fast reactor NPP – up to 40%; 5) Subcritical-pressure coal-fired thermal power plants – up to 38%; and 6) Modern water-cooled reactors – 30 – 36%.
3. In spite of advances in coal-fired thermal power-plants design and operation worldwide they are still considered as not environmental friendly due to producing a lot of carbon-dioxide emissions as a result of combustion process plus ash, slag and even acid rains.
4. Combined-cycle thermal power plants with natural-gas fuel are considered as relatively clean fossil-fuel-fired plants compared to coal and oil power plants, but still emits a lot of carbon dioxide due to combustion process.
5. Nuclear power is, in general, a non-renewable source as the fossil fuels, but nuclear resources can be used significantly longer than some fossil fuels plus nuclear power does not emit carbon dioxide into atmosphere. Currently, this source of energy is considered as the most viable one for electrical generation for the next 50 – 100 years.
6. However, all current and oncoming Generation III+ NPPs are not very competitive with modern thermal power plants in terms of thermal efficiency, the difference in values of thermal efficiencies between thermal and nuclear power plants can be up to 20 – 25%.
7. Therefore, new generation (Generation IV) NPPs with thermal efficiencies close to those of modern thermal power plants, i.e., within a range of 45 – 50% at least, should be designed and built in the nearest future.

Nomenclature

P, *p* pressure, Pa

H specific enthalpy, J/kg

m massflow rate, kg/s

T temperature, °C

V specific volume, m^3/kg

Greek letters

ϱ density, kg/m^3

Subscripts

cr critical

el elctrical

in inlet

pc pseudocritical

s, sat saturation

th thermal

Abbreviations

ABWR Advanced Boiling Water Reactor

AECL Atomic Energy of Canada Limited

AGR Advanced Gas-cooled Reactor

BN Fast Neutrons (reactor) (in Russian abbreviation)

BWR Boiling Water Reactor

CANDU CANada Deuterium Uranium

DOE Department Of Energy (USA)

EGP Power Heterogeneous Loop reactor (in Russian abbreviations)

EU European Union

GCR Gas-Cooled Reactor

GFR Gas Fast Reactor

HP High Pressure

HTR High Temperature Reactor

ID Inside Diameter

IP Intermediate Pressure

KAERI Korea Atomic Energy Research Institute (South Korea)

LFR Lead-cooled Fast Reactor

LGR Light-water Graphite-moderated Reactor

LMFBR Liquid-Metal Fast-Breeder Reactor

LP Low Pressure

LWR Light-Water Reactor

MOX Mixed OXides

NIKIET Research and Development Institute of Power Engineering (in Russian abbreviations) or RDIPE, Moscow, Russia

NIST National Institute of Standards and Technology (USA)

NPP Nuclear Power Plant

NRC National Regulatory Commission (USA)

NRU National Research Universal (reactor), AECL, Canada

PCh Pressure Channel

PHWR Pressurized Heavy-Water Reactor

PT Pressure Tube

PV Pressure Vessel

PWR Pressurized Water Reactor

RBMK Reactor of Large Capacity Channel type (in Russian abbreviations)

RPV Reactor Pressure Vessel

SC SuperCritical

SCW SuperCritical Water

SCWR SuperCritical Water Reactor

SFR Sodium Fast Reactor

UK United Kingdom

USA United States of America

VHTR Very High Temperature Reactor

VVER Water-Water Power Reactor (in Russian abbreviation)

Author details

Igor Pioro*

Faculty of Energy Systems and Nuclear Science, University of Ontario Institute of Technology, Oshawa, Ontario, Canada

References

[1] Grigor'ev, V.A. and Zorin, V.M., Editors, 1988. Thermal and Nuclear Power Plants. Handbook, (In Russian), 2nd edition, Energoatomizdat Publishing House, Moscow, Russia, 625 pages.

[2] Gupta, S., McGillivray, D., Surendran, P., et al., 2012. Developing Heat-Transfer Correlations for Supercritical CO_2 Flowing in Vertical Bare Tubes, Proceedings of the 20th International Conference On Nuclear Engineering (ICONE-20) – ASME 2012 POWER Conference, July 30 - August 3, Anaheim, California, USA, Paper #54626, 13 pages.

[3] Hewitt, G.F. and Collier, J.G., 2000. Introduction to Nuclear Power, 2nd ed., Taylor & Francis, New York, NY, USA, 304 pages.

[4] Kirillov, P.L., Terentieva, M.I. and Deniskina, N.B., 2007. Thermophysical Properties of Materials for Nuclear Engineering, 2nd edition augmented and revised, Edited by P.L. Kirilov, Publishing House IzdAT, Moscow, Russia, 200 pages.

[5] Kruglikov, P.A., Smolkin, Yu.V. and Sokolov, K.V., 2009. Development of engineering solutions for thermal scheme of power unit of thermal power plant with supercritical parameters of steam, (In Russian), Proc. Int. Workshop "Supercritical Water and Steam in Nuclear Power Engineering: Problems and Solutions", Moscow, Russia, October 22–23, 6 pages.

[6] Mokry, S., Pioro, I.L., Farah, A., et al., 2011. Development of Supercritical Water Heat-Transfer Correlation for Vertical Bare Tubes, Nuclear Engineering and Design, Vol. 241, pp. 1126-1136.

[7] National Institute of Standards and Technology, 2010. NIST Reference Fluid Thermodynamic and Transport Properties-REFPROP. NIST Standard Reference Database 23, Ver. 9.0. Boulder, CO, U.S.: Department of Commerce.

[8] Nuclear News, 2012, March, A Publication of the American Nuclear Society (ANS), pp. 55-88.

[9] Nuclear News, 2011, March, A Publication of the American Nuclear Society (ANS), pp. 45-78.

[10] Oka, Yo., Koshizuka, S., Ishiwatari, Y. and Yamaji, A., 2010. Super Light Water Reactors and Super Fast Reactors, Springer, 416 pages.

[11] Peiman, W., Pioro, I. and Gabriel, K., 2012. Thermal Aspects of Conventional and Alternative Fuels in SuperCritical Water-Cooled Reactor (SCWR) Applications, Chapter in book "Nuclear Reactors", Editor A.Z. Mesquita, INTECH, Rijeka, Croatia, pp. 123-156.

[12] Pioro, I., 2011. The Potential Use of Supercritical Water-Cooling in Nuclear Reactors. Chapter in Nuclear Energy Encyclopedia: Science, Technology, and Applications, Editors: S.B. Krivit, J.H. Lehr and Th.B. Kingery, J. Wiley & Sons, Hoboken, NJ, USA, pp. 309-347 pages.

[13] Pioro, I.L. and Duffey, R.B., 2007. *Heat Transfer and Hydraulic Resistance at Supercritical Pressures in Power Engineering Applications*, ASME Press, New York, NY, USA, 328 pages.

[14] Pioro, I. and Mokry, S., 2011a. Thermophysical Properties at Critical and Supercritical Conditions, Chapter in book "Heat Transfer. Theoretical Analysis, Experimental Investigations and Industrial Systems", Editor: A. Belmiloudi, INTECH, Rijeka, Croatia, pp. 573-592.

[15] Pioro, I. and Mokry, S., 2011b. Heat Transfer to Fluids at Supercritical Pressures, Chapter in book "Heat Transfer. Theoretical Analysis, Experimental Investigations and Industrial Systems", Editor: A. Belmiloudi, INTECH, Rijeka, Croatia, pp. 481-504.

[16] Pioro, I., Mokry, S. and Draper, Sh., 2011. Specifics of Thermophysical Properties and Forced-Convective Heat Transfer at Critical and Supercritical Pressures, Reviews in Chemical Engineering, Vol. 27, Issue 3-4, pp. 191–214.

[17] Pioro, L.S., Pioro, I.L., Soroka, B.S. and Kostyuk, T.O. 2010. Advanced Melting Technologies with Submerged Combustion, RoseDog Publ. Co., Pittsburgh, PA, USA, 420 pages.

[18] ROSENERGOATOM, 2004. Russian Nuclear Power Plants. 50 Years of Nuclear Power, Moscow, Russia, 120 pages.

[19] Ryzhov, S.B., Mokhov, V.A., Nikitenko, M.P. et al., 2010. Advanced Designs of VVER Reactor Plant, Proceedings of the 8th International Topical Meeting on Nuclear Thermal-Hydraulics, Operation and Safety (NUTHOS-8), Shanghai, China, October 10 – 14.

[20] Saltanov, Eu. and Pioro, I., 2011. World Experience in Nuclear Steam Reheat, Chapter in book "Nuclear Power: Operation, Safety and Environment", Editor: P.V. Tsvetkov, INTECH, Rijeka, Croatia, pp. 3-28.

[21] Schulenberg, Th. and Starflinger, J., Editors, 2012. High Performance Light Water Reactor. Design and Analyses, KIT Scientific Publishing, Germany, 241 pages.

[22] Shultis, J.K. and Faw, R.E., 2008. Fundamentals of Nuclear Science and Engineering, 2nd ed., CRC Press, Boca Raton, FL, USA, 591 pages.

Chapter 11

Nanostructured Materials and Shaped Solids for Essential Improvement of Energetic Effectiveness and Safety of Nuclear Reactors and Radioactive Wastes

N.V. Klassen, A.E. Ershov, V.V. Kedrov, V.N. Kurlov, S.Z. Shmurak, I.M. Shmytko, O.A. Shakhray and D.O. Stryukov

Additional information is available at the end of the chapter

http://dx.doi.org/10.5772/55205

1. Introduction

It is generally considered that nuclear energetics started with the discovery of nuclear fission in 1939 [1,2]. But much earlier, in 1913, H.Moseley, a prominent pupil of E. Rutherford [3], analyzed the possibilities of direct conversion of ionizing radiation to electricity and was the first to experimentally demonstrate that ionizing radiation did produce electricity [4,5]. Henry Moseley is known among most physicists as the author of "Moseley's law", connecting the energy characteristics of X-Ray emission of chemical elements with their atomic numbers [1]. On the other hand, his experiments in 1913 on creating an electrostatic potential between two insulated electrodes subjected to ionizing irradiation can be considered as the starting point for nuclear power engineering. At least his electric circuits are widely applied in modern nuclear reactors for detection of ionizing radiation by means of direct charging devices and separation of electric charges by ionizing irradiation for direct conversion of radiation to electricity is the subject of wide speculation [4–7]. These considerations seem to be quite natural because ionization means separation of positive and negative electric charges which is the basis of any kind of generation of electrical energy. In the middle of the 20th century when Nuclear Power Plants (NPP) started their scheme of electric energy production using a fairly long chain of "nuclear fission – heating energy – water vapor – vapor electric generator" transformations, discussions about direct "radiation – electricity" conversion continued [6,7]. Nowadays, after two global disasters at NPP (1986 – Chernobyl, 2011 – Fukushima), the issue of direct generation of electricity from nuclear and radiation processes has become extremely urgent because this way of energy production would convert nuclear power engineering from a hazardous phenomenon to the most economic and ecologically safe technology.

The abovementioned disasters at nuclear power plants with significant radioactive contamination of the environment and a large amount of human victims induced a severely negative attitude of the officials and human society to nuclear power engineering in most civilized countries. Besides, the society is highly anxious about the ecological danger of radioactive wastes because their amount will grow inevitably with increased production of nuclear energy. These two negative factors connected with nuclear power plants induce serious obstacles for further development of nuclear power engineering. Moreover, several governments have decided to gradually close their NPP (e.g. Germany). But objective comparative analysis of the ecological dangers of nuclear-based and carbon-based energy production shows that the latter is much more harmful to human life due to continuous accumulation of carbon dioxide in the Earth atmosphere and the related green house effect. Its negative influence on our climate is growing inevitably resulting in weather instabilities with extreme hazardous phenomena. Therefore, an intensive search for ways to improve the safety of Nuclear Power Plants in order to make them more acceptable to our society is a burning problem. In addition to ecological problems, the competition between nuclear, carbon, solar, hydro and other techniques of electrical energy production involves their economical effectiveness. Our studies of the prospects of application of nanostructures and shaped solids demonstrate that these two kinds of materials are capable to improve essentially either the ecological safety or the effectiveness of NPP as well as radioactive waste storage. Such improvements are based on development of units and devices for: 1) operative differential monitoring of radiation flows inside active reactors zones for instantaneous detection of damages of uranium oxide rod cladding inside nuclear reactors and optimization of regulation of reactor functioning; 2) direct conversion of radiation flows to electrical energy inside the reactors, in water pools with depleted nuclear fuel rods pulled out from the reactors and in radioactive waste stores anywhere.

The next part of this chapter is devoted to analysis of the current situation with active zone control, energy production effectiveness, resource electricity supplies, utilization of radioactive wastes at nuclear power plants around the world. The drawbacks of these systems as well as possible ways of elimination of these drawbacks are discussed. Two subsequent parts describe the properties of nanostructures and shaped sapphire which can be useful for nuclear power engineering. Manifold enhancement of radiation hardness in nanostructures and possibilities of direct transformation of radiation to electricity connected with active migration of electron excitations between nano-grains are described. A set of experimental facts confirming good radiation hardness of shaped sapphire is presented. The subsequent part deals with application of nanostructures and shaped solids to continuous and informative monitoring of radiation flows, nuclear fuel elements and other constituents inside active reactor zones. The prospects of improving the safety and economic characteristics of NPP connected with these innovations are described. The sixth part is devoted to practical ways of direct conversion of radiation to electricity at NPP, radioactive waste storage and other objects of nuclear engineering based on nanostructures and shaped solids. The influence of application of direct radiation-to-electricity conversion on safety and effectiveness of NPP is discussed.

The conclusion of this chapter summarizes the improvements in safety and effectiveness of nuclear power engineering resulting from application of nanostructures and shaped crystals. The terms of their possible realization are estimated.

2. Current situation with control of active zones, effectiveness and safety of nuclear energy production

In spite of the wide variety of designs of nuclear reactors, their operation is described with a set of typical processes. Uranium or plutonium nuclear fission produces two secondary nuclear splinters with the total kinetic energy of about 80 % of the total energy of the process (Fig. 1) [1, 2]. The rest of the fission energy is distributed between the secondary neutrons, alpha and beta particles and gamma radiation. The fissile nuclear fuel in the form of enriched uranium oxide pellets is contained inside the cylindrical fuel elements with metal cladding made of either zirconium alloy or stainless steel. Usually, when the cladding is not damaged, the neutron and gamma particles escape from the fuel element to the outer space whereas other fission products remain inside the rod. The major part of the energy is converted to heating directly inside a rod. Water or some other cooler is circulating between the assemblies of fuel rods, absorbs their thermal energy and delivers it from the active zone of the reactor to the outer space where heating is converted to electricity.. The gamma and neutron radiation emitted from the fuel rods is absorbed either inside the active zone or in the reactor shielding. Monitoring of the reactor operation inside the active zone is usually performed with radiation detectors registering neutron and gamma flows and with thermoelectric devices measuring the distribution of the temperature inside the reactor [1–7]. The rate of the nuclear fission is regulated with absorbers of neutron flows (cadmium, boron, etc.) by means of variation of their content inside the active zone (in the most cases by means of introduction of rods with absorbing material into the active zone or dissolving of the absorbing elements in the water used for cooling).

Typical nuclear reactors contain thousands of fuel rods with fissile material (enriched with U^{235} or Pu^{239}), collected to assemblies containing several tens of rods. The temperature of the cooling liquid when it leaves the active zone and its flow intensity are the main factors determining the electric power that can be achieved by the electric converters installed outside. The liquid is heated as it passes by the fuel rods where the nuclear fission takes place. The specific fission power is determined by the total amount of the fissile nuclei and the flow of fast and thermal neutrons irradiating the rod. On the other hand, these flows are determined by the intensities of the fission in other rods, the processes of moderation and absorption of neutrons, etc. Thus, the set of the fuel rods is a multi-component system with a huge amount of links and feedbacks that are flexible, so it is natural to expect local fluctuating instabilities of the fission rate and corresponding oscillations of temperatures within separate rods and their assemblies. For example, one can assume that thermal growth of a certain fuel rod will

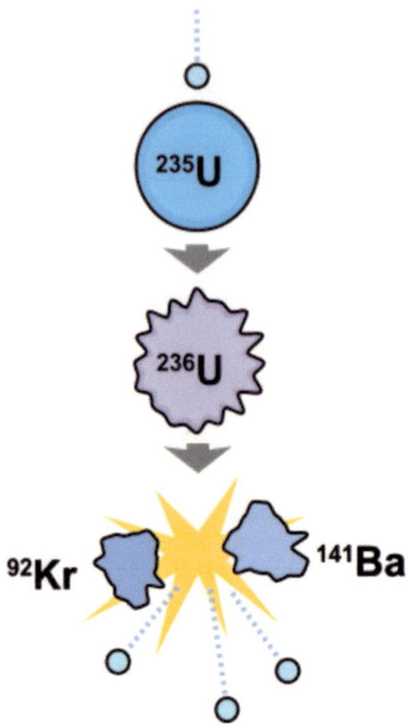

Figure 1. Scheme of ^{235}U nuclear fission induced by a thermal neutron.

produce a vapor bubble in its vicinity, resulting in local focusing or defocusing of the neutron flows related to their additional scattering produced by water nonhomogeneities and further increase or decrease of the fission rate and the corresponding variations of the local temperature and radiation instabilities. Severe fluctuations of the fission rate and local temperatures can develop in fractions of seconds. For instance, the fission rate and corresponding power of the Chernobyl reactor before its disaster increased many fold in a second due to the drawbacks of the regulation system, boiling of water and the human errors. The fission rate is determined mainly by absorption of thermal neutrons. But the presence of the Pu^{239} isotope (about 1%) in the depleted fuel rods after their removal from the active zones reveals the participation of fast neutrons also in the nuclear fission processes. The exchange with the fast neutrons occurring between the neighboring rods provides a positive feedback as an additional factor for local fast oscillations of the reactor operation. These temperature oscillations create oscillating internal stresses in the claddings of the rods which are capable of inducing their cracking with subsequent penetration of the radioactive splinters from the rod interior to the cooling liquid and then to the outside of the reactor core.

Thus, observations of the amplitudes and characteristic times of fast instabilities of fissile rates, radiation flows and temperatures in separate rods seem to be important for better understanding of the internal processes in reactors and further improvement of their operation control, regulation and energy effectiveness. From this viewpoint, monitoring of the averaged values of these parameters commonly done by means of measurements with thermoelectric and direct charging devices seems to be too slow (the characteristic times of these devices are of the order of several tens of seconds [1,2]). The averaging of the temperature and radiation flows proceeds inevitably due to the very long characteristic times of these devices. So, we must confess that up to now there is no available information on faster instabilities. We suppose that this is due to the absence of fast radiation detectors and temperature sensors with satisfactory radiation hardness for stable functioning within intense gamma and neutron flows inside active zones of nuclear reactors.

Our studies on nanocrystalline scintillators show that their radiation hardness can be improved up to the level required for measurements inside active zones of nuclear reactors (see part 3 of this chapter). In Part 4 we describe radiation hard light guides made of (based on) shaped sapphire and glass fibers with a complex cross-section in the form of photonic crystals. Part 5 describes the experimental schemes of local measurements of radiation flows and temperatures by means of nano-scintillators and radiation hard light guides with a temporal resolution better than one millisecond. It should be emphasized that radiation flows can be measured differentially, i.e. separate sets of data on gamma-, alpha-, beta-particle flows, fast and thermal neutrons can be registered. The intensities of scintillation flashes can be used for estimation of fast temperature variations of the rod claddings. On the other hand, the data on the oscillations of the cladding temperatures enable determination of the amplitudes of the thermo-elastic stresses and strains of the rods. By these means much more adequate information on the working resources of the claddings can be obtained.

It is worth noting that differential measurements of intensities of beta- and alpha- flows provide reliable information on local cladding damages, because these particles are usually localized inside the rods and their presence in the external region points to perforation of the cladding. Hence, the damage of the cladding can be revealed much earlier than the moment when the radioactive splinters washed out by the cooling liquid from the rods appear outside the active zone. Prompt detection of such perforations ensures fast removal of the damaged fuel assembly before the total radioactive contamination of the cooling chain. Thus, shutdown of the reactor can be avoided resulting in improvement of its ecological safety and economic effectiveness.

On the other hand, operative information on the neutron and gamma flows from definite fuel rods would result in better optimization of the fuel usage, because the degree of fissile material depletion will be determined for every separate fuel rod (or at least for separate assemblies of rods). So, more exhausted assemblies will be substituted earlier and the total coefficient of fuel usage will be increased.

Generally speaking, the data on differentiated flows of various ionizing radiations obtained with good temporal and spatial resolutions could ensure deeper understanding of the physical processes in nuclear reactors providing their better safety and effectiveness.

The Fukushima NPP disaster showed that the resource electrical supplies of the reactor cooling system based on diesel engines are not reliable because their moving parts can be broken by a water flow or any other accident, whereas fuel delivery can become problematic due to destruction of roads. Our experiments on propagation of ionizing radiation through triple-layer nanostructures including a strongly absorbing electrical conductor, an insulator and a weakly absorbing conductor showed that such adequately constructed structures can provide direct, sufficiently efficient conversion of neutron and gamma radiations to electricity. These devices could be attached to the assemblies of the depleted fuel rods taken out of the active zones and stored in cooling pools until exhaustion of their radiation. On the other hand, besides essential improvement of reliability of resource electrical supplies of NPP these converters could change (transform) the situation with nuclear wastes. Nowadays they are the cause of social tensions due to their radioactivity, because such radioactive materials will become low cost sources of electricity and a source of commercial profit (see part 6). Annual exploitation of a nuclear power plant results in 1.500 cubic meters of radioactive waste per one GWt of electrical power which is the power of an average nuclear reactor [1,2]). Considering the fact that the Nuclear Power Plants around the world have now more than 400 nuclear reactors, the annual amount of radioactive wastes can be estimated as 600,000 cubic meters As it will be seen from the arguments presented below, direct conversion of their radiation to electricity can produce the amount of energy equivalent to construction of several tens of new nuclear reactors.

3. Unique features of nanostructures applicable to nuclear power engineering

The current situation with monitoring operation of nuclear reactors, utilization of depleted fuel rods and radioactive wastes described in the previous part shows that the safety and effectiveness of nuclear power engineering could be essentially improved with application of much faster detectors for separate control of different kinds of ionizing radiation inside the active zones and direct radiation-electricity converters for application at NPP as reliable sources of reserve electricity supplies as well as in nuclear waste stores for production of electrical energy for NPP own needs and supply of the surrounding areas. These problems have been pressing since the very beginning of nuclear engineering (see, for instance, books [4,6,7]). Moreover, nuclear reactions inside fuel rods are connected with the electrical charge separation process that is the main constituent of any generator of electrical energy. So, direct conversion of nuclear fission energy to electricity would be the most natural decision. But very short distances of charged particle propagation inside condensed materials require nanoscopic dimensions of construction elements of such converters [7–14]. So, when the main construction elements of experimental converters had thicknesses exceeding tens of micrometers, the energy effectiveness of the radiation – electricity converters was extremely low (less than 0.01 % with respect to the energy of absorbed ionizing radiation, see, for instance, one of the patents on this issue [16]). Hence, prior to active studies of technologies based on nano-dimensional materials and their properties, development of fast and radiation hard detectors as well as sufficiently effective radiation- electricity converters remained problematic.

A great body of the experimental and theoretical results on interaction of ionizing radiations with nanostructures demonstrated the prospects for developing either fast and radiation hard detectors for selective registration of gamma, neutron, alpha, beta and proton flows or direct converters of such types of radiation to electricity [10 – 14, 17 - 20]. Among the main features of nano-dimensional materials which provide successful development of their based devices

http://dx.doi.org/10.5772/55205

are the high probability of structural defects or electron excitation to reach quickly the external boundary of the nano-dimensional grain as well as a significant ratio of the amount of surface atoms to the total amount of the atoms in the grain. The surface -to-volume ratio varies inversely with nanoparticle size. On the other hand, absorption of any kind of external irradiation by a nanoparticle is proportional to its volume whereas re-emission of the absorbed energy to the external space is proportional to the surface. Hence, the absorption rate decreases in accordance with the third power of the particle radius whereas the decrease of the re-emission rate is much slower, i.e. proportional to the square of the radius. So, at a certain radius value the re–emission rate becomes equal to the rate of external pumping. This means that particles with smaller radii will not change their stationary state when subjected to this kind of irradiation because the absorbed energy returns to the external space by means of re-emission via the surface (by surface re-emission). This consideration belongs to the arguments confirming the increase of radiation stability of nano-particles following the decrease of their dimensions.

It is (rather) easy to show that the stability of nano-particles with respect to accumulation of radiation defects should be immediately enhanced with the decrease of their dimensions. Surfaces of solids are well known channels of annihilation of structural defects [22]. Thus, intensive irradiation-induced migration of point defects towards the external grain boundaries is able to enhance manifold radiation hardness of materials with respect to their bulk radiation hardness. This effect is explained by fast annihilation of radiation-induced defects as compared to the rate of irradiation pumping of the material. The time interval between the moment of creation of a point defect by radiation and its annihilation at the surface is inversely proportional to the square of the radius of the particle divided by the coefficient of diffusion of defects of this kind [20]:

$$\frac{1}{\tau_a} \approx \frac{D}{R^2} \tag{1}$$

On the other hand, the frequency of absorption of the ionizing radiation particle quanta by this particle is proportional to its volume multiplied by the intensity of the radiation flux and the absorption coefficient of the material (2)

$$\nu_{\ddot{\imath}} = F \cdot k \cdot R \cdot \pi R^2 = \pi F \cdot k R^3 \tag{2}$$

Comparison of expressions (1) and (2) enables to conclude that the critical intensity of radiation flux F_c that the particle subjected to irradiation can withstand without accumulation of defects and the corresponding degradation of its structure and properties is inversely proportional to the fifth power of the particle radius, i.e. it increases fast with decreasing dimensions:

$$F_c = D / \pi\, k\, R^5 \tag{3}$$

Therefore, it is clear that in this version of determination of radiation hardness for any irradiation intensity it is possible to select the critical nanoparticle radius when particles with smaller radii will withstand such radiation intensity, because annihilation of the point defect created by radiation due to the particle diffusion to the surface will proceed faster than production of a new one.

Besides, surface atoms belong to the category of structural defects themselves, because the atomic structure around them does not resemble the bulk structure. When the particle dimensions become less than 100 nm, the averaged concentration of these defects (i.e. the number of the surface atoms with respect to the total amount of the atoms in the particle) exceeds 1 %. This value is much higher than the averaged concentration of point defects which can be created in the bulk by radiation. So the influence of radiation-induced defects on the electron properties of the particles will be negligible with respect of the influence of the surface. Hence, either the atomic or electron structure of the nanoparticles subjected to irradiation should preserve their initial parameters even with high intensities appropriate for active zones of nuclear reactors.

The good stability of nanoparticles observed during their studies in scanning and transmission electron microscopes when they are subjected to irradiation of electron beams with high energy flow density shows that most of the incident energy is scattered outside. The possible scattering channels are secondary electrons and X-rays, thermal and optical radiation. Our studies of X-ray excited luminescence of composites composed of inorganic polymer (polystyrene, etc.) - bound nanoparticles showed that in this case most absorbed X-ray energy is not stored in the absorbing particle [21, 23-26,28]. Other authors confirm enhancement of the radiation hardness of the materials transforming ionization radiations when the dimensions of their grains become nanoscopic [29, 30]. The energy is transferred to the surrounding organic molecules by means of secondary electrons and soft X-rays, excitons, light photons. These phenomena can be used for direct conversion of radiation energy to electricity in composite structures when one of the components is characterized by much stronger absorption of ionizing radiation than the other. In this case the radiation will produce separation of electric charges between these two components resulting in generation of an electromotive force. The idea of direct conversion of radiation to electrical energy has been discussed in scientific publications since the beginning of the 20th century [3-5]. But as it mentioned above, the experiments provide too low effectiveness of this conversion (not more than 0.01%) [16]. The abovementioned direct charging detectors can be classified as a type of low-efficiency radiation-electricity converter [2] which is too small for practical energy generators. But application of layered structures of radiation-electricity converters with nanoscopic ranges opens up new prospects for increasing conversion effectiveness. Theoretical estimations show that about 80% of radiation energy can be transformed to electricity by means of nano-dimensional converters with optimized structure, see, for example, patent application and publications [13,14, 17-20]. But the scheme of nano-structural converters described there implies artificial constructions made of nano-tubes and nano-layers which can be rather expensive to produce. Below we will present one of the possible and more economic ways for construction of radiation-electricity converters with an effectiveness sufficient for practical applications and a relatively low production cost [45, 49-51].

Up to now a wide variety of different techniques of preparation of nanoscintillators and other nanoscopic solids applicable to nuclear engineering have been developed [41, 42, 44–48].

4. Shaped sapphire and glass fiber radiation hard light guides with hollow cores

The considerations presented above show that applications of scintillating nanoparticles are rather promising for fast monitoring of radiation fluxes inside cores of nuclear reactors, because they possess sufficient radiation hardness for stable work within high intensity radiation fluxes. But the scintillation light emitted by these particles should be delivered to the outside of the core where the radiation intensity is much lower and photodetectors transforming the scintillation pulses to electric signals can be installed.

So radiation hard light guides are quite necessary. Usual silica fibers applied widely as light guides do not possess sufficiently high radiation stability [34]. We propose two other kinds of light guides that are radiation hard for effective transfer of light signals from cores of reactors to photodetectors. The first kind uses shaped sapphire fibers grown by the EFG technique. Several papers of independent authors devoted to shaped sapphire single crystals show that this material is able to withstand intensive radiation flows inside active zones of nuclear reactors [31–34]. Moreover, the color centers produced by irradiation deteriorate the optical transparency of sapphire in the ultraviolet region whereas we can choose nanoscintillators emitting light in the visible range. On the other hand, the color centers can be bleached by a laser beam with a wavelength in their absorption band. It should be emphasized that single crystalline sapphire tubes manufactured at the Institute of Solid State Physics (ISSP) RAS (Chernogolovka, Russia) were successfully used as guides of neutron flows inside the core of nuclear reactor in Grenoble and demonstrated a high radiation stability of their properties and structure [38, 39]. Another original technique for preparation of sapphire fibers with a modulated core of lateral structures made by regulation of the chemical composition of the melt delivered to the growth region have been developed [35–37]. Modulation of the chemical compositions of the fibers by varying the content activators of light emission along or across the fiber ensures control of spectral compositions of light emission from different fiber regions. Below we will describe the possibilities of differential radiation flow monitoring achieved by modulation of local chemical compositions of sapphire fibers doped by titanium. The examples of sapphire fibers grown in Chernogolovka are presented in Figs. 2–5.

The core of another radiation hard fiber light guide does not contain any solid, i.e. it is empty [43]. So, nothing can be damaged by irradiation of the core. The localization of the light flow along the fiber axis is achieved by the specific geometry of the micro-channels in the fiber cross – section. The channels are parallel to the fiber axis whereas in the transverse direction they form a photon crystal limiting light propagation in radial directions (Fig. 6). The design was patented by ISSP [46].

Figure 2. Non-doped single crystalline sapphire fibers grown by EFG technique at ISSP [35-37].

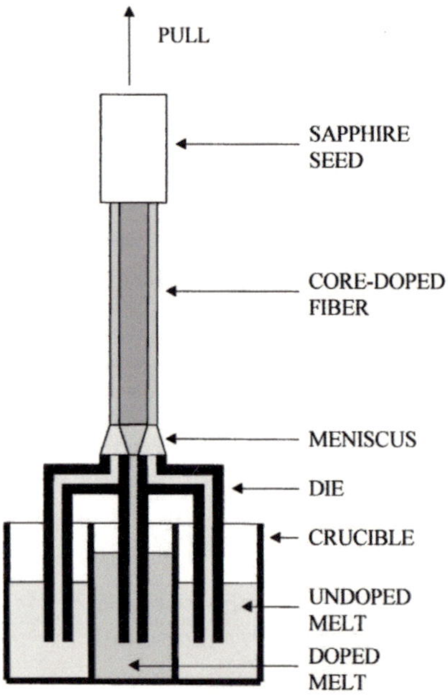

Figure 3. Scheme of growth of single crystalline sapphire fiber with doped core.

5. Differential and fast monitoring of radiation flows inside active zones for better safety and optimized operation of nuclear reactors

Shaped sapphire nanocrystalline scintillators and fibers possess high radiation hardness and can be used effectively for fast differential control of radiation fluxes in active zones of nuclear reactors. Due to their small dimensions, the composition and structures of nanoscintillators

http://dx.doi.org/10.5772/55205

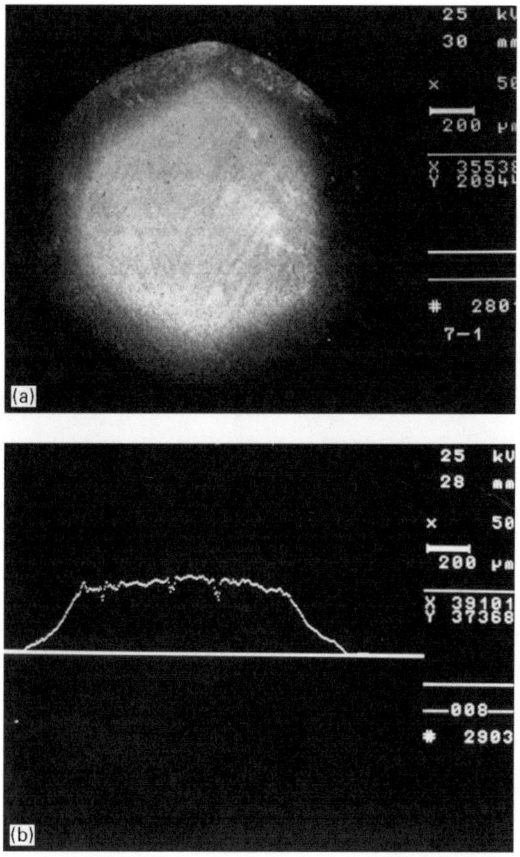

Figure 4. Cathode luminescence micrograph of the cross-section of the sapphire fiber with doped core (upper photo) and radial distribution of the luminescence intensity across the radius (below). Titanium was used as the dopant.

are regulated over a wide range providing excellent opportunities for optimized selection of materials for registration of specific types of radiation. Nanocrystalline particles are placed at lateral surfaces of fiber light guides which also possess high radiation hardness.. The light signals delivered to the outside of the active zone are characterized by millisecond duration and transfer separate information on the intensities of the radiation flows of fast and thermal neutrons, alpha and beta particles, gamma – quanta in definite points of the reactor core to the entrance of the control system. Such a design provides well – timed preventing of accidents and optimized control of the reactor operation. For instance, fiber light guides placed along the claddings of the fissile rods enable timely control of the wall tightness. With intact walls the scintillators register neutron and gamma radiation only, because alpha and beta particles cannot penetrate through the cladding. But as soon as the wall has become too thin, beta emission occurs and with thinning of the wall there also appears alpha – emission. In this case the light guide transfers a warning signal to the regulation system much earlier than the radioactive substances reach the outside of the active zone. Thus, fiber- scintillator detectors ensure timely preventing of diverse accidents which increases significantly the reliability of the reactor operation.

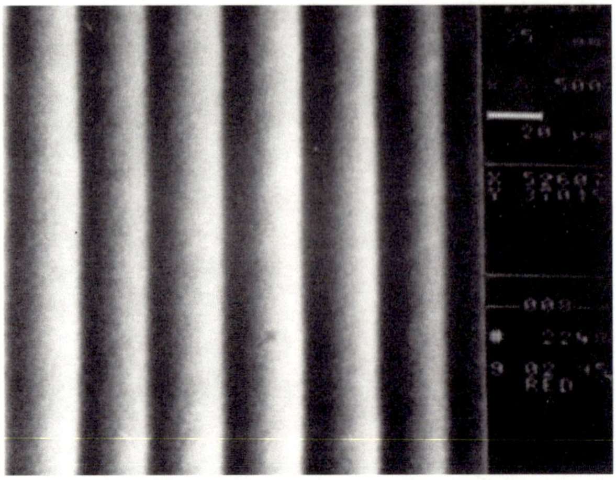

Figure 5. Cathode luminescence of shaped sapphire fiber with modulation of activator doping along the fiber axis. Titanium was used as the dopant.

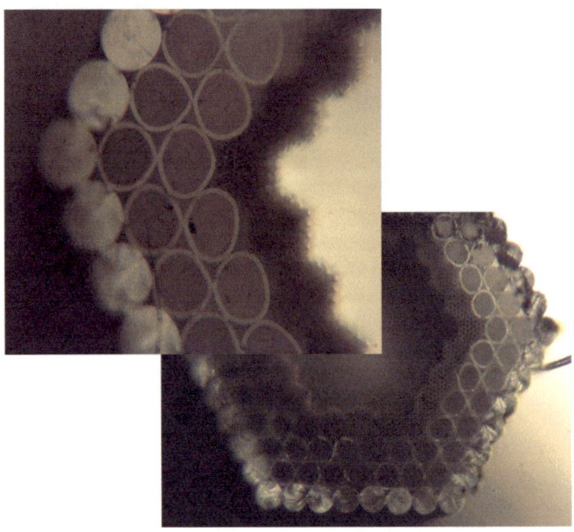

Figure 6. Radiation hard light guide with empty core, transverse propagation of light limited by photon crystal geometry [43, 46].

As it was mentioned above, development of differential monitoring of radiation fluxes inside active zones of nuclear reactors with temporal resolution much better than a second is desirable for deeper understanding of the physical processes inside nuclear reactors, preventing of leakages of radioactive wastes, more complete burning of fission fuel, etc., in other words, for general improvement of reactor safety and energy effectiveness. Application of nano-dimensional scintillators and radiation hard light guides which can work reliably inside active zones of reactors allow for resolving this problem. High radiation hardness of nano size scintillators as well as other nano-dimensional objects was discussed in the previous part. Scintillation light registration of the level of radiation flow in a certain point of the active zone will be transferred by of the light guides to the outside of the active zone where photodetectors can be installed for transformation of light to electric signals for further computer processing (Fig. 7).

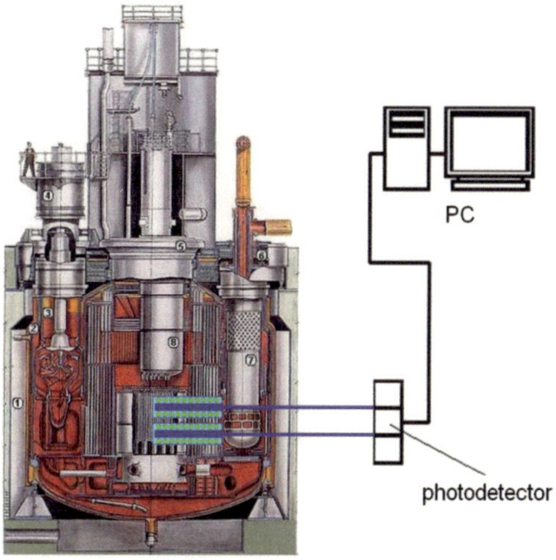

Figure 7. Scheme of transfer of scintillation signals to photodetectors outside the reactor core.

At least two kinds of radiation hard light guides can used for the delivery of scintillation signals to the outside of the active zone: shaped sapphire and glass fibers (Fig. 2) with empty cores with cross-sections in the form of photonic crystals (Fig. 6). According to the measurements made by V.V. Nesvizhevskiy at the Institut Laue-Langevin (Grenoble, France) [38, 39], the products made from shaped sapphire crystals at ISSP RAS invariably retain their parameters in the active zone of the reactor. The data are supported by the direct experiments on the transparency of sapphire fibers subjected to ionizing irradiation [31-34]. On the other hand, light guides with transversal geometry of photonic crystals contain empty cores whose transparency cannot be deteriorated by radiation defects [30, 46]. Based on the above, the researchers at the ISSP RAS are developing various combinations of nanocrystal scintillators and light sapphire fiber waveguides for rapid differential control of radiation flows inside nuclear reactors [40 – 42, 47]. Radiation hard nano-scintillators are placed at the lateral guide surfaces (Fig. 8).

As one of the nano-particles absorbs ionizing radiation it produces a light outburst of a definite color. Here differentiation means to obtain separate information about the intensities of the fluxes of the fast and slow neutrons, alpha-, beta-, and gamma radiation in specific points of the active zone. Such information appears quite useful for improving the safety of the reactor operation and optimizing its operational parameters (for example, for improve the burnup fraction of the fissile materials in the fuel elements).

The information on the intensities of different ionizing radiations is differentiated with the aid of the nanoscintillators on the basis of a wide variation of their structure. The ISSP RAS has developed the techniques of synthesis of nanocrystal scintillators with largely varied compo-

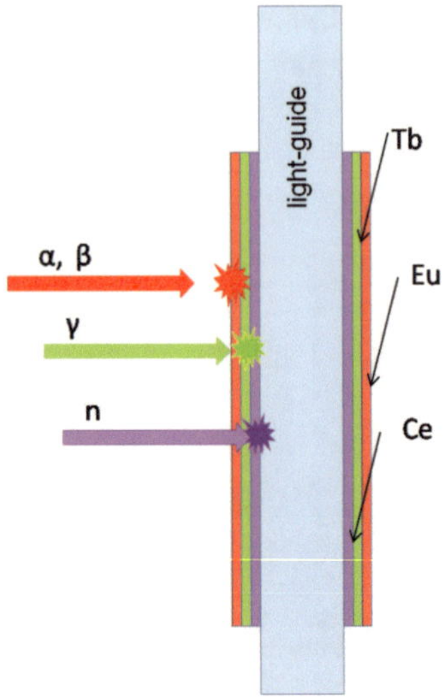

Figure 8. Deposition of nanoscintillators with various dopants at lateral surfaces of radiation hard light guides.

sitions and sizes to ensure selective information on any specific type of radiation (Fig. 9) [21, 23-25, 44].

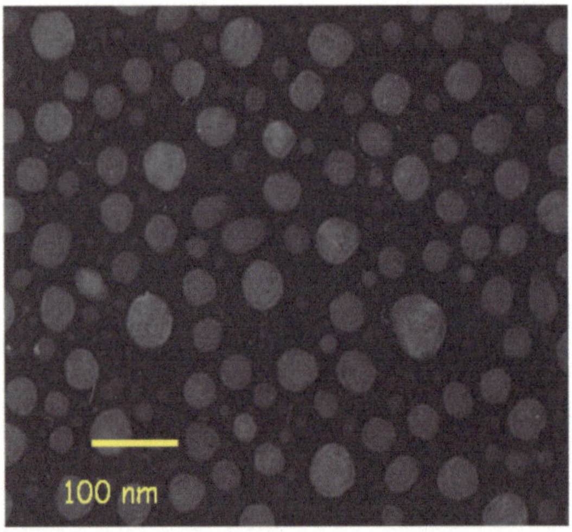

Figure 9. Lutetium-sodium fluoride nanoscintillators doped with cerium activator.

Thus, the prevalence of heavy elements (tungsten, lutecium, lead, etc.) enhances sensitivity to hard gamma-radiation. When alloying with gadolinium, lithium, boron increases the role of scintillations induced by thermal neutrons. Hydrogen saturation of nanoparticle allows receiving scintillation signals from the fast neutrons. The selective sensitivity of the radiation detectors to fast neutrons can be ensured by making compositions of inorganic nanoscintillators and organic molecules [23, 24, 41, 47, 48]. When fast neutrons collide with the protons of

http://dx.doi.org/10.5772/55205

the organic matter, the protons penetrate fast into the inorganic particles contacting with these molecules and exciting their scintillation radiation. For selectivity in relation to alpha-ray and beta-ray radiation the depth of its penetration into the material may be strongly limited (alpha-rays: by units of microns, beta-rays: by tens of microns). Thus, for selective recording of the charged particles, it is sufficient to place the relevant nanoscintillators in a thin layer on the outer surface of the light waveguide. The scintillation signals from different radiations may be of different spectral compositions. For example, nanoscintillators activated with cerium will produce blue flashes, green flashes will be produced by terbium and red flashes by europium.

As stated above, nanoscintillators may be fixed on the lateral surface of the same sapphire light waveguide in order to register different radiations. Separating the scintillators by the spectral compositions of the light emission at the light waveguide output enables to determine the specific radiation in the relevant point of the active zone. When nanoscintillators are uniformly positioned along the whole light waveguide length, the coordinate of the scintillation radiation point may be determined from the ratio of the signal amplitudes at the opposite ends of the light waveguide. As the light passing along the light waveguide reduces exponentially, the difference of the distances between the point of radiation of scintillation flash and the ends of the light waveguide is determined as the logarithm of the ratio of intensities of the relevant signals (Fig. 10).

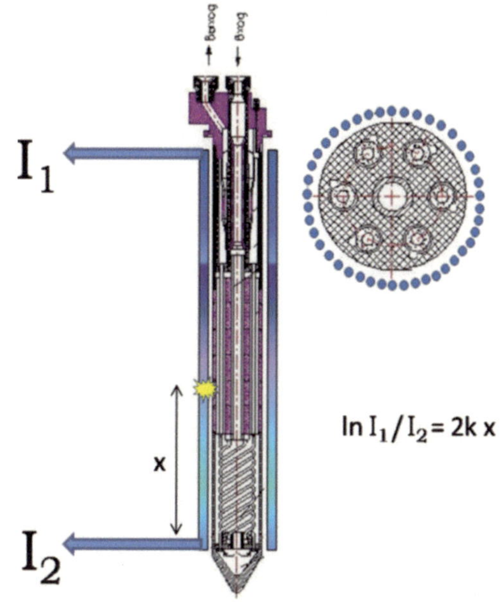

Figure 10. Scheme showing localization of point of damage of fuel rod cladding.

The control of impermeability of the fuel-element cladding may serve an example of the efficiency of separate recording of ionizing radiations for improving the reactor safety. When a fiber sapphire light waveguide with fixed nanoscintillators is positioned along its outer wall, as described above for specific radiations (see Fig. 8), and the cladding is fully impermeable, only gamma and neutron radiation signals will be recorded from the light waveguide (as the alpha and beta particles generated by the nuclear reactions cannot pass through an impermeable wall). With thinning and further perforation of the cladding wall the alpha and beta radiations will start. Hence, the sapphire fiber located near the fuel element will send a signal "colored" for the alpha and beta radiations to the control equipment (first, the beta-ray signal will appear; then, as the cladding "perforation" develops, the alpha particle signal will occur). As nanoscintillators are quite fast radiation detectors, and the signals go along a light waveguide at the speed of light, the response time of the control system to a cladding defect will not exceed milli-seconds. This is much faster than the moment when a radioactive leak appears at the active zone outlet. Hence, efficient emergency actions can be taken much faster. Hence, fixing a sapphire light waveguide with specially adjusted nanoscintillators to every fuel element will significantly improve the reactor safety. Moreover, this method of control ensures, without extra risk, the service life of a fuel element in the active zone, improving the fuel burn up fraction and the reactor efficiency as a whole. It is important to note that beta radiation control ensures detection of corrosion-induced thinning of the fuel-element cladding down to several tens of microns prior to perforation. Thus, active zone emergencies can be prevented as such, increasing, in parallel, the service life of the fuel assembly.

Composite scintillators made of a combination of inorganic nanoparticles and organic luminofors efficiently record fast neutrons, and, due to the fast transfer of the electronic excitations from the inorganic nanoparticle to the organic phosphor increase considerably the response speed of the radiation detectors.

6. Practical development of direct radiation-electricity converters based on nanostructures

In the formation of nanostructures, due to the small sizes of their grains, the role of the interactions of the atomic and electronic excitations with the interface surfaces is extremely great. The radiation defects diffuse rapidly to the surface and annihilate on it. The nonequilibrium electrons formed upon interaction of the ionizing radiations with the nanoparticles collide repeatedly, with their edges during their lifetime. Bremsstrahlung and transition radiations are generated during these collisions and electrons are also emitted.. The above processes rapidly carry the consequences of the radiation effects outside; hence, the radiation strength of the nanostructures increases manyfold as their sizes decrease. On this basis, several new devices are proposed; which could to stabilize work in a field of intense radiation, ensuring efficient transformation of ionizing radiations into electric power, based on the external photoelectric effect and thermal electromotive force. For example, in a nanostructure of alternating "light weight conductor — insulator — heavy metal" layers gamma radiation is absorbed, mainly, in the heavy metal. This induces emission of photoelectrons; some of them coming, via the insulator layer, to the light weight conductor (these may be, for example, thin aluminum or graphite layers of). Thus, the heavy metal gets a positive charge while the light weight conductor acquires extra electrons and gets a negative charge, i.e., an electromotive

http://dx.doi.org/10.5772/55205

force is generated. When the electrodes interlock via a current receiver, work is produced in the latter. Thus, the energy of the ionizing radiation is directly transformed into electric energy. The produced power is determined by the difference of the potentials between the electrodes and summed current which is directly related to the number of the electrons transferred from one electrode to the other. Meanwhile, this number depends upon the configuration of the heavy metal surface because only the electrons whose normal to the surface component of the momentum exceeds a certain threshold will take part in the charge transfer. From this viewpoint, the flat geometry of the electrode is far from being advantageous because only a relatively small number of the electrons can move in the direction required. The coaxial configuration with an insulator layer and a tubular light weight conductor around the heavy metal axis (this configuration is close to that of the direct charge transducers used to record the ionizing radiations [1, 2] is much closer to the optimum. When the thickness of the heavy metal conductor is comparable to the track length of the hot photoelectrons, a significant share of them will take part in the electrode charge exchange, and, hence, in the transformation of the absorbed radiation into electric energy.

From the viewpoint of the above, the honeycomb structure of the cylindrical cells with conductive walls when the core of the cells is filled with heavy metal separated from the walls with a thin insulator layer seems appropriate. At the ISSP RAS we use biomorphic silicon carbide structure as the basis for such a device (see Fig.11). Such structure is obtained as follows: a cross-cut piece of wood (in this case spruce was used) was subjected to pyrolysis, i.e., thermal disintegration in an oxygen-free medium. The pyrolysis yielded a carbonic frame reproducing the structure of the channels parallel to the tree growth axis (upper Figure). Following this, the carbonic frame was saturated with silicon yielding a silicon carbide frame reproducing the original morphology of the wood [49-51]. The silicon carbide surface was oxidized to obtain a silicon dioxide insulating layer. Next, a metal layer was applied to the inner surfaces of the channels. In such a configuration, the total current-forming surface is fairly large: with the spruce cross-section surface of 1 cm^2 and thickness of 1 cm the inner surface of the cells is about 3000 cm^2. The Figures 11 and 12 show the inner configuration of the cell with visible nanograins of the silicon carbide (that make the surface even larger). The carbon of the pyrolyzed wood frame forming the integrated structure for the whole frame works as a light weight conductor. In this case, the silicon carbide plays the role of a mechanical frame. It is made sufficiently thin to allow transfer of photoelectrons emitted from the heavy metal to the light weight conductor.

In the second variant of thermoelectric generator the silicon carbide formed as described above plays the key role in transforming the heat into electricity. In one of the variants of the generator the contact surface of the silicon carbide with the carbon frame acts as the double charge layer. On thermal generation inside this layer of nonequilibrium electrons and holes the inner difference of potentials carries them in opposite directions generating electric current. The intensity of the current is determined by the surface of the silicon carbide – carbon contact that is, as it has been shown above, quite large.

A set of tubular radiation-electric and thermal-electric converters that may be coaxially put onto a cylindrical assembly of fuel cells will, at a 100 kW total flow of energy from the assembly, generate at least 20 kW of electric energy. As the fuel containment sump of the reactor can contain several hundred assemblies of the kind (depending upon the reactor type) the total power output from the sump will be several megawatts. Compared with the working reactor power this makes only fractions of a percent, and, hence, is insignificant. However, the Fukushima 1 accident has shown the critical need, in an event of emergency, for a redundant and absolutely reliable source of electric power. For this purpose NPPs usually have diesel generators. However, their start requires at least a minute, and they need fuel to be supplied. Moreover, a serious accident may damage the moving parts of the diesel generators. Meanwhile, radiation and thermoelectric generators need no fuel, have no moving parts, and work nonstop; hence, they can be much more reliable. The key issue of reliability of such generators is the radiation strength of their generating units. As the main elements of these generating units are nanocrystals, a high radiation strength of the generating units may be expected.

As for nuclear power engineering, it is important to point out that the silicon carbide is a wide zone semiconductor, with a high thermochemical stability, and capable to retain its high semiconductor capacities up to 600÷800ºC. Thus, generally speaking, it can work not only in the sumps of the used fuel assemblies but in the active zones of reactors, as well.

There is another prospect for the nuclear power engineering use of the above radiation-electric and thermal-electric generators based on biomorphic silicon carbide: standalone electricity sources for radioactive waste storage sites. As such storage sites are arranged, as a rule, in back lands, the possibility to get the electric energy without fuel supplies or power transmission lines seems extremely attractive..

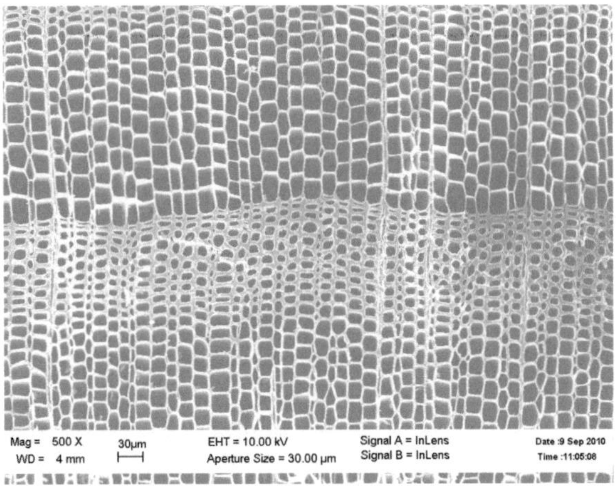

Figure 11. Electronic microscopy image of cross cut surface of spruce after pyrolysis.

http://dx.doi.org/10.5772/55205

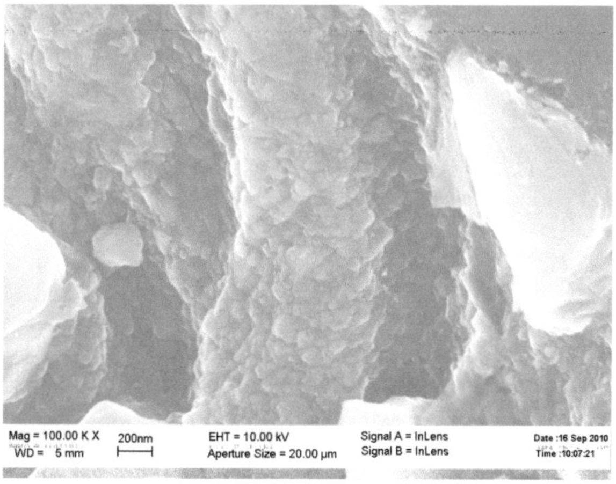

Figure 12. High magnification image of morphology of inner wall of channel in pyrolyzed wood after synthesis of silicon carbide by siliconizing.

γ-Ray irradiation of layered structures consisting of interleaving layers of heavy metal, insulator, and lightweight metal produces electromotive power [49-51]. This occurs due to the prevailing absorption of γ-radiation with heavy metal, thus resulting in emission of fast electrons. Some fraction of them penetrates the lightweight metal making it negatively charged. The contact of the lightweight and heavy metals under an external load makes these electrons return through it into the heavy metal, thus performing useful work. The ratio of this work to the radiation energy absorbed by the layered structure determines the efficiency of radiation-electricity conversion in devices of this kind.

Ideally, all electrons dislodged from the heavy emitter reach the lightweight conductor and are absorbed there. If each of these electrons returns from the collector to the emitter through the external circuit, electric current I will occur in the circuit.

$$I = e \cdot N = e \cdot \gamma \cdot \hbar \cdot \omega / (K + W), \tag{4}$$

where e is elementary charge; $\hbar$ is Plank's constant; ω is circular wave frequency of electromagnetic field; K is electron kinetic energy; γ is photon flux absorbed in the emitter material; W is electronic work function.

In the external circuit with electrical resistance R, electric power will occur. It is possible to estimate the efficiency (η is the efficiency factor) of the system, which is determined as the ratio of power generated in the external circuit W_{ext} to the power of the absorbed γ-rays W_{abs}:

$$\eta = W_{ext} / W_{abs} = I^2 \cdot R / \gamma \cdot \hbar \cdot \omega = \left[\left[e \cdot \gamma \cdot \hbar \cdot \omega / (K + W) \right]^2 \cdot R \right] / \gamma \cdot \hbar \cdot \omega, \tag{5}$$

From formula (5) one can see that as the electron kinetic energy K in the denominator decreases, the efficiency factor is expected to increase monotonously until the energy reaches zero; however, in order to have current I in the external circuit, there should be potential difference $\phi = I \cdot R$ between the emitter and the collector.

For the electrons escaping from the emitter to overcome this potential difference and to reach the collector, their kinetic energy should be not lower than this difference multiplied by elementary charge:

$$K \geq e \cdot \phi = e \cdot I \cdot R = e^2 \cdot \gamma \cdot \hbar \cdot \omega \cdot R / (K + W). \tag{6}$$

From (6) there follows a quadratic equation for the minimum acceptable value of kinetic energy of emitted electrons, by solving which we obtain:

$$K = -W/2 + \sqrt{(W^2/4 + e^2 \cdot \gamma \cdot \hbar \cdot \omega \cdot R)}. \tag{7}$$

As one can see from expression (7), in the ideal case, when the system is pumped with gamma-rays so much that the second term under the square root $e^2 \cdot \gamma \cdot \hbar \cdot \omega \cdot R$ is much higher than the first one $W^2/4$ (therefore, power consumption of the work function can be neglected), the efficiency factor value in expression (5) approaches unity within the accuracy of the energy spent on the work function.

The physical sense of this derivation is as follows: when the kinetic energy of electrons emitted from the emitter to the collector becomes equal to their electrostatic energy between the emitter and the collector, the electrons enter the collector with zero kinetic energy, thus eliminating heat release losses. Therefore, the efficiency factor approaches unity, i.e. the ideal value. But, if kinetic energy exceeds the value

$$K = e \cdot \sqrt{\gamma \cdot \hbar \cdot \omega \cdot R}, \tag{8}$$

then, according to expression (5), the efficiency factor starts to decrease in inverse proportion to kinetic energy.

With the opposite approach, i.e. when the first term under the square root in equation (7) is much higher than the second one, the kinetic energy of emitted electrons only slightly exceeds the work function; therefore, a sufficiently high charge cannot be accumulated on the collector, and the efficiency of the system will be always low.

Despite the fact that the efficiency of an idealized system could reach 100 %, the experimental data reported in the patent [16] have shown an efficiency factor of ~ 10^{-5} %. Below we analyze the reasons of such a considerable discrepancy, and discuss the ways to improve the efficiency of conversion of ionizing-radiation energy into electricity up to a practically significant level.

http://dx.doi.org/10.5772/55205

1. *Non-optimal geometry of converters.* In the experiments mentioned above, the layered converters were of planar geometry. At photoelectric absorption and Compton scattering of gamma-rays, the directions of outgoing electrons receiving their energy do not coincide with the initial direction of radiation propagation; hence, the planar geometry means that traveling of the predominating fraction of these electrons is strongly deviated from the normal line to the outflow surface. Due to this, the distance to the collector increases and can considerably exceed the full range of the fast electron. Accordingly, such electrons fail to leave the plate and participate in formation of radiation EMF. From this point of view, spherical geometry for the heavy metal fragments absorbing gamma-rays seems to be optimal. But for these photoelectrons to travel into the lightweight metal, the size of the spheres should not exceed their ranges. Besides, all spheres should be electrically-connected to each other. Since the ranges of fast electrons in heavy metals are of the micron or even submicron scales, manufacture of such a device is a complicated engineering problem. As an intermediate stage, the geometry of biomorphic silicon-carbide or carbon matrices described earlier [49 - 51] can be used, where heavy metal (lead or tin) fills parallel micro channels formed by means of pyrolysis and siliconizing of wood sections (see figure below). Although the geometry of parallel metal micro fibers is far from the geometry of spheres, it is much more beneficial for photoelectron emission in comparison with the planar geometry.

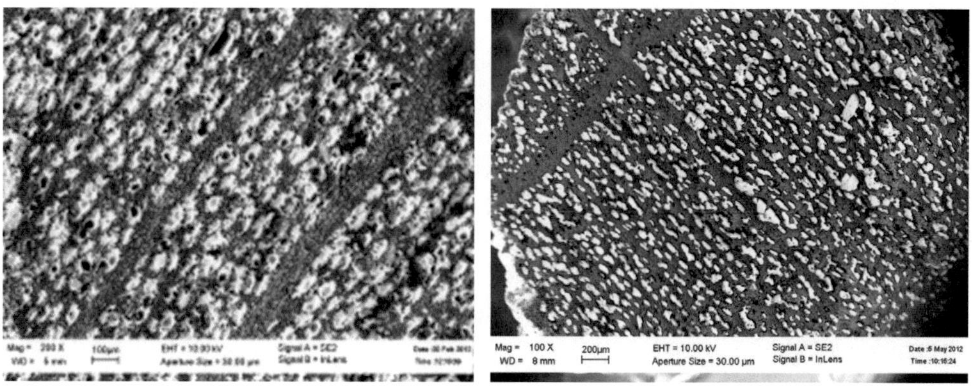

Figure 13. Matrix of silicon carbide microchannels obtained by wood pyrolysis filled with tungsten (left picture); and with lead (right picture).

2. *Excessive thickness of heavy metal layers.* In the patent [16], the planar structure "heavy metal – insulator – lightweight metal" is formed by layers, each of tens of micrometers thick. This thickness is acceptable for the light metal (collector) that is intended to capture emitted electrons; but for the heavy metal (emitter) such thicknesses significantly exceed the optimum values and contribute to a considerable decrease of the effect of radiation separation of charges. A considerable fraction of both primary and secondary electrons (occurring in collisions with the primary ones via collision ionization) fail to travel such long distances and stop in the heavy metal. Naturally, these electrons do not participate in the generation of useful current in the external circuit.

3. *Excessive thickness of the insulator.* In the case of excessive insulator thickness, many of the electrons emitted by heavy metal are stopped in the insulator layer. Apart from the fact that these electrons do not reach the light metal and accordingly drop out of the process of formation of useful current, they considerably decrease the electrical resistance of the insulator, contributing to the return of the emitted electrons from the lightweight metal to the heavy metal not via the external circuit, but immediately back through the insulator layer. Reduction of the insulator thickness will allow increasing the conversion efficiency considerably. For example, with the insulator thickness reduced from 200 μm to 1 μm in the papers [49 - 51], the conversion efficiency increased by 3 orders. At the same time, this value should be sufficient to prevent breakdown, as well as large leakage current.

The most preferable insulator is a vacuum layer, where emitted electrons make no impact such as collision ionization, and consequently, the probability of backflow is minimized, being determined only by the tunnel process. With the purpose to create an insulator with properties close to vacuum, it is possible to use a layer of aerogel (material formed by fibers of such dielectrics as aluminum oxide or silicon oxide, the porosity of which exceeds 95 % [15]), and mechanical strength allows holding the layers of heavy and lightweight metals at a certain distance preventing their direct contact.

4. *Non-optimal energy of electrons leaving the emitter*. The optimal energy of electrons leaving the emitter corresponds to the situation, when by the moment of entering the collector, all the kinetic energy has already been spent on overcoming the repulsive potential between the emitter and the collector. Besides, the electron needs a certain kinetic energy store for losses when travelling in the insulator. Note that in a real situation, electrons leaving the emitter have certain statistical spread of energy; therefore, it is almost impossible to achieve an exact situation, when all electrons would precisely spend their all kinetic energy before reaching the collector. The energy required for electrons to exit from the emitter is largely determined by the range of the electron in the absorbent material and by the energy of initial radiation. Therefore, adjustment of exit energy primarily implies selection of the optimal size and shape of the emitter surface, and this adjustment should be performed with reference to the technical parameters of particular operating conditions.

Conversion of energy of fast neutrons created in fission processes to electricity needs preliminary transfer of a neutron kinetic energy to kinetic energy of a proton. This necessity is explained by the too small cross - section of a neutron interaction with electrons and a much bigger corresponding cross - section of a proton. The micro-channels of biomorphic matrices can be filled with material containing a high percentage of hydrogen atoms. For example, pure water H_2O can be used as a preliminary converter of this kind. Parrafin and other organic materials are acceptable as well. Protons generated by fast neutrons will produce showers of fast electrons which will penetrate through the insulator nano-layer to the light conductor as in the case described above.

It should be noted that sources of ionizing radiation can be placed just inside the micro-capillaries. For example, it can be metal uranium or uranium oxide. The experimental studies of nuclear energy conversion to electricity in plane layered structures with U^{235} [13, 14] revealed

http://dx.doi.org/10.5772/55205

a conversion efficiency of 25 %. Microcapillary structures based on biomorphic matrices provide much better geometry for penetration of electrical charges generated in nuclear reactions through the insulating layer to the collector. So we can assume that filling of biomorphic matrices described above with nuclear fuel will increase the conversion efficiency to higher values.

7. Conclusion

Energy effectiveness and ecological safety of nuclear power engineering can be improved essentially with application of nanoparticles, layered nanostructures, shaped sapphire and microporous light guides as well as microcapillary biomorphic matrices. On the one hand, usage of these materials can improve essentially the situation at operating NPP and at stores of radioactive wastes. Nanoscintillators fixed at lateral surfaces of radiation hard light guides are capable to deliver instantaneous information about local situations at separate fuel rods and their assemblies with temporal resolution of milliseconds and spatial resolution of centimeters. This information will form the basis for much deeper understanding of operation of rather complicated multi-component objects such as NPP. Besides, this information will improve essentially the safety and effectiveness of performance of nuclear reactors.

Direct converters of radiation energy to electricity will improve significantly the safety of operation of nuclear reactors and transform radioactive wastes from the object of continuous anxiety of the society to ecologically safe and low cost fuel.

Moreover, qualitatively new designed of nuclear reactors can be developed based on nano-structured fissile fuel deposited inside micro-capillary matrices. Reactors of this kind will produce most energy by direct conversion of nuclear energy to electricity. Due to this radical improvement their construction will be more simple, economic and safe. The volume of radioactive wastes will also be reduced.

Author details

N.V. Klassen, A.E. Ershov, V.V. Kedrov, V.N. Kurlov, S.Z. Shmurak, I.M. Shmytko, O.A. Shakhray and D.O. Stryukov

Institute of Solid State Physics Russian Academy of Sciences, Chernogolovka, Russia

References

[1] DOE Fundamental Handbook . Nuclear Physiscs and Reactor Theory. Vol. 1 of 2. DOE-HDBK-1019/2-93, January 93

[2] DOE Fundamental Handbook . Nuclear Physiscs and Reactor Theory. Vol. 2 of 2. DOE-HDBK-1019/2-93, January 93

[3] H.G.J. Moseley (1913) "The Attainment of High Potentials by the Use of Radium", Proc.Roy.Soc.(London), A88, 471-

[4] S.L. Soo, Direct Energy Conversion, Prentice-Hall, Englewood Cliffs, NJ, 1968.

[5] J. L. Heilbron, (1974). H.G.J. Moseley: The Life and Letters of an English Physicist, 1887–1915. Berkeley: University of California Press.

[6] W.R. Corliss, D.G. Harvey, 1964, Radioisotopic Power Generation

[7] G.H. Miley, "Direct Conversion of Nuclear Radiation Energy", American Nuclear Society, 1970.

[8] G. Safanov, Direct Conversion of Fission to Electric Energy in Low temperature reactors, RAND, Research Report No. RM-1870, 1957.

[9] F.W. Krieve, JPL Fission-Electric Cell Experiment, JPL, Technical Report No 32-981, Pasadena, CA, 1966.

[10] M.S. Dresselhous, Nanostructures and Energy Conversion, Proceedings of 2003 Rohsenow Symposium on Future trends in Heat Transfer, May 16, Massachusetts Institute of Technology, Cambridge, MA, USA, pp. 1 - 3

[11] Y. Ronen, A. Hatav, N. Hazenshprung, "242 m Am-Fueled Nuclear Battery, Nuclear Instruments and Methods in Physics Research A, vol. 531, Oct. 2004, pp. 639-644.

[12] Y. Ronen, M. Kurtzland, L. Drizman, E. Shwargeraus "Conceptual Design of Americium Nuclear Battery for Space Power Applications" - J. of Propulsion and Power, vol. 23, No 4, July – August 2004.

[13] V.B. Anufrienko, V.P. Kovalev, A.V. Kulikov, B.A. Chernov "Converters of nuclear energy to electricity by secondary electrons" - Russian Chemical Journal, 2006, v. 1, No5, pp. 120 – 125.

[14] V.B. Anufrienko, B.A. Chernov, A.M. Mikhailov et. al. "Applcation of Super-Multilayered Nanostructures for Direct Conversion of Nuclear Energy to Electrical" - Journal for Nano-and Micro-system technique (in Russian), № 8, 2008, p.30 – 39.

[15] J. Friske, Aerogels, Ed. J. Friske – Berlin; Heidelberg; New York, Springer – Verlag, 1986.

[16] T. Yoshida, T. Tanabe, A. Chen "Method for Generating Electrical Power and Electric battery" United States Patent Application, Pub. No. US 2005/0077876 A1, Apr. 14, 2005.

[17] L. Popa-Simil "Pseudo-Capacitor Structure for Direct Nuclear Energy Conversion", United States Patent Application, № US2010/0061503 A1, Mar. 11, 2010.

[18] L. Popa-Simil, Advanced Nano-Nuclear Program Proposal, LAVM LLC, Los Alamos, 2011.

[19] L. Popa-Simil, C. Muntele, Direct Conversion Nano-hybrid Fuel, MRS Spring Meeting Proceedings, vol. 1104, 2008.

[20] N. Klassen, O. Krivko, V. Kedrov, S. Shmurak, I. Shmyt'ko, E. Kudrenko, E. Yakimov, V. Beloglasov, Yu. Skibina, "Significant Enhancement of Radiation Hardness of Nanodimensional Scintillators and their Application for High Rate Control of Radiation Flows inside nuclear Reactors", Proceedings of 10 th International Conference on Inorganic Scintillators and Their Applications SCINT 2009, Korea, June 8 – 12, 2009, p. 134,

[21] Krivko O., N. Klassen, V. Kurlov, V. Kedrov, I. Shmyt'ko, S. Shmurak, E. Kudrenko, A. Orlov, "Prospects of Application of Record Characteristics of Nanocrystalline Scintillators for Radiation Monitoring", IEEE Nuclear Science Symposium (Dresden), (2008) Abstract Book, p. 27

[22] R. Morrison, The Chemical Physics of Surfaces, Plenum Press, New York and London, 1977.

[23] N. Klassen, V. Kedrov, S. Shmurak, I. Shmyt'ko, A. Ganin, D. McDaniel, J. Voigt, "Development of Fast Radiation Detectors Based on Nanocomposites form Inorganic and Organic Scintillators", Proceedings of 10 th International Conference on Inorganic Scintillators and Their Applications SCINT 2009, Korea, June 8 – 12, 2009, p. 128.

[24] N. Klassen, O. Krivko, V. Kedrov, S. Shmurak, I. Shmyt'ko, E. Kudrenko, V. Dunin, V. Beloglasov, Yu. Skibina, "Essential Improvement of Sensitivity, Kinetics, Temporal and Spatial Resolutions of Radiation Detectors by Nanoscintiollators", Proceedings of 10 th International Conference on Inorganic Scintillators and Their Applications SCINT 2009, Korea, June 8 – 12, 2009, p. 131.

[25] N. Klassen, O. Krivko, V. Kedrov, S. Shmurak, I. Shmyt'ko, E. Kudrenko, E. Yakimov, V. Beloglasov, Yu. Skibina, "Significant Enhancement of Radiation Hardness of Nanodimensional Scintillators and their Application for High Rate Control of Radiation Flows inside nuclear Reactors", Proceedings of 10 th International Conference on Inorganic Scintillators and Their Applications SCINT 2009, Korea, June 8 – 12, 2009, p. 134,

[26] N.V. Klassen, V.V. Kedrov, Y.A.Ossipyan, S.Z. Shmurak, I.M. Shmyt'ko, O.A. Krivko, E.A. Kudrenko, V.N. Kurlov, N.P. Kobelev, A.P. Kiselev, S.I. Bozhko "Nanoscintillators for Microscopic Diagnostics of Biological and Medical Objects and Medical Therapy", IEEE Transactions on Nanobioscience, Vol. 8, No 1, March 2009, p. 20 – 32.

[27] I.M. Shmytko, E.A. Kudrenko, G.K. Strukova, V.V. Kedrov, N.V. Klassen. Anomalous structure of nanocrystallites of rare-earth compounds produced by sol-gel methods", Z. Kristallorg. Suppl., 27 (2008), p. 211-218

[28] N.V. Klassen, V.V. Kedrov, I.M. Shmyt'ko, S.Z. Shmurak, O.A. Krivko, E.A. Kudrenko, A.P. Kiselev, A.V. Shekhtman "Essential improvement of spatial and temporal resolutions of radiation detectors by application of nanoscintillators" Proceedings of 5th International Symposium on Laser, Scintillator and Non Linear Optical Material ISLNOM -5, Pisa, 3 – 5 September 2009. p. 90,

[29] V. Nesvizhevsky, Application of diamond nanoparticles in low energy neutron physics, Materials, 2009, 2.

[30] A.I. Fedoseev, M. Turovski, Q, Shao, and A. A. Balandin, Solar Cell Nanotechnology for Improved Efficiency and Radiation Hardness, Photonics for Space Environments XI, edited by Edward W. Taylor, Proc. of SPIE, Vol. 6308, 630806, (2006) 0277-786X/06/

[31] E.R. Dobrovinskaya et al., Sapphire: Material, Manufacturing, Applications, 177, DOI: 10.1007/978-0-387-85695-7_3, © Springer Science + Business Media, LLC 2009

[32] D. Sporea, A. Sporea "Radiation effects on sapphire optical fibers" - Phys. Stat. Sol. (C), 2007, vol. 4, No. 3, pp. 1356-1359. / DOI 10.1002/pssc.200673709,

[33] Gui-Gen Wang, Jie-Cai Han, Hua-Yu Zhang, Ming-Fu Zhang, Hong-Bo Zuo, Zhao-Hui Hu, and Xiao-Dong He, Radiation resistance of synthetic sapphire crystal irradiated by low-energy neutron flux, Cryst. Res. Technol., 2009, vol. 44, No. 9, pp. 995-1000 / DOI 10.1002/crat.200900243.

[34] A.Kh. Islamov, E.M. Ibragimova, I. Nuritdinov, Radiation-optical characteristics of quartz glass and sapphire, Journal of Nuclear Materials, 2007, vol. 362, pp. 222–226.

[35] V.N. Kurlov, S.V. Belenko "In Situ Preparation of Bulk Crystals with Regularly Doped Structures" - Adv. Mater., 1998, vol. 10, No. 7, pp. 539-541.

[36] V.N. Kurlov, S.N. Rossolenko, S.V. Belenko "Growth of sapphire core-doped fibers" - Journal of Crystal Growth, 1998, vol. 191, pp. 520-524.

[37] V.N. Kurlov, S.V. Belenko "Growth of sapphire shaped crystals with continuously modulated dopants" - Journal of Crystal Growth, 1998, vol. 191, pp. 779-782.

[38] V.V. Nesvizhevsky "Near-surface quantum states of neutrons in the gravitational and centrifugal potentials" - Physics-Uspekhi, 2010, vol. 53(7), pp. 645-675.

[39] V.V. Nesvizhevsky, R. Cubitt, K.V. Protasov, A.Yu. Voronin "The neutron whispering gallery effect in neutron scattering" New J. Phys., 2010, vol. 12, pp. 1130-1050.

[40] N.V. Klassen, V.N. Kurlov, S.N. Rossolenko, O.A .Krivko, A.D Orlov., S.Z. Shmurak "Scintillation fibers and nanoscintillators for improvement of spatial, spectral and temporal resolution of radiation detectors" – Bull. Russ. Acad. Sci., 2009, vol. 73, No 10, pp. 1541 – 1456.

[41] N.V. Klassen, V.V. Kedrov, I.M. Shmyt'ko, S.Z. Shmurak, E.A. Kudrenko, E.B. Yakimov, "Anomalous spatial correlation of borate nanoscintillators grown from melt

solutions leading to essential improvement of scintillation parameters", Proceedings of 5th International Symposium on Laser, Scintillator and Non Linear Optical Material ISLNOM -5, Pisa, 3 – 5 September 2009, p. 89.

[42] N.V. Klassen, O.A. Shakhray, V.V. Kedrov, V.N. Kurlov, D.O. Stryukov, I.M. Shmytko, S.Z. Shmurak "Nanocrystalline scintillators and fiber sapphire for radiation control" - Journ. of Nuclear Measuring and Information Technologies, 2011,Moscow, No. 4, pp. 30 – 35.

[43] Yu.S. Skibina, A.B. Fedotov, L.A. Melnikov, V.I. Beloglazov, Tuning the photonic bandgap of sub 500 nm –pitch holey fibers in the 930 – 1030 nm range, Laser Physics 2010, vol. 10, pp. 723 – 726.

[44] N.V. Klassen, O.A. Krivko, V. V. Kedrov, S.Z. Shmurak, A.P. Kiselev, I.M. Shmyt'ko, E.A. Kudrenko, A.V. Shekhtman, A.V. Bazhenov, T.N. Fursova, "Laser and Electric Arc Synthesis of Nanocrystalline Scintillators" IEEE Transactions on Nuclear Science, v. 57, No 3, June 2010, p. 1377 – 1381

[45] N. Klassen, V. Kedrov, S. Shmurak, O. Krivko, N. Prokopiuk, I. Shmyt'ko, E. Kudrenko, S. Koshelev, "Electric Signals from Nanoscintillators and Layered Nanostructures Subjected to Ionizing Irradiation", Proceedings of 10 th International Conference on Inorganic Scintillators and Their Applications SCINT 2009, Korea, June 8 – 12, 2009, p. 151,

[46] Beloglazov V.I., V.V. Kedrov, N.V. Klassen, Smurak S.Z. et. al. "Radiation hard detector" - Patent of Russian Federation", № 85669 from 10.08. 2009.

[47] N.V. Klassen, V.V. Kedrov, S.Z. Shmurak, I.M. Shmyt'ko, O.A. Shakhray, V.N. Kurlov, D.O. Striukov "Differential registration of various radiation flows by means of nanocrystalline scintillators and sapphire fibers" - Abstracts of 11-th International Conference on Inorganic Scintillators and their Applications SCINT2011, Germany, 11-16 Sept. 2011. P.1.4

[48] N.V. Klassen, V.V. Kedrov, S.Z. Shmurak, I.M. Shmyt'ko, O.A. Shakhray, P.A. Rodnyi, S.D. Gain "Essential improvement of scintillation parameters in composites from nanocrystalline dielectrics and organic phosphors", Abstracts of 11-th International Conference on Inorganic Scintillators and their Applications SCINT2011, Germany, 11-16 Sept. 2011. P 1.3.

[49] A.E. Yershov, V.V. Kedrov, N.V. Klassen, S.Z. Shmurak, I.M. Shmytko, O.A. Shakhray, V.N. Kurlov, S.V. Koshelev "New Radiation Hard Devices for Control of Ionizing Radiations and Direct Conversion of Radiation to Electrical Energy Based on Nanostructured Materials" - Problems of Nuclear Science and Engineering, series Applied Physics and Automation, 2011, vol. 66, pp. 132-139.

[50] N.V. Klassen, A.E. Ershov, V.N. Kurlov,et. al. "Particularities and applications of interactions of ionizing radiation with nanostructures" - Abstracts of German-Russian

Conference on Fundamentals and Applications of Nanoscience, Berlin, May 2012, p. 52.

[51] A.E. Yershov, V.V. Kedrov, N.V. Klassen, V.N. Kurlov, I.M. Shmytko, S.Z. Shmurak , Efficiency of direct converters of radiation to electric power, Journ. of Nuclear Measuring and Information Technologies, 2012, No. 4, pp. 21-26.

Chapter 12

Multilateral Nuclear Approach to Nuclear Fuel Cycles

Yusuke Kuno

Additional information is available at the end of the chapter

http://dx.doi.org/10.5772/53178

1. Introduction

The Fukushima Daiichi Nuclear Power Plant (Fu-NPP) accident, which occurred in March 2011, has significantly influenced the recent trend in the growing interest in nuclear reactor deployment. On the other hand, it is common knowledge that there remain problems in global climate changes and energy security in the long term view that human-beings are obliged to solve. Meanwhile, the issue on treatment of spent fuels (SFs) has become remarkably recognized. These facts should result in the continuous needs of nuclear energy and proper management of SFs, i.e., well-organized nuclear fuel cycle (NFC) services, including uranium mining, refining, conversion, enrichment, reconversion, fuel fabrication, spent fuel treatment such as storage, reprocessing, and repository.

The concerns for the nuclear proliferation of so-called "Sensitive Nuclear Technologies (SNTs)", and weapon-use nuclear materials, namely, enrichment technology (frontend), spent fuel reprocessing technology (backend) and fissile materials will increase. The latter includes concern with worldwide increase in the amount of SFs, which may have to be stored in individual states. Namely, there will be growing concerns from the nuclear non-proliferation and security perspectives that plutonium may globally proliferate as a form of SFs.

Measures for nuclear non-proliferation have so far been taken mainly by the combination of institutional systems and supply side approaches (see Fig.1). International society has been responding to the above-mentioned concerns by strengthening schematic measures centered around the safeguards under the NPT and Convention on the Physical Protection of Nuclear Material, etc. Bilateral agreements represent the latter one (supply side approach) particularly those between the US and individual states that have been functioning strongly. However, increase in the supply of fuel source materials from the Eastern Block has been remarkable in recent years, as shown in Fig.2, which may potentially weaken the influence of the Western Block on nuclear non-proliferation.

The measures for enhancement of nuclear non-proliferation on the supply side that mainly consist of the nuclear power technology advanced countries may interfere with the inalienable right of peaceful uses of nuclear power that is guaranteed by Article 4 of the NPT. Thus, there is a need to develop nuclear non-proliferation measures with high non-proliferation capacity based on new concepts which are completely different from the conventional ones. In addition, as for the nuclear security for handling SNTs and nuclear materials as well as safety management of nuclear facility operations, the conventional state-by-state efforts have limitations from the viewpoints of effectiveness, efficiency, and economic reasonability.

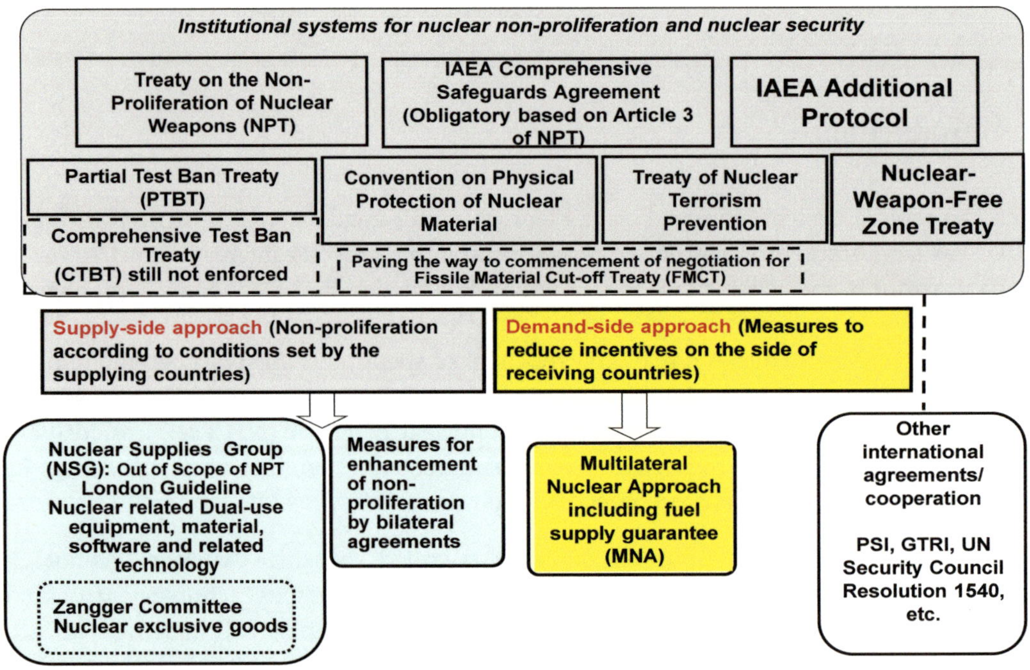

Figure 1. International Effort for Nudear Non-proliferation

Demand side approach represented by Multilateral Nuclear Approach (MNA), where services on the frontend and the backend are provided to the states possessing nuclear power plants without interfering with the inalienable right in NPT and measures for nuclear non-proliferation properly function, may be one of the most effective and efficient goals to solve all the problems discussed above. Originally the MNA was proposed as an idea to reduce the possibility of nuclear proliferation of sensitive technologies by supplying enrichment and reprocessing services to newcomer countries [1].

Regional MNA, e.g., in Asian regions, may also complement or reinforce the weakened non-proliferation regime of Western Block-based regions.

http://dx.doi.org/10.5772/53178

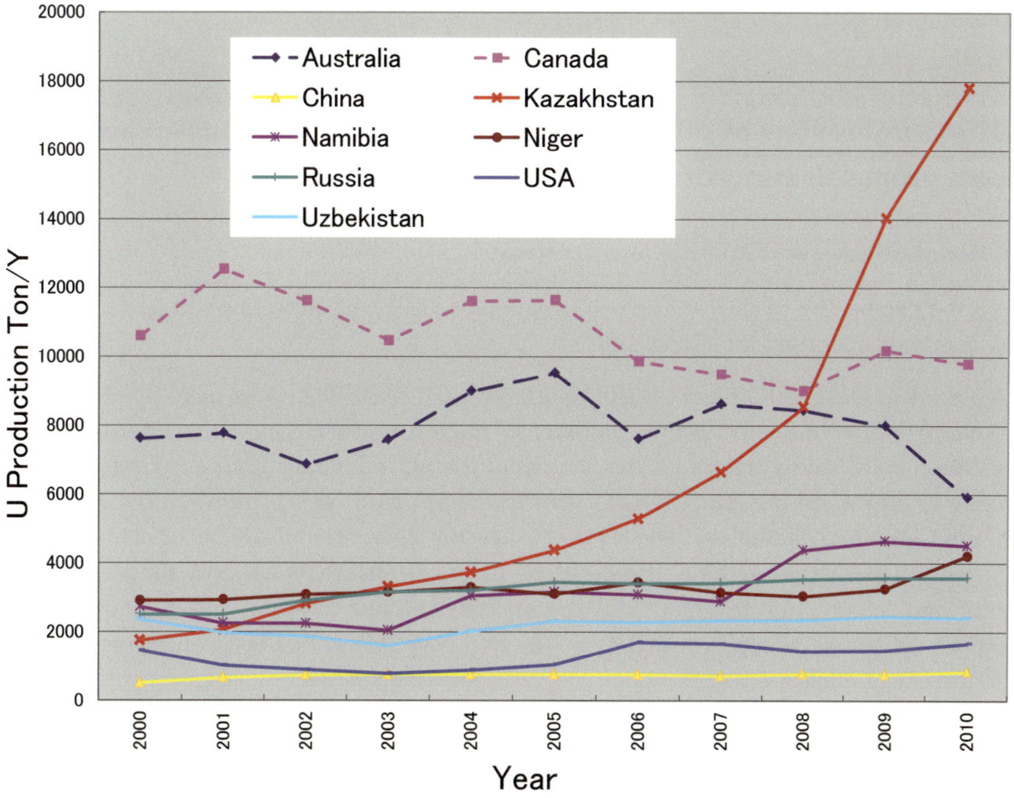

Figure 2. Diversification of Uranium Resource Supply. Data source: http:/www.world-nuclear.org/info/uprod.html

In the foreseeable future, most of newcomer countries utilizing nuclear energy would like to have a reliable system of fresh fuel supply and spent fuel treatment services, free of any political disruptions to fuel their nuclear reactors.

Several proposals [2,3] on the Multilateral Nuclear Approach (MNA) have recently been studied and a few are now ready to be implemented, in which no restraint of the peaceful use of nuclear energy due to the issues of proliferation of sensitive technologies is taken into account. Recent discussions, however, tend to focus on reliable fuel supply, namely front-end of NFC, where proliferation of uranium enrichment can be deterred. At the same time, the MNA capability to provide assurance/service that the SFs be managed properly is actually more important [4].

Storing SFs as well as possessing those in power reactors in individual countries remains problematic not only for Safety but also the risks in nuclear proliferation/Safeguards and Security (3S), due to the presence of large amounts of imbedded plutonium (Pu). Although Pu in SFs is protected by its high radiation dose rate, the technology to separate the plutonium from SFs (reprocessing) is not as difficult as uranium enrichment technology. It is therefore important to address the issues associated with the backend of the nuclear fuel cycle and to propose to properly manage/treat SFs. In this context, MNA may also be beneficial from the viewpoints of 3S.

2. Historical review of international framework discussions – Past and recent proposals concerning MNA

2.1. Historical review of international framework [5]

From the perspective of preventing proliferation of SNTs, the concept of "international control" has been proposed for a long time.

The old one is the international control of nuclear materials, which was developed under the Truman Administration in 1946 (i.e. pooling all nuclear materials, etc. in an international organization and lending them to states that want them). In January 1946, the United Nations Atomic Energy Commission (UNAEC) was founded based on the proposal by the US, the UK and Canada in November 1945 [6] that asked the international control of atomic energy, i.e., "control of atomic energy to the extent necessary to ensure its use only for peaceful purposes" and "elimination from national armaments of atomic weapons and of all other major weapons adaptable to mass destruction". In this way, MNA has been encouraged in line with the use of nuclear energy.

In the US, "A Report on the International Control of Atomic Energy" (also known as the "Acheson-Lilienthal Report") was prepared for discussion in the UNAEC. Based on the Declaration on the Atomic Bomb, the Report proposed to establish a new international organization called the "Atomic Development Authority (ADA)" which owns all fissionable material and controls them under effective leasing arrangements [7]. The report also proposed that the ADA would be in charge of all "dangerous" activities relating to raw materials, construction and operation of production plants, and the conducting of research into explosives, while "non-dangerous" nuclear activities, such as the construction and operation of power-producing piles, "would be left in national hands". It is interesting to note that in the report, all nuclear fuel cycle activities, except nuclear reactors, were categorized as "dangerous activities" and should not be conducted by an individual state.

In 1946, Bernard Baruch, the US representative to the UNAEC, submitted his plan for the international control of nuclear energy based on the Acheson-Lilienthal Report [8]. However, he modified the report by inserting the prohibition of the development of nuclear-weapons capability by new states and punishment for violations of such prohibition. The plan was not accepted by the Soviet Union and as a result, the international control of nuclear energy did not bear fruit in the 1940s.

This plan was later put on the table of the UNAEC in the form of "Baruch Plan" by UN Representative B. Baruch. However, the Plan did not take off successfully because it was contradicting with the US's free enterprise system of that time as it was promoting international ownership of the US technology. It also reached a deadlock in the negotiations between the US and the Soviet Union. However, the Plan triggered the "Age of International Collaboration for Peaceful Use of Nuclear Energy" in the "Atoms for Peace" speech by US President Eisenhower in 1953 at the UN. In this initiative, the uranium bank (reserve) with an intension of international management of fissile materials was proposed. After these debates, the International Atomic Energy Agency (IAEA) was established in 1957. Provision of nuclear materials, etc. became one of the missions of the IAEA. However, the uranium bank plan was eventually abandoned because a) uranium supply was not as limited as was initially envisioned, and b) competition of commercial nuclear energy technology/supply of nuclear materials in the major supplying states based on the above speech was intensified.

In post-war Europe, the European Atomic Energy Community (EURATOM) was established to promote nuclear energy development. The most important requirement of the EURATOM Convention was "to guarantee nuclear materials supply" by the member states [2]. At the same time, the Convention had safeguard systems to ensure that the nuclear materials within EURATOM were to be used only for peaceful uses.

International debate with regards to exporting nuclear technology and material/equipment promoted another international framework concerning the supply. In 1971, the Zangger Committee was established. The member states shall apply the IAEA's safeguards to the exported "nuclear materials" when exporting them to the non-NPT member states without nuclear weapons as well as when transporting them from these non-nuclear weapons states. The Committee also created a list of equipment as subjects of the regulation. Meanwhile, after the first nuclear test by India, the Nuclear Suppliers Group (NSG) was established in 1974 for a similar purpose. The NSG controls exports based on the so-called "NSG Guidelines", the guidelines designed for the states which export nuclear energy related equipment, material and technologies (it is a "gentleman's agreement" without any legal binding power).

In 1975, the IAEA began the exploration of the first Regional Nuclear Fuel Cycle Center (RFCC) [9] and assessed the advantages of applying backend to the RFCC. The RFCC report examined and presented basic research from international and regional approaches regarding the backend of fuel cycle in various geographical sites. From 1977 to 1980, the International Nuclear Fuel Cycle Evaluation (INFCE) [10] was conducted, and the effectiveness of nuclear fuel cycle was thoroughly evaluated by 8 working groups (WGs). Through this activity, many WGs picked up "fuel cycle center" and described it as a systematic arrangement to strengthen nuclear non-proliferation. Furthermore, for the spent fuel issues, they considered the fuel cycle as a solution that includes legal framework and multinational arrangement. Based on the results of the INFCE, the IAEA supported the experts group to examine the concept of international plutonium storage (IPS) [9], established the Committee for Assurance of Supply (CAS) [9] in 1980 and continued the deliberation until 1987. The experts' examination concluded that the multilateral approach was technically and economically feasible but there were still issues in terms of difficulty in prerequisites for participation and transfer of rights towards nuclear non-proliferation. Most of those activities, initiated by the US-initiatives, could not agree on the non-proliferation commitments and conditions that would entitle states to participate in the multilateral activities" [11], because nuclear developed states in Western Europe and Japan had already engaged in the development of their own sensitive capabilities. Therefore, they tried to maintain their activities, and not let them be interfered with by such initiatives. Developing states, especially NAM states, argued that any requirements for the non-proliferation commitment of not engaging in sensitive nuclear activities were against Article IV of the NPT. They also insisted such a requirement would discriminate between the "haves" and "have-nots" of sensitive capabilities, in addition to there being an existing discrimination by the NPT between "NWS" and "NNWS". The US was thus left alone and could not gain enough support for promoting its initiatives any further. Together with Cold War tensions, a decline in the growth rate of the US economy and a decrease of energy demand due to the second oil shock in 1979, and discouragement of nuclear energy use after TMI and the Chernobyl accidents in 1979 and 1986, the US itself lost its motivation for MNA.

At GLOBAL 93, an international conference, the "International Monitored Retrievable Storage System (IMRSS)" [12] was proposed by Dr. Häfele from Germany. IMRSS proposed that spent fuel and plutonium shall be stored in a retrievable condition under monitoring by an international entity. It chose the IAEA as a desirable entity to lead the initiative. Although it was considered a temporally measure to buy some time until the conclusion of whether SFs would be directly disposed or plutonium would be retrieved, there was no development thereafter. Dr. A. Suzuki of the University of Tokyo made a proposal for spent fuel storage in the East Asia region, and Dr. J-S. Choi of CISAC/Stanford University made a proposal for the regional treaty including regional spent fuel storage. Their proposals show the significance of the systems in which the host states offer interim storage of SFs for a limited time (40 to 50 years), even though the handling of SFs from other states is not easy.

In 1994, the US and Russia agreed that the US would purchase 500 tons of highly-enriched uranium (HEU) from Russia, convert it to low-enriched uranium (LEU) and make peaceful uses of it. Furthermore, both states agreed that each state would declare 50 tons of excess plutonium to be used for defense purposes, dismantle and retrieve 34 tons from nuclear weapons, and convert it to power generating fuel as MOX. For the purpose of nuclear non-proliferation, the US also began the "Foreign Research Reactor Spent Nuclear Fuel Acceptance Program (FRRSNFA)" in 1996 to accept the US-origin spent HEU and LEU fuels from foreign research reactors by May 2009. Furthermore, under the Russian Research Reactor Fuel Return (RRRFR) Program, some 2 tons of HEU and some 2.5 tons of LEU SFs, which were previously supplied by Soviet Union/Russia to foreign reactors, were shipped to the Mayak reprocessing complex near Chelyabinsk. The US and the Russian Federation cooperated in several repatriation projects for Russian-origin HEU fuels.

Based on the recognition that SFs and high level waste (HLW) are the common critical issues which could be factors to hinder nuclear energy promotion in the East Asia region, the Pacific Nuclear Council (PNC) began deliberation to promote understanding and cooperation for the management of SFs and HLW among the PNC members and to investigate possibilities of the International Interim Storage Scheme (IISS) in 1997. The IISS is managed at national, regional, or international levels and is to augment (not to replace) the national system. The IISS operates during the contract period from the time when SFs and HLW are deposited to the storage facility in the host state till the time when "they are returned to the originating state". The host state would be responsible for safety and safeguards of the storage facility and receive financial compensation from the contact member state, which is the owner of the SFs and HLW.

In reality, the interim storage of SFs, a part of a reprocessing contract, had been offered by reprocessing operators such as the BNFL and the AREVA. With this system, the state which makes a reprocessing contract can store SFs as long as it is stored in the reprocessing facility; however the separated plutonium and HLW at the time of reprocessing would be returned to the state. On the other hand, the concepts of the IMRSS, the RSSFEA, regional treaty and the IISS demand the host state to store or dispose of other state's SF. However, this is not easy in reality.

2.2. Recent proposalsp [13, 14]

The concerns about nuclear proliferation by states and the acquisition of nuclear weapons by terrorists has grown after nuclear testing by India/Pakistan in 1998 and the terrorist attack on September 11, 2001. The nuclear weapons black market network issues by Democratic People's Republic of Korea (DPRK, hereafter referred to as North Korea), Libya, Iran and A.Q. Khan are driving the international society to make efforts through various trials and proposals in preventing proliferation of the SNT related to fuel cycle such as isotope separation and reprocessing.

The proposals made by Ex-Director General of the IAEA, Dr. M. ElBaradei, in October 2003 presented that (1) reprocessing and enrichment operations must be restricted under the multinational control, (2) nuclear energy system shall have nuclear non-proliferation resistance, and (3) multinational approaches shall be considered for the management and disposal of SFs and radioactive waste. However, it was anticipated that his idea of a multilateral system of SNTs and radioactive substances would take a long time to overcome the issues.

Former US President G.W. Bush strongly demanded in his speech at the Defense University in February 2004 that exporting SNTs should be limited to the states which were already using them on a full scale and respecting the Additional Protocol. However, this proposal may lead to international cartels and may split the member states into the states with SNTs and without SNTs. The "Nuclear Fuel Leasing" proposal by V. Rice, et al. and "Nuclear Fuel Service Assurance Initiative" proposal by E. Moniz, et al. expect the improvement of nuclear non-proliferation though institutionalization. However, the proposals still contain a concern over supply assurance to the user states as well as a concern over the dichotomization of the member states, similar to the other proposals.

Later, a group of experts for multinational nuclear (fuel cycle) approaches (MNA) was formed (ElBaradei Commission). The group was assigned to (1) identify and provide an analysis of issues and options relevant to multilateral approaches to the frontend and back-end of the nuclear fuel cycle, (2) provide an overview of policy, legal, security, economic, institutional and technological incentives and disincentives for cooperation in multinational arrangements, and (3) provide a brief review of the historical and current experiences and analysis relating to multinational fuel cycle arrangements. In the report, MNA was assessed based on two primary factors, namely, assurance of supply and services, and assurance of nuclear non-proliferation. Furthermore, 3 potential MNA options were presented.

- **i.** To strengthen existing market mechanisms case by case with assistance from governments through long-term and transparent arrangement;
- **ii.** To establish an international supply assurance such as fuel bank in collaboration with the IAEA as an organization to assure fuel supply; and
- **iii.** To promote voluntary transformation of existing facilities of member states to MNA (including regional MNA by collaborative ownership and collaborative administration)

The study results by the expert group at the IAEA are summarized in INFCIRC/640, which give an impact on the successive examination of multinational approach framework. After this report, a number of proposals related to supply assurance and multilateral approaches were put forward. The following are some of these proposals/approaches [15]:

1. In order to achieve "Reliable Fuel Supply (RFS) Initiative", announced by former Secretary of the US Department of Energy (DOE), Bodman in September 2005, the US is in the process of down-blending about 17.4 tons of HEU to about 290 tons of LEU (4.9%) within 3 years and storing them. The RFS Initiative was later renamed to the American Assured Fuel Supply (AFS) and it will be operational in 2012.

2. During the discussion of fuel supply assurance at the Global Nuclear Energy Partnership (GNEP), the US, in collaboration with the partner states, declared that it would aim at establishing a fuel service mechanism including fuel supply at frontend and SF disposal at backend to achieve international nuclear non-proliferation. In the Non-proliferation Impact Assessment (NPIA) presented by DOE in January 2009, the importance of maintaining advanced reprocessing capacity including minor actinide recycling was insisted. It also emphasized the significance of the US's participation in the overall fuel services including backend service in order to suppress incentives for the emerging states to individually develop enrichment and reprocessing technologies. Later, being influenced by political regime change, the GNEP terminated its domestic activities (i.e. cancellation of prompt construction of commercial reprocessing facility and fast reactor) and decided that they would maintain international collaboration framework as International Framework for Nuclear Energy Cooperation (IF-NEC) only for international activities from 2010. The fuel supply working group at IFNEC expressed its willingness to support collaborative actions among member

http://dx.doi.org/10.5772/53178

states and organizations towards establishment of an international fuel supply framework. It would also provide trustworthy and worth-the-cost fuel services/supply to the global market and provide options relating to the development of nuclear energy usage in accordance with reductions of nuclear proliferation risks. The new director expressed his speech its willingness to achieve so-called "from cradle to graveyard".

3. World Nuclear Association (WNA) proposed a three-level assurance mechanism: 1) basic supply assurance provided by the existing market, 2) collective guarantees by enrichment operators supported by relevant governmental and the IAEA commitments, and 3) government stocks of enriched uranium product. According to them, it is necessary to promote the idea of an international reprocessing recycling center when nuclear energy usage is expanded in the future.

4. Reliable Access to Nuclear Fuel (RANF) (nuclear fuel supply assurance concept by 6 states): Similar to the above, this proposal contains a three-level mechanism: 1) supply through market, 2) system in which enrichment operators would substitute for each other based on the collaboration with the IAEA, and 3) virtual or physical low-enriched uranium banks by a state or the IAEA.

5. Japanese proposal: The states willing to participate shall voluntarily register at/notify the IAEA of their capacities (current stockpiles and supply capacity), and the member states shall notify the IAEA of their service provision capacity in accordance with the availability of service utilization capability by three levels (Level 1: provision of service on the domestic commercial basis – no exporting on a commercial scale, Level 2: international provision on a commercial basis, Level 3: storage that can be exported for a short time). The IAEA would make an agreement of standby-arrangement with member states and manage the system. If the fuel supply actually gets confused in a state, IAEA will play a role as a mediator. This proposal is to improve market transparency, prevent supply termination, and augment the RANF proposal.

6. UK Enrichment Bond proposal: Enrichment tasks shall be carried out by domestic enrichment operators. The supplying state, the consuming state and the IAEA will make a treaty in advance. The IAEA shall approve commitment of the consuming state for nuclear non-proliferation. If assurance is activated by bonding, the supplying state would not be prevented from supplying enrichment services to a consuming state. This proposal is to enhance credibility of supply assurance mechanisms and augment the RANF proposal. The Bond proposal was later renamed the Nuclear Fuel Assurance (NAF) proposal and was approved by the IAEA Board of Governors in March, 2011.

7. The Nuclear Threat Initiative (NTI) proposal [16]: This is a storage system for LEU stockpiles possessed and controlled by the IAEA, and it is the anchor proposal for actual realization. For the activity of the NTI, the US pledged $50 million, Norway $5 million, the United Arab Emirates $10 million, the EU $32 million, and Kuwait offered $10 million. The total pledge has reached $107 million. Furthermore, in April 2009, Kazakhstan's President Nazarbayev announced that the country was ready to receive the IAEA nuclear fuel bank and officially announced its willingness to be a host state in January

2010 (INFCIRC/782). In May 2009, the IAEA presented a proposal for deliberation at the Board of Governors meeting held in June 2009. The proposal included consuming state's requirements in relation to the IAEA nuclear fuel bank, supply processes, contents of model agreement (e.g. supply price of LEU, safeguards, nuclear material protection, nuclear liability), etc. Later, at a regular Board of Governors meeting on December 3, 2010, the establishment of "nuclear fuel bank" which will internationally manage and supply LEU to be used as fuel for nuclear energy generation was agreed on. If the IAEA receives a request from a state which cannot purchase LEU due to exceptional circumstances impacting availability and/or transfer and is unable to secure LEU from the commercial market, state-to-state arrangements, or by any other such means, the IAEA will supply LEU to the state at the market price under the guidance of the Director General of IAEA. Through this agreement, the first system in which LEU would be controlled by an international organization began. The IAEA owns the bank based on the contributions from the member states. The Board of Directors will later deliberate the location of the bank. Kazakhstan is already declaring its candidacy to be a host state. The resolution was proposed collaboratively by over 10 states including the US, Japan and Russia and was adopted with 28 states voting in favor. The developing countries which were planning to have nuclear energy later had been insisting that the bank would lead to a monopoly of nuclear technology by developed countries and "right for peaceful use of nuclear energy" stipulated by the NPT would be threatened. To address this issue, the resolution clearly stated that it would not "ask for abandoning" nuclear technology development by each state and obtained understanding from the developing countries.

8. International Uranium Enrichment Center (IUEC) [17]: The IUEC was established in Angarsk, Russia, with investment by Russia and Kazakhstan. The IUEC is not only to assure supply but to provide uranium enrichment services. Thus, this proposal is more realistic than the others. The proposal states that the uranium enrichment technology will be black-boxed, namely, the investing states will not be informed, and the technology will be under the control of the IAEA. Other than Russia and Kazakhstan, Armenia and Ukraine are now members of the IUEC, while Uzbekistan is expressing its intention of participation. It will have the LEU reserve of two 1000MW-level cores. In May 2009, for the deliberation at the IAEA Board of Governors meeting held in June, Russia submitted the proposal including the summary of agreement for LEU storage between the IAEA and Russia and summary of agreement for the LEU supply between the IAEA and the consuming states. In November 2009, being led by Russia, the nuclear advanced states submitted a resolution to the IAEA Board of Governors in November. The resolution was to seek approval of two agreement plans: 1) agreement plan between the IAEA and Russia to establish the LEU reserve under Russian IUEC, and 2) a model agreement plan between the IAEA and the LEU recipient states concerning the LEU supply from the reserve. The resolution was approved by a majority. In March 2010, the IAEA's Director General, Amano, and Director General of Rosatom Nuclear Energy State Corporation, Kiriyenco, sign-

http://dx.doi.org/10.5772/53178

ed on the agreement for the establishment of the LEU reserve under Russian IUEC, and the LEU storage was established in December, 2010.

9. Multinational Enrichment Sanctuary Project (MESP) (proposed by Germany): This proposal is for the IAEA to manage enrichment plants and exportation on an extra-territorial basis in a host state. The SNT will be black-boxed.

10. The Science Academies of the US and Russia presented analysis and proposals for nuclear fuel assurance as a measure to prevent proliferation of nuclear weapons under the title of "Internationalization of Nuclear Fuel Cycle – Goals, Strategies, and Challenges"13. In its report, the options and technological issues for the future international nuclear fuel cycle are presented. The report also contains the analysis of the incentives for the states that opt for accepting fuel supply assurance and developing enrichment or reprocessing facilities and do not opt for it. Furthermore, they examined new technologies for reprocessing/recycling and new reactors and made various proposals to the governments of the US and Russia and other nuclear supplier states to stop proliferation of SNTs and contribute to reduction in the risk of nuclear weapons proliferation. The report analyzed and summarized critical issues and presented several standards for assessing the options.

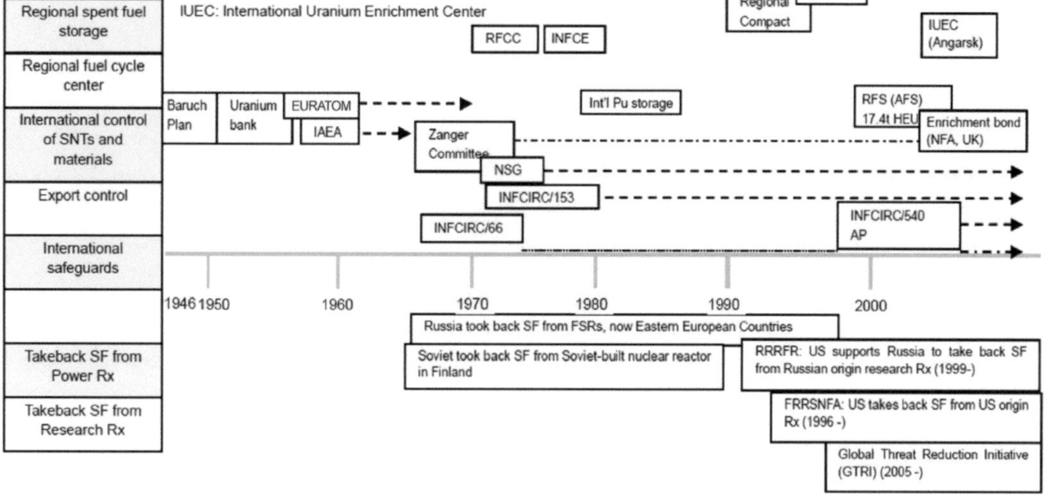

Figure 3. Transition of Proposals/Initiatives for International/Regional Management of Nuclear Fuel Cycle Relevant to Nuclear Non-proliferation

Figure 3. summarizes the flow of nuclear non-proliferation measures centered on multilateral approach/supply assurance in the past. As shown, the debates have become more and more active in recent years, and the need for internationalization of fuel cycle, which was not very realistic until now, is gradually becoming a reality. As described above, as of December 2011, the IAEA nuclear fuel bank, LEU reserve in Angarsk, Russia, and the UK's NFA proposal were approved by the IAEA Board of Governors, and the US's AFS begin its operation in 2012.

2.3. Issues with the past and current proposals

Most of the past proposed MNAs had never been implemented in any form until the nuclear fuel bank 7) and the LEU-IUEC storage 8) were approved by the IAEA Board of Governors. It was probably because nuclear proliferation was not recognized as a sufficiently serious issue and there was not a very strong economic motivation. Many proposals included unfair double standards, i.e., "have" and "have not", and inconsistency with Market Mechanism. Also need of MNA may not have matured, or become critical yet.

However, as explained above, the situation has been changing in the last few years. Despite the Fukushima Nuclear Power Plant accident as well as the actual global concern over nuclear non-proliferation, the expansion of peaceful uses of nuclear energy in the world is unavoidable in the a long term. In that sense, the role of supply assurance was reviewed, and some of the above mentioned proposals have been approved by the IAEA Board of Governors.

3. Significance of MNA

Significance of NMA, namely, MNA's benefits and incentives of individual stakeholders may be summarized as follows:

New nuclear non-proliferation regimes based upon mutual confidence and transparency, including regional safeguards, can be established, which can strengthen the function of nuclear non-proliferation.

Formulation of no discriminatory framework can be the primary incentive to make states join MNA. Recent criteria-based approaches of export of sensitive technologies in NSG [18] would help create a framework taking into account NPT Article IV.

Nevertheless, the number of enrichment and reprocessing facilities can be limited from the viewpoints of their needs (capacities) and nuclear nonproliferation, although every participating country can formally have the right to possess such SNTs.

Services on spent nuclear fuels, take-back, take-away, storage, reprocessing etc, can systematically be assured. Recipient countries can enjoy such services in NMA framework.

It is also expected that the host country in MNA would be discouraged to divert nuclear materials and to misuse related technologies because of the multilateral control of the fuel cycle facilities.

To minimize proliferation risks on SFs: accumulation of SFs, e.g., in power reactor user countries, has become serious issue in the world. By leaving such spent fuel in individual countries, there is also a certain level of risk to make such countries change the policy, i.e., to have an incentive to try reprocessing.

Improvement in 2S (safety/security) can be expected if for NMA framework systems to deal with such issues within the framework can be included, e.g., application of international standards among the participating countries.

Host countries may be able to expand their nuclear fuel cycle business capabilities further although facilities are expected to be controlled under/by MNA.

4. Prerequisites/features for establishing MNA

INFCIRC/640 (Pellaud Report) [19] proposed 7 elements of assessment called "Label" as prerequisites/features. In INFCIEC/640a variety of different issues are included altogether and the importance of each issue, such as nuclear security and safety, and political and public acceptance, is not focused individually, even though these are contemporary topics particularly following the Fukushima nuclear accident. Therefore, the following 12 elements, namely with 5 additional Labels, can individually be described as a full set of prerequisites or features to be considered for the formulation of a new framework of MNA as discussed elsewhere [20].

Label a: Nuclear non-proliferation

This includes safeguards, nuclear security and export control. If a state meets certain criteria (e.g. regional safeguards under MNA, nuclear security, export control), it is considered that the state can adequately maintain nuclear non-proliferation resume. Thus, the possession of sensitive nuclear technologies (SNTs) (i.e. uranium enrichment and spent fuel reprocessing), which is one of the measures for nuclear non-proliferation, would not necessarily be limited (criteria-based approach).

Label b: Fuel cycle service

An appropriate state becomes a host state or siting state (that provides/lets site) and offers fuel cycle service, based on the above-mentioned criteria-based approach. It includes uranium fuel supply service and services on spent fuel treatment. The latter should be made with a clear plan/agreement for long-term spent fuel treatment (storage / recycling / direct disposal), e.g., reduction of nuclear waste toxicity (HLW to medium level), individual member states to receive final waste, and use of MOX, in order not to bring concern to host/siting states.

Label c : Selection of a host state (siting state)

The state that meets all the criteria can be a host state or siting state. The specific criteria to participate in the multinational framework or to be host/siting state are, for instance, to satisfy conditions almost equivalent to the "objective criteria" described in INFCIRC 254 part 1-6, 7 (NSG guideline revised) 18, that is, member states are in full compliance with its obligations under NPT/safeguards agreement, are adhering to the NSG Guidelines, apply agreed standards of physical protection and have committed to IAEA safety standards.

Label d : Access to technology

Particularly, the access to SNTs should be strictly controlled under the MNA Framework.

Label e : Multilateral involvement

This includes 1) having multilateral cooperative system on e.g., safeguards, safety, and security, and 2) provision of services with or without transfer of facility ownership to MNA.

Label f : Economics

MNA, as a whole, increases economy when compared with management by individual states.

Label g : Transport

Member states should cooperate and maintain international standards for nuclear material transportation beyond borders.

Label h : Safety

International safety standards should be met within MNA.

Label i : Liability

MNA should cover a certain level of liability.

Label j : Political and public acceptance

Individual host or siting states should obtain political and public acceptance in corporation with MNA.

Label k : Geopolitics

Practically it should be taken into account if the stat is geopolitically stable.

Label l ; Legal aspect

Table 1 summarizes the existing treaties and agreements that correspond to each Label to be considered for MNA.

The gap between the new-MNA and existing related laws and agreements, which may conflict in some cases, should be adjusted. In particular, new MNA framework must have equal or higher capability on nuclear non-proliferation (Label a), e.g., in order to adjust the existing bilateral agreement that may be one of the strongest measures among the existing non-proliferation systems. In other words, the MNA member states must basically assure conditions set forth in the international treaties and agreements.

http://dx.doi.org/10.5772/53178

Evaluation element (label) and its contents			Related treaties, agreements, etc.
a	Nuclear non-proliferation	General	Treaty on the Non-Proliferation of Nuclear Weapons (NPT)
		Safeguards	Comprehensive Safeguards Agreement (INFCIRC/153)
			Additional Protocol (AP)(INFCIRC/540)
			Regional safeguards agreement (e.g. EURATOM, ABACC)
		Nuclear material protection, nuclear security	The Physical Protection of Nuclear Material and Nuclear Facilities (INFCIRC/225 Rev.5)
			Convention on the Physical Protection of Nuclear Material (INFCIRC/274)
			International Convention for the Suppression of Acts of Nuclear Terrorism
		Export control	Nuclear Suppliers' Group Guideline for Nuclear Transfers (INFCIRC/254 Part 1) Nuclear Suppliers' Group Guideline for Transfers of Nuclear-Related Dual-Use Equipment, Material and Related Technologies (INFCIRC/254 Part 2)
			United Nations Security Council Resolution 1540
		Bilateral nuclear energy cooperation agreement	e.g. Bilateral nuclear cooperation agreement, particularly with the US
b	(Nuclear fuel) Supply assurance		e.g. IAEA nuclear fuel bank, LEU storage at International Uranium Enrichment Center (IUEC) in Angarsk, Russia.
c	Selection of host states (Only the case where Asian states are the member states)		Southeast Asian Nuclear Weapon Free Zone Treaty (Bangkok Treaty) [21]
			Treaty on a Nuclear Weapon Free Zone in Central Asia (Treaty of Semei) [22]
			Mongolia Nuclear Weapons-Free Zone
			Korean Peninsula Non-Nuclear Weapon Declaration
d	Access to technologies		NA
e	Degree of involvement in multinational initiative		NA
f	Economics		NA
g	Transportation		IAEA recommendation regarding physical protection of nuclear material (INFCIRC/225 Rev.5)
			Convention on the Physical Protection of Nuclear Material (INFCIRC/274)
			Regulations for the Safe Transport of Radioactive Material (TS-R-1, IAEA Transport Regulations)
			A code of practice on the international transboundary movement of radioactive waste (INFCIRC/386)
h	Safety		Convention on Nuclear Safety
			Convention on Early Notification of Nuclear Accidents
			Convention on Assistance in the Case of a Nuclear Accident or Radiological Emergency
			Joint Convention on the Safety of Spent Fuel Management and on the Safety of Radioactive Waste Management
i	Liability		Vienna Convention on Civil Liability for Nuclear Damage
			Paris Convention on Nuclear Third Party Liability
			Convention on Compensation for Nuclear Damage
j	Political and social acceptance		NA
k	Geopolitics		NA
l	Legal regulations		As shown in this table

Table 1. Existing Treaties and Agreements to be Considered for MNA

5. An example of specific MNA framework study [21, 22, 23, 24, 25]

The author's group has been studying an example of MNA framework, where strengthening of international non-proliferation scheme and provision of stable energy/nuclear fuel cycle services in a region are discussed. It contributes to enhancement of transparency and trust-building in the region. The study investigated the schematic issues and the countermeasures concerning the specific measures to achieve stable maintenance of the multilateral international nuclear fuel cycle including stable uranium supply system, spent fuel treatment system, usage of plutonium, establishment of regional safeguards scheme for the international nuclear fuel cycle, requirements for an organization that carries out international nuclear fuel cycle, and roles of industry in the international nuclear fuel cycle scheme. An image of framework scope is given in Fig.4. Outline of the study is shown below,

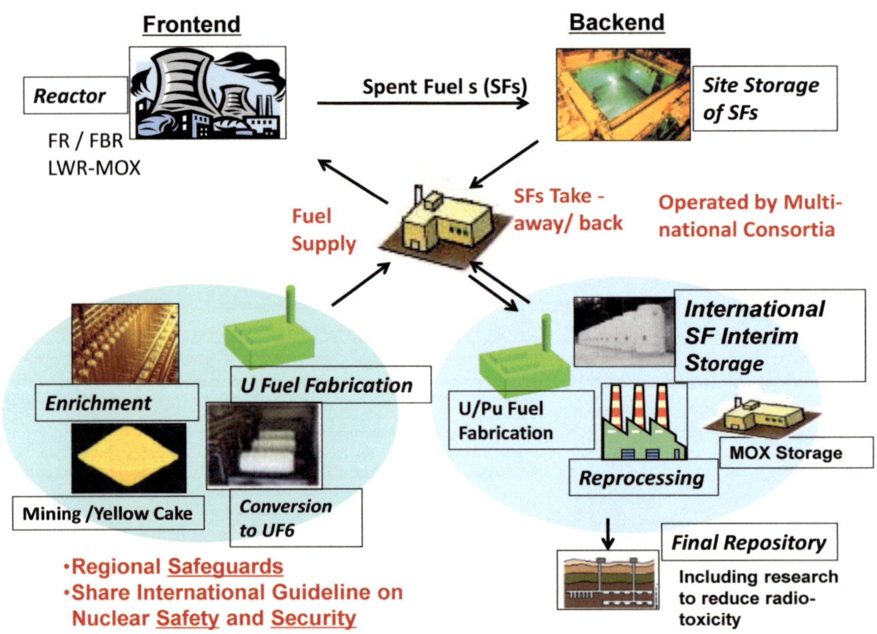

Figure 4. Possible framework of Future Nuclear Fuel Cycle

Three options on MNA system, Type A, B and C as shown below are defined.

Type A: No involvement of services (assured) of fuel supply, spent fuel storage and reprocessing, but regional framework for 3S.

Type B: Provision of services (assured) of fuel supply, SF storage and reprocessing without transfer of ownership of facilities; including regional framework for 3S.

Type C: Provision of services (assured) of fuel supply, spent fuel storage and reprocessing, MOX storage with ownership transfer of facilities to MNA; regional framework for 3S (with IAEA - arrangement).

http://dx.doi.org/10.5772/53178

(Specific framework proposed)

It is targeted to the Asian region

It establishes MNA Operating Organization as the core of the framework function

Conclude Treaty on Regional NFC and related Agreements between States and the Organization.

In the multilateral framework, the system/facilities are divided into the Type A, Type B and Type C. Plutonium-handling facilities such as reprocessing, MOX fuel fabrication, Fast Reactors, and MOX storage facilities should be controlled under type C, whereas uranium enrichment facility and spent fuel storage can probably be categorized as Type B or C depending on the siting countries. LWR MOX reactor would be Type A, while direct disposal should be Type B.

Regarding SFs, the MNA consisting of host, siting, and recipient states has clear plan for long-term spent fuel treatment, i.e., recycling / direct disposal, reduction of nuclear waste toxicity and use of MOX, within a specific certain period in order not to bring concern to host/siting states.

MNA develops technologies and services of reprocessing to reduce radio-toxicity of HLW (e.g., HLW to medium level) that would make an individual member receive final disposal waste easier.

It establishes Regional Material Accounting and Safeguards system within the MNA Framework to implement the nuclear non-proliferation regime, as described in Fig 5. The MNA agreement contains high level of nuclear non-proliferation capability, equivalent to the existing bilateral agreements (e.g. one with the United States).

The MNA has function to attain the international level on nuclear safety and security for facilities within the Framework (not only for fuel cycle facilities but also nuclear power reactors); criteria and inspection system.

The MNA has agreement with technology holders to precisely manage and control the SNTs (limited to technology-holding operators only).

Obligation with regards to nuclear non-proliferation is performed equally by the member states, while it is guaranteed that the right of peaceful uses of nuclear energy pursuant to Article 4 of NPT is not interfered with.

The specific requirement to participate in the multilateral framework is to satisfy conditions almost equivalent to the "objective criteria" described in INFCIRC 254 part 1, 6-7 (NSG Guidelines revised in 2011, see below*).

The MNA Framework is to be more economically advantageous than the fuel service on per state basis.

Framework member states cooperate on and agree to "transport" with regards to the nuclear fuel cycle service.

Liability for compensation of damages at a possible level is agreed to within the Framework.

The member states cooperate in efforts to obtain public consensus in host and siting states. Any legal regulation to be inconsistent with, or antagonistic to, existing international rules, bilateral agreements, etc. is cleared.

Framework meets the recent Nuclear Industries' recognition on Safety, Health and Radiation Protection; Physical Security; Environmental Protection and Handling of Spent Fuel and Wastes; Compensation for Nuclear Damage; Nuclear Non-Proliferation and Safeguards; and Ethics, as described in "Principle of Conduct" for nuclear power plant exporters [26], where the plant manufactures (=Venders) should consider when exporting nuclear products.

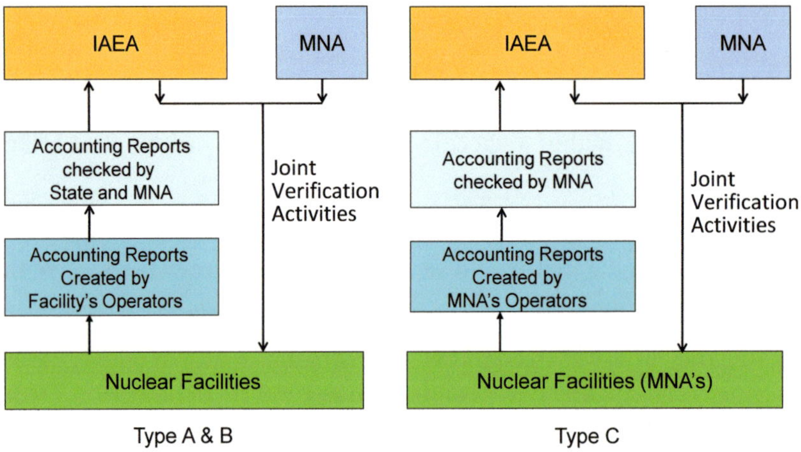

Figure 5. Regional Safeguard System

[NSG Guidelines* (INFCIRC/Rev.10/Part.1, 26 July 2011)[18]

NSG members would not authorize transfers of enrichment and reprocessing technology unless the intended recipient met certain "objective" criteria:

Be party to the NPT and in full compliance with the treaty,

Has not been identified in a report by the IAEA Secretariat which is under consideration by the IAEA Board of Governors, as being in breach of its obligations to comply with its safeguards agreement, nor continues to be the subject of Board of Governors decisions calling upon it to take additional steps to comply with its safeguards obligations or to build confidence in the peaceful nature of its nuclear program, nor has been reported by the IAEA,

Is adhering to the NSG Guidelines and has reported to the Security Council of the United Nations that it is implementing effective export controls as identified by Security Council Resolution 1540,

Has concluded an inter-governmental agreement with the supplier including assurances regarding non-explosive use, effective safeguards in perpetuity, and retransfer,

Has made a commitment to the supplier to apply mutually agreed standards of physical protection based on current international guidelines, and

Has committed to IAEA safety standards and adheres to accepted international safety conventions.

The outline of the management structure for the framework to meet the proposed concept is given in Fig. 6, where a basic treaty and many agreements on export, fuel supply, safety, security, safeguards, transfer of facility, SF services, sensitive technology control needs to be concluded among member states, and between state and AMNAO, technology holder and AMNAO, IAEA (and/or other international organization) and AMNAO. Also support of IAEA is essential for establishment of such international framework and system.

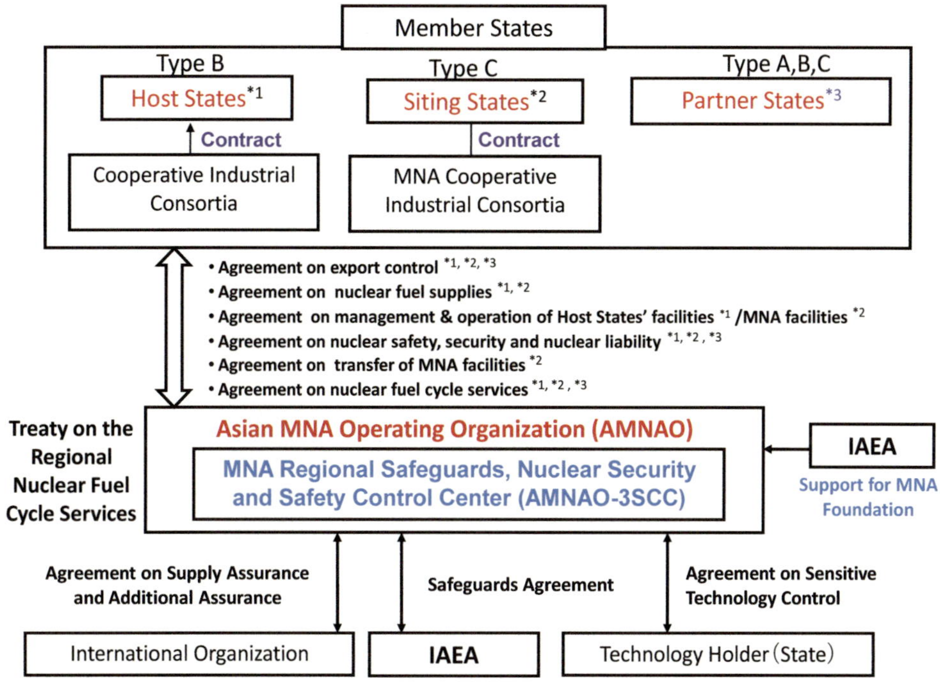

Figure 6. Treaty/Agreements Needed for the Foundation of MNA Model Structure

The treaty on the Regional Nuclear Fuel Cycle may consist of the following articles:

Preamble: Objective, Scope, Definition, Rights and Obligation, Signors

Article 1: Promotion of Foundation and Operation of Cooperative Industrial Consortia

Article 2: Limitation of commercial use, Obligation for Cooperative Industrial Consortia

Article 3: Industrial Rights Article 4: Sensitive Technology

Article 5: Safeguards Article 6: Resolution of Dispute

Article 7: Nuclear Security and safety Article 8: Withdrawal

Article 9: Licensing Article 10: Transport

Ending Part Protocol, Attached Documents

An example of specific Asia region MNA cooperation is shown in Fig. 7, which may complement or even reinforce the present non-proliferation regime. At the same time, the new proposed scheme may enable the region to have international storage of SFs and MOX.

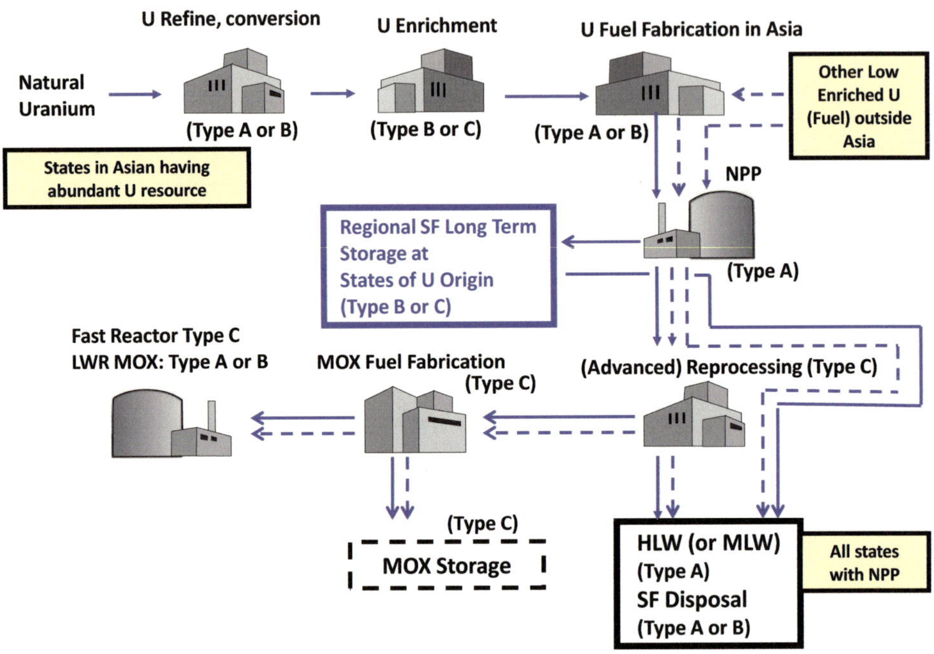

Figure 7. An Example of Potential MNA Cooperation in Asia Region

6. Future perspective and challenges on MNA

As discussed in Section 5, many prerequisites/features (Label a-l) have to be studied in order to establish a feasible and sustainable "Multilateral Approach Framework of Nuclear Fuel Cycle". Even if a proposal enables the fulfillment of all the prerequisites/features including equal right for peaceful use of nuclear energy, furthermore, remaining challenges may be how to make states have incentives toward or be attracted to the proposed MNA.

As shown in Section 2, EURATOM as a regional MNA was successfully established in the post-war. An incentive for European states to participate in EURATOM was to securely acquire source materials for nuclear energy development [2] due to the presence of the risk of shortage of energy source, although their Convention included safeguards systems to ensure that the nuclear materials to be used only for peaceful purpose.

The priorities for the above-shown example study are the following: "to eliminate inequality from the perspective of peaceful uses of nuclear power", "involvement of industry", "to have nuclear non-proliferation capacity of the current or higher level (including political and geopolitical perspectives)", "to realize international standards for safety and security", "to have higher economic potential for fuel cycle than a single state approach", "to eliminate conflicts/inconsistency with existing laws and regulations", and "to solve transport issues of nuclear fuel, etc". Particularly, involvement of industry would be a key issue when such incentive or attractiveness of the new proposal is discussed. Probably, nuclear societies including industry have internationally received greater recognition of the importance in Safety, Health and Radiation Protection; Physical Security; Environmental Protection and Handling of Spent Fuels and Wastes; Compensation for Nuclear Damage; Nuclear Non-Proliferation and Safeguards and Ethics, as described by Principle of Conduct [28], since the Fukushima Power Plant Accident.

Taking into account the example of EURATOM, the need of MNA for participants is the overriding issue in having "incentive" towards establishment of MNA. The author would like to note that the environment is getting ripe for the need of MNA, in terms of SF and waste treatment, maintenance and improvement in safety, security and nuclear non-proliferation-safeguards (3S).

There are still many challenges, pursuing incentives on economic efficiency, 3S, finding the solutions for nuclear material transportation within MNA, effective and efficient organizational management with not only member states but industries, and legal conflicts between new MNA's treaty/agreements and existing ones.

7. Conclusion

Even after Fu-NPP accident, use of nuclear reactors may be expanded particularly in emerging countries, where reliable systems of fresh fuel supply as well as proper management of spent fuels, free of any political disruptions to their nuclear reactors, are highly desirable. Establishment of international cooperative systems, which includes services for fresh fuel supply, spent fuel take-back/take-away, interim storage, reprocessing, and possibly repository disposal, may be able to contribute to a) enhancement of 3S, nuclear non-proliferation (Safeguards), Safety, and Security, b) economic rationality, c) promotion of confidence-building, and d) prevention to the occurrence of unfair business such as government-to-government transaction based on cradle-to-grave service that nuclear weapon state's privilege enables. This kind of internationally cooperative framework may become essential for future sustainable utilization of nuclear power.

Acknowledgements

The author should like to express sincere gratitude to Ms. M.Tazaki, Dr. M.Akiba, Dr. T.Adachi, Dr.T.Oda, Dr. R.Takashima, Dr. J-S.Choi, Professor A.Omoto and Professor S.Tanaka for their great contribution to this study.

A part of this study is the result of "Study on establishment and sustainable management of multi-national nuclear fuel cycle framework" carried out under the Strategic Promotion Program for Basic Nuclear Research by the Ministry of Education, Culture, Sports, Science and Technology of Japan.

Author details

Yusuke Kuno

Department of Nuclear Engineering and Management,University of Tokyo, Japan

References

[1] U.S. Committee on the Internationalization of the Civilian Nuclear Fuel Cycle; Committee on International Security and Arms Control, Policy and Global Affairs; National Academy of Sciences and National Research Council http://www.nap.edu/catalog/12477.html September 30, 2008

[2] Y.Kuno, J-S.Choi: Nuclear Eye 59-62, Vol.55, No.5, (2009) http://www.carnegieendowment.org/files/fuel_assurances_rauf.pdf

[3] http://www.carnegieendowment.org/files/fuel_assurances_rauf.pdf Individual Countries'Proposals at IAEA: INFCIRC/ 659(USA), INFCIRC/708(RUS), INFCIRC/707(UK), INFCIRC/704(GER), INFCIRC/683(JAP)

[4] Individual Countries' Proposals at IAEA: INFCIRC/ 659(USA), INFCIRC/708(RUS), INFCIRC/707(UK), INFCIRC/704(GER), INFCIRC/683(JAP)

[5] http://www.jaea.go.jp/04/np/activity/2008-07-10/2008-07-10-9.pdf

[6] Declaration on Atomic Bomb By President Truman and Prime Ministers Attlee and King, Washington", November 15, 1945

[7] Chester I. Barnard; Dr. J. R. Oppenheimer, Dr. Charles A. Thomas, Harry A. Winne, David E. Lilienthal, Chairman, "A Report on the International Control of Atomic Energy", Prepared for the Secretary of State's Committee on Atomic Energy by a Board of Consultants, Department of State Publication 2498, Washington, D.C., March 16, 1946, pp. 25-43

[8] The Baruch Plan, presented to the United Nations Atomic Energy Commission, June 14, 1946, atomicarchives.com, http://www.atomicarchive.com/Docs/Deterrence/BaruchPlan.shtml

[9] Tariq Rauf, "Perspectives on Multilateral Approaches to the Nuclear Fuel Cycle", 30 April 2004, http://www.iaea.org/newscenter/focus/npt/npt2004_3004_mnfc_npt.pdf

[10] Tariq Rauf; Fiona Simpson, "The Nuclear Fuel Cycle: Is it Time for a Multilateral Approach?", Arms Control Today, December 2004

[11] Yury Yudin, Multilateralization of the Nuclear Fuel Cycle: Assessing the Existing Proposals, UNIDIR/2009/4, United Nations Institute for Disarmament Research (UNIDIR), Geneva, Switzerland; p.6

[12] Costing of Spent Nuclear Fuel Storage", IAEA Nuclear Energy Series, No. NF-T-3.5, IAEA, Vienna, 2009, p42

[13] http://www.jaea.go.jp/04/np/activity/2008-07-10/2008-07-10-9.pdf & U.S. and Russian Committees on Internationalization of the Nuclear Fuel Cycle, National Research Council and Russian Academy of Sciences: Internationalization of the Nuclear Fuel Cycle : Goals, Strategies, and Challenges, September 30, 2008,

[14] Y.Kuno, J-S.Choi: Internationalization and regional administration of nuclear fuel cycle – Why internationalize nuclear fuel cycle. Nuclear Eye 59-62, Vol.55, No.5 (2009)

[15] Tariq Rauf; Zoryana Vovchok, IAEA Bulletin 49-2, Vienna, March 2008, pp.62-63

[16] IAEA Board of Governors Conclude December Meeting", IAEA Top Stories & Features, December 3, 2010, IAEA homepage, http://www.iaea.org/newscenter/news/2010/bog031210.html

[17] Russia Inaugurates World's First Low Enriched Uranium Reserve", IAEA Top Stories & Features, December 17, 2010, IAEA homepage, http://www.iaea.org/newscenter/news/2010/leureserve.html

[18] INFCIRC/254/Rev.10/Part 1, IAEA, Vienna, 26 July 2011

[19] INFCIRC/640, IAEA, 22 February 2005, pp. 42-44

[20] M.Tazaki, Y.Kuno: "The Contribution of Multilateral Nuclear Approaches (MNAs) to the Sustainability of Nuclear Energy", Sustainability, ISSN 2071-1050, 4, 1755-1775 doi:10.3390/su4081755 (2012)

[21] Signed in 1995 and became effective in 1997. The member states are: Laos PDR, Myanmar, Malaysia, Brunei, Viet Nam, Thailand, Cambodia, Singapore, Indonesia, and the Philippines (10 ASEAN states)

[22] Y.Kuno, M.Tazaki, M.Akiba, T.Adachi, R.Takashima, A.Omoto, T.Oda, J-S.Choi, S.Tanaka: Study on Sustainable Regional Nuclear Fuel Cycle Framework from Nuclear Non-Proliferation Viewpoint - (I) Historical Review and Basic Concept to Propose New Framework, Institute of Nuclear Material Management (INMM) Annual Meeting, July 15-19, 2012, Orlando, USA (On-line Proceedings, 10 pages)

[23] Mitsunori Akiba, Takeo Adachi, Makiko Tazaki, Ryuta Takashima, Yusuke Kuno and Satoru Tanaka : Study on Sustainable Regional Nuclear Fuel Cycle Framework from Nuclear Non-Proliferation Viewpoint- (II) Prerequisites for formulation of the Framework, Institute of Nuclear Material Management (INMM) Annual Meeting, July 15-19, 2012, Orlando, USA (On-line Proceedings)

[24] T.Adachi, M.Akiba, M.Tazaki, R.Takashima, A.Omoto, S. Hoshiba, Y.Kuno, S.Tanaka: Study on Sustainable Regional Nuclear Fuel Cycle Framework from Nuclear Non-Proliferation Viewpoint - (III) Proposal of Specific Agreements for Multilateral Nuclear Fuel Cycle Approach, Institute of Nuclear Material Management (INMM) Annual Meeting, July 15-19, 2012, Orlando, USA (On-line Proceedings)

[25] M.Tazaki, M.Akiba and Y.Kuno, "The Legal Aspects of Internationalization of Nuclear Fuel Cycle", Paper No. 392033, Proc., GLOBAL 2011, Makuhari, Japan, Dec. 11-16, 2011

[26] J-S.Choi, Takuji Oda, S.Tanaka, and Y.Kuno, "The Roles of Industry for Internationalization of Nuclear Fuel Cycle", Paper No. 392343, Proc., GLOBAL 2011, Makuhari, Japan, Dec. 11-16, 2011

[27] http://carnegieendowment.org/publications/special/misc/nppe/

Chapter 13

The Fukushima Disaster: A Cold Analysis

Cristian R. Ghezzi, Walter Cravero and
Nestor Sanchez Fornillo

Additional information is available at the end of the chapter

http://dx.doi.org/10.5772/54262

1. Introduction

The accident of Fukushima Daiichi Nuclear Power Plant on March 11, 2011, followed by an earthquake and tsunami at the Honshu island of Japan was one of the worst accidents in the history of mankind. It was classified as a level 7 nuclear accident, comparable to the Chernobyl accident in 1986. There was a general feeling of dissatisfaction about the information provided by the company Tokyo Electric Power Company (TEPCO), that operated the plant, and there were criticisms about the government decisions and on how the data about the measured dose rates in the island were interpreted. It is understood that in such a huge crisis, the people at charge in the company and in the government should not contribute to spread the panic, and it is useful to keep control of the situation. However, a flood of data invaded the news media and it turned difficult to harvest the truth among several contradictions, and to know about the destiny of a good portion of Japan and its population. Moreover, we scientist must feel compelled to clarify that truth, no matter if not directly working in the nuclear industry. We must feel an ethical and moral commitment to understand the situation in such a human calamity, because we are contributing directly or indirectly to the development of new technologies. It is necessary to learn from the errors of the past to plan carefully the highway towards the real progress. Our knowledge must be not abandon in hands of irresponsible people, but it must be driven to make our existence in this planet safer and happier.

This chapter is a technical report that exposes some basic concepts about nuclear physics, gives a concise chronology of the events at the Fukushima nuclear disaster, and analyzes a large database of dose measurements with a code specifically designed for it. It is not a document in favor or against the nuclear energy.

2. Chronicle of the disaster of Fukushima I

On March 11 of 2011, at 14:46 of Japan standard time, the earthquake Tōhoku impacted at the Honshu island of Japan. The earthquake had its epicenter approximately 70 kilometers east of the Oshika Peninsula of Tōhoku (coordinates 38º5′36″ N 142º6′15″E) at the north of the main island of Japan, with a moment magnitude scale of 9 (equivalent to 8.9 in the Richter scale). For comparison, a ten kilotons nuclear explosion produces roughly a seismic wave of moment magnitude 4.8 [1]. The earthquake produced large accelerations on the ground of Fukushima I Dai-ichi nuclear power plant, at southwest of the epicenter (coordinates 37°25′17″N 141°1′57″E). The plant had six boiling water reactors, each one in separated containment buildings, and operated by the Tokyo Electric Power Company (TEPCO). The plant suffered accelerations in excess of 5 m/s^2, which is larger than the designed tolerances for the installations 2, 3 and 5, but within the tolerances of units 1 and 6 [2]. When the earthquake strikes, the reactors 4, 5, and 6 were in cold shutdown for maintenance, while reactors 1, 2, and 3 were in operation. However, reactors 1, 2, and 3 scrammed (shutdown automatically, see below) after the quake.

The plant lost its electric power, while the connection with the offsite power electric grid was lost due to the damages in the transmission lines by the earthquake. Thirteen electric diesel generators automatically came into operation to continue with the cooling of the six reactors. A giant tsunami wave followed approximately forty minutes after the earthquake, and surpassed the seawall of the plant of 5.7 meters, although the plant was about ten meters above the sea. The tsunami flooded the plant, and most of the diesel generators stopped working approximately one hour after the quake [3]. However, a diesel generator remained in operation and maintained the cooling capabilities of the cores and spent fuel pools of units 5 and 6 (units 5 and 6 are at 13 meters over the sea level).

After the diesel generators failure, an emergency power system sourced by electric batteries came into operation. But the batteries lasted eight hours, and although the replacement batteries arrived on time, it was impossible to connect the generators to the water pumps, since the connection point was flooded and was difficult to find the cables [4].

Therefore, TEPCO concentrated their efforts on establishing electric power from the offsite grid, and there was a long delay in connecting them to the cooling system.

2.1. Description of the reactors

A boiling water reactor (BWR) is a nuclear reactor that uses fresh water as moderator and coolant. The fuel rods are located in a pressurized container: the reactor pressure vessel (RPV). The control bars are inserted in from the bottom of the vessel, and together with the water are responsible for slowing down neutrons. The steam dryers are at the top of the RPV, and prepare the working fluid to enter the turbine. The RPV is itself contained in another container which is called drywell (DW). That container prevents that vapor vented from the reactor core enter into contact with the environment. In fact, the drywell contains (or is connected to) another vessel that surrounds the RPV bottom called the pressure suppression chamber (SC) (also known as the wet well). During normal operation of the reactor there is a significant pressure difference between the RPV and the DW. If the RPV is vented to the DW, the condensate water is collected in the pressure suppression chamber. All this is in turn housed in a concrete structure, called the primary containment, which is intended to protect the indoor mechanically and to isolate the reactor.

The reactors are called based on core related systems, and that is done by a slash and a number after the acronym BWR: BWR/2 (/ 3, / 4 or / 5). It also makes the designation of the type of reactors the DW and SC that surrounds the RPV with the indicative Mark I or Mark II. The Mark I type has the DW electric bulb-shaped having the receiving SC -with toroidal shape- surrounding its base. The DW has an interconnecting vent network with the SC. The Mark II type has a DW in the form of a frustum of a cone or a truncated cone, surrounded around its base by a cylindrical SC separated from de DW by reinforced concrete slabs. At Fukushima Daiichi were located 6 units, the number 1 a BWR/3 Mark I. From 2 to 5 BWR/4 Mark I and number 6 was a BWR/5 Mark II[1]. The units involved in the accident were BWR/3 and /4, all Mark I (see Table 1).

BWR/3 and BWR/4 have some similar features as the forced circulation systems with two recirculation loops and twenty single nosed-variable speed jet pumps. In both the "emergency core cooling system" (ECCS) high pressure pumping delivery point is in the vessel annulus via feedwater sparger, and for the same system the "low pressure core spray system" (CS) comprises two core spray (independent) loops. Also in both models the "hydrogen control short term" is provided by nitrogen inerting during normal operation.

Some pertinent differences between BWR/3 and BWR/4 are in the "reactor isolation pressure control". Both have an "isolation condenser" (IC) and "safety relief valves" (SRV), but BWR/4 has steam condensing mode of the "residual heat removal system" (RHR). A remarkable difference is in the "reactor isolation inventory control": BWR/3 has an isolation condenser while BWR/4 has a "reactor core isolation cooling system" (RCIC). While in the BWR/3 the shutdown cooling is performed by a "shutdown cooling system" or the residual heat removal system, in the BWR/4 only the last system is available. Something similar happens with the "containment spray and cooling", BWR/3 has the option between "low pressure cooling injection" (LPCI) or RHR while BWR/4 only the last one. The high pressure pumping for the "emergency core cooling system" (ECCS) of the BWR/3 disposes of "feedwater pumps" or "high pressure cooling injection system" (HPCI). Only the last mentioned system is available in the BWR/4. A detailed description of the differences among reactors is shown in Table 6.0.1 of reference [6].

The accidents that occurred in all the reactors that were in operation followed an analogous path, as will be described in the following sections.

3. Main events after the earthquake

In the unit 1, the isolation condensers (IC) automatically actuated on the high reactor pressure after the scram, but were shutdown manually to prevent the cooling rate of the reactor pressure vessel from exceeding a design limit (55 K/hour) [3,7]. The reason is because in the boiling water reactor the water acts as a neutron moderator for the nuclear reactions. Thus, a correct flow rate must be maintained to control the power of the reactor, because the flux of water allows fine reactivity adjustments. Depending on the quantity of vapor bubbles, the power of the reactor can vary. For example, if there are fewer bubbles, i.e., if the flux of water is greater, the neutrons are moderated to energies at which can be absorbed to produce fission. Thus, the power of the reactor increases when the flux of water is greater.

Unfortunately, when the tsunami struck, all the electric power was lost and the valves of the isolation condensers remained closed. Thus the IC of the unit 1 became useless. At units 2 and 3, the reactor core isolation cooling system (RCIC) started automatically after the earthquake and continued in operation for many hours after the tsunami. The unit 1 has an older reactor without a RCIC system.

	Unit 1	**Unit 2**	**Unit 3**	**Unit 4**	**Unit 5**	**Unit 6**
Reactor type	BWR-3	BWR-4	BWR-4	BWR-4	BWR-4	BWR-5
PCV type	Mark I	Mark I	Mark I	Mark I	Mark I	Mark II
Max. pressure of RPV (MPa)	8.24	8.24	8.24	8.24	8.62	8.62
Max. press. of PCV (MPa)	0.43	0.38	0.38	0.38	0.38	0.28

Table 1.

Next it is given a short review of the events that lead to hydrogen explosions in three of the reactors within four days after the tsunami. The main events are condensed in the Appendix A.

3.1. Unit 1

All the valves of the isolated condenser were motor operated and were left in the closed position, as explained above, and this led to an increasing pressure inside the reactor 1 [7].

The pressure build up in the primary containment vessel was due to the liberation of steam - produced in the reactor pressure vessel- through the safety relief valves.

The measured pressure at the dry well was 0.84 MPa at 2:30 on March 12. This value doubles the pressure design for the dry well. It was argued that this pressure could not be due to steam alone, but built up from a mixture of steam and hydrogen [7].

The pressure in the reactor pressure vessel was as high as 6.9 MPa at 20:07 on March 11, but fell to 0.8 MPa at 2:45 on March 12. It is not clear if this depressurization occurred through the safety relief valves or if there was some cracking in the reactor pressure vessel within that interval [7]. If there were a meltdown of the fuel, and some portion fell to the bottom of the vessel, it is possible that it has damaged the reactor pressure vessel at the places where the control rods are inserted. Thus, a leakage of the molten fuel to the primary containment vessel is possible [7]. This could be the cause of the depressurization at 2:45 on March 12. However, this is still a conjecture.

The injection of water to the reactor resumes at 5:46 on 12 March -using fire engines- 14 hours after the tsunami. But in this case, it is not clear if the pumps delivered the water at a pressure higher than the pressure in the reactor pressure vessel at that time (0.75 MPa). If the pressure of the pumps were lower, the water would have been supplied to the core about 22 hours after the tsunami, when the reactor pressure vessel was vented and depressurized [7].

As the fuel rods overheated, it was produced a water-Zirconium reaction that liberates hydrogen. The Zirconium of the zircaloy cladding of the fuel rods oxidizes with the steam according to the reaction:

$$Zr + 2H_2O \rightarrow ZrO_2 + 2H_2 \quad (1)$$

The hydrogen is highly explosive and a deflagration or a detonation is easily triggered when mixed with air at a high pressure.

As the hydrogen had been vented to the secondary containment, an explosion occurs at 15:36 on 12 March 2011. The explosion blew the secondary containment without visibly damaging to the primary containment vessel. The sea water injection began at 19:04 on 12 March 2011.

3.2. Unit 2

The reactor core isolation cooling system (RCIC) operated continuously during three days and the water level on the reactor was maintained at its normal value. The RCIC stopped at 13:25 on March 14, 2011, by unknown causes. The pressure at the reactor pressure vessel began to rise, and the water level decreased.

The reactor was depressurized at 18:00 hs, through the safety relief valves. The seawater injection started at 19:54, and the water level recovery was confirmed at 22:00 on 14 March. An explosion occurs at 6:00 on 15 March. The suppression pool was assumed damaged, because of the large amount of contaminated water at the turbines building [7].

3.3. Unit 3

The reactor core isolation cooling operated normally for 20 hours after the tsunami, but failed at 11:36 on 12 March 2011. The operators started soon the high pressure core injection (HPCI), but it failed at 2:42 on 13 March 2011. It is yet not clear why the HPCI stopped. But a probable cause was the depletion of water in the condensate storage tank [7]. The reactor was left without water injection for more than six hours. The fresh water injection starts at 9:25 and changed to seawater injection at 13:12 on March 13, 2011. The pressure start to rise at the dry well and a series of venting operations followed to relief the pressure at the reactor pressure vessel. The depressurization process led -like in the other reactors- to the accumulation of hydrogen at the secondary containment. So, an explosion blew up the secondary containment of unit 3 at 11:00 on March 14, 2011.

There is an interesting observation made by Gundersen [9] related with the hydrogen explosions in units 1 and 3. The explosion in unit 1 occurred mainly sidewise, while explosion in unit 3 occurred visibly in the vertical direction and apparently the explosion was much more violent. Gundersen claims that explosion in unit 3 was a hydrogen detonation, while explosion in unit 1 was a hydrogen deflagration. A detonation is a supersonic combustion wave (respect to fuel), while a deflagration is a subsonic combustion. Gundersen noticed that this could be an indication of prompt criticality in the spent fuel pool of reactor 3 [9]. Although this is a possible explanation for the difference in the explosions, it must be noticed that there are other factors that could alter the development of the explosion, i.e., the preheating of the gas before the ignition of the combustion can lead to a detonation. The development of a detonation depends on the temperature gradients of the gas. Thus, the combustion mode depends on the complex history in which the hydrogen was released to the second containment vessel. If the hydrogen was promptly released the gradient could be too steeply to develop a detonation. On the other hand, turbulence in the fuel mixture can also lead to a deflagration to detonation transition, since the turbulence can induce a shallow temperature profile, mixing hot fuel regions with cold ones.

4. Basic concepts

Radioactivity is the spontaneous disintegration of an atomic nucleus. In the radioactive decay the nucleus decay in lighter nuclei, which are the fragments of the nuclear disintegration, in addition there could be an emission of neutrons, gamma rays, electrons and neutrinos.

Several radioactive isotopes are man-made, but there are also radioactive traces found in some minerals, in soils; in the plants; in the animal's tissue; in the air; and in the water.

Uranium -as other heavy elements- is radioactive and its main isotope decays according with the reaction:

$$^{238}_{92}\mathrm{U} \rightarrow {}^{234}_{90}\mathrm{Th} + {}^{4}_{2}\mathrm{He}, \tag{2}$$

That is the uranium-238 decay into thorium-234, emitting an alpha particle α ($^{4}{}_{2}$He). The daughter isotope thorium is also radioactive and decays as:

$$^{234}_{90}\mathrm{Th} \rightarrow {}^{234}_{91}\mathrm{Pa} + e + \nu, \tag{3}$$

Where the first element on the right side is protactinium, the second particle is an electron (also called β^- beta particle), and the last particle is a neutrino.

These two reactions are the start for a long chain of reactions that ends at the stable isotope of lead $^{206}_{82}Pb$. There are similar chains for uranium-235 and thorium-232.

Element	Mean life
caesium-137	30.3 years
radon-222	3.82 days
iodine-129	1.7×10^7 years
iodine-131	8.04 days
strontium-90	29.1 years
uranium-235	7.04×10^8 years
uranium-238	4.46×10^9 years
plutonio-239	2.41×10^4 years

Table 2.

The mean life of a radioactive nucleus t_h (half-life) is the time in which half of the radioactive sample decays. The radioactive decay is essentially probabilistic, and its law is a negative exponential of time:

$$N(t) = N_0 \, e^{-t/tm}, \tag{4}$$

where t_m is the mean-lifetime. The meaning of this law is that the number of parent nucleus in a sample at a given time, N(t), diminishes exponentially with time from the initial number N_0. Some mean lifetimes for known radioactive nucleus produced in a nuclear reactor are given in Table 2. The activity (A) of a radionuclide is the number of decays per unit time:

$$A = N / tm. \tag{5}$$

Thus, the larger is the radionuclide sample, the higher will be its activity. For the same amount of nuclide the activity is higher for the radionuclide with lower mean lifetime. The activity is measured in becquerel (Bq), honoring the scientist who discovered radioactivity. A Bq is equivalent to one disintegration per second. The other commonly used unit is the curie (Ci), equal to 3.7×10^{10} disintegrations/second, which is equivalent to the activity of one gram of radio.

The fission of the uranium-235 occurs by means of the absorption of thermal neutrons produced in other radioactive decays. The most probable reaction gives krypton and barium:

$$^{235}_{92}\mathrm{U}+^{1}_{0}\mathrm{n}\rightarrow\,^{90}_{36}\mathrm{Kr}+^{144}_{56}\mathrm{Ba}+2\,^{1}_{0}\mathrm{n}+\mathrm{energy}, \tag{6}$$

The total energy release on the induced fission of $^{235}{}_{92}U$ is on average 205 MeV [10]. From this total the neutrinos escape unimpeded with 12 MeV, while the remaining energy is distributed in the kinetic energy of the fission fragments, the neutrons, the photons, the betas, plus the delayed gammas and betas. Thus, the fission of uranium produce a useful energy of 190 MeV, which is equivalent to 3.04 × 10^{-11} J. The number of fissions to produce a watt-second is:

$$1/(3.04\ 10^{-11})=3.3\ \times 10^{10}\,decays,$$

which is roughly the number of decays in one gram of radio during one second. Thus, in order to produce a megawatt of thermal energy per day is required, roughly:

$$10^{6}\,W\,(3.3\,10^{10}\,decays\,/\,W\cdot s)\,86400\,s\,/\,day=3.3\,10^{21}\,decays\,/\,day.$$

But, a mol of uranium-235 has the Avogadro number of atoms (6 10^{23}). So, the consumption of uranium per megawatt is:

$$3.3\,10^{21}\,(decays\,/\,day)\,235\,grams\,/\,(6\ 10^{23})=1.3\,grams\,/\,day,$$

without considering the efficiency of the chain reactions. This is equivalent to 1.3 kg/day of uranium-235 for a 1000 MW power plant, or 474 kg of uranium-235 per year (or equivalent 27 kW/kg, taking into account an enrichment of 1%). If reactor 1 of Fukushima has 78 ton of fuel in its core, and assuming the fuel is new and enriched at 1.5 %, there is enough fuel for 2.5 years. The quantity of spent nuclear fuel in the pool of each reactor is [11]:

- Reactor No. 1: 50 ton of nuclear fuel
- Reactor No. 2: 81 ton
- Reactor No. 3: 88 ton
- Reactor No. 4: 135 ton
- Reactor No. 5: 142 ton
- Reactor No. 6: 151 ton

Each reactor at Fukushima has less than 100 ton of nuclear fuel.

4.1. Why the fossil fuel is less viable and produces a large damage to the environment?

It can be shown that is required millions of times more mass of carbon than uranium to produce the same amount of energy. The carbon gives 32 kilojoules per gram of energy. In order to produce 1000 MW of energy it is required [12]:

$$(1000\,10^{6}\ joules\,/\,s\ \times 86400\,s\,/\,day)\,/\,(32000\ joules\,/\,gr)=2.7\,millions\ of\ ton\ \,/\,day.$$

http://dx.doi.org/10.5772/54262

Thus, the fossil fuel is highly contaminant to our atmosphere. However, on the view of the Chernobyl and Fukushima nuclear accidents, we must know reconsider which energy source is more dangerous for human life.

4.2. Induced fission

The objective of a nuclear reactor is to produce a self-maintained and controlled chain reaction of a fissionable element.

There are two classes of fissionable materials, depending on how they behave when absorbing a neutron [13]. A fissile material is one that will undergo a nuclear fission when bombarded by neutrons of any energy. A fertile material is one which after capturing a neutron, will transmute by radioactive decay into a fissile material. The uranium-235 is the only fissile material found in nature. The fertile material ^{238}U becames radioactive and decays in series of beta decays into ^{239}Pu after capturing a neutron:

$$^{238}_{92}U{+}n \rightarrow {}^{239}_{92}U \xrightarrow{\beta} {}^{239}_{93}Np \xrightarrow{\beta} {}^{239}_{94}Pu. \tag{7}$$

The plutonium-239 is a fissionable material, i.e., it will decay if a neutron strikes it. The ^{239}Pu has a very large half-life, so it can be stored and used as a reactor fuel (see the table 1).

The thorium-232 is another fertile material found in nature. It decays in a series of beta decays after absorbing a neutron according to the reaction:

$$^{232}_{90}Th{+}n \rightarrow {}^{233}_{90}Th \xrightarrow{\beta} {}^{233}_{91}Pa \xrightarrow{\beta} {}^{233}_{92}U, \tag{8}$$

where the uranium-233 is a fissile material.

In general, when a neutron impinges on the nucleus of uranium-235, it breaks into fragments and there are a mean of 2.4 neutrons emitted, gamma rays, and neutrinos:

$$^{235}U{+}\text{neutron} \rightarrow \textit{f}\text{ission fragments}{+}2.4\text{ neutrons}{+}\text{gamma rays}{+}\text{neutrinos}{+}193\text{ MeV}. \tag{9}$$

The uranium-235 fission leads to a neutron multiplication; each neutron emitted in the fission has the potential to lead to a new fission in the fuel sample. The reactions will continue until all the uranium-235 decay, or until no free neutrons are left in the sample, i.e., until all the neutrons escape from the boundaries of the fuel or are absorbed by another non-fissionable material.

A typical fission reaction for uranium-235 is:

$$^{235}_{92}U{+}n \rightarrow {}^{140}_{54}Xe{+}{}^{94}_{38}Sr{+}2\,n{+}200\text{ MeV}. \tag{10}$$

The fission fragments in this reaction are also unstable because their neutron to proton ratio is too large. The fission fragments decay mostly by beta particle emission accompanied by gamma rays. The decay occurs in several stages until reaching the stable nuclei branch, for example:

$$^{140}_{54}Xe \xrightarrow{\beta} {}^{140}_{55}Cs \xrightarrow{\beta} {}^{140}_{56}Ba \xrightarrow{\beta} {}^{140}_{57}La \xrightarrow{\beta} {}^{140}_{58}Ce, \qquad (11)$$

the reaction for strontium is:

$$^{94}_{38}Sr \xrightarrow{\beta} {}^{94}_{39}Y \xrightarrow{\beta} {}^{94}_{40}Zr. \qquad (12)$$

The energy release from the nuclear fission is distributed into the kinetic energy of the fission fragments, beta particles and neutrons, and in the energy of the gamma rays and neutrinos.

	MeV
Fission fragment kinetic energy	166
Neutrons	5
Prompt gamma rays	7
Fission products gamma rays	7
Beta particles	5
Neutrinos	10
Total	200

Table 3.

The energy balance for the decay of a uranium nucleus is given in the Table 3 [10, 12].

A chain reaction is self-maintained when the fuel do not need to be bombarded by external particles to maintain the fission process [12].

The chain reaction will be self-sustained and maintained in time if the mass of the nuclear fuel is larger than a critical mass value. The critical mass is the quantity of matter necessary to compensate the neutrons loss across the boundaries of the fuel, in such a geometric configuration that allows the chain reactions.

Fissile materials can be produced in the core of a reactor if fertile material is mixed with the nuclear fuel. In the decay of ^{235}U there are enough neutrons to sustain a chain reaction and to convert fertile to fissile material. If a nuclear reactor produces more fissile material than it consumes, it is said that the reactor is a breeder. As most reactors use natural uranium, or uranium with low enrichment, there are enough ^{238}U -a fertile material- in the fuel to be converted into plutonium [10,13].

However, the design of a breeder is not easy due to the absorption of neutrons by other materials. Moreover the neutrons can leak through the boundaries of the fuel. Thus, most reactors are not breeders [13].

4.3. Neutron cross-section for fission

The cross section of ^{235}U and ^{238}U, as a function of the energy of the incident neutrons is of great relevance to sustain a chain reaction.

Since neutrons are uncharged they do not find a coulomb barrier to reach the nucleus; thus the neutrons can interact with the nuclei even if they have very low energy [10,12]. At low neutron energies the exothermic nuclear reaction rate is roughly independent of the energy and proportional to the density of the incident flux of neutrons [10]. The reaction rate for nuclear fission induced by neutrons can be written as [10,12]:

$$R = \rho_n \, v \, \sigma \tag{13}$$

where R is the reaction rate; ρ_n is the neutron density; v is the mean speed of the neutrons and σ is the cross section of the target nucleus. Thus, the cross section at low energies is proportional to the inverse of the speed of the neutrons, or equivalently, inversely proportional to the square root of their energy:

$$\sigma \propto const / v. \tag{14}$$

This law was experimentally observed [10,12,13]. The dependence of the total cross section with energy changes very fast for neutron energies close to the resonances of the compound nucleus. In this case, the cross section is characterized by a series of narrow resonance peaks, and is expressed by the Breit-Wigner formula [10]. These peaks are due to the exited states of the compound nucleus (see fig 9.1 in ref [10]).

The principal elements in the fuel of a reactor are uranium-235 and uranium-238, which are found in the nature in the proportions 0.72 % and 99.27 % respectively [12].

The neutron absorption cross section can be divided into two parts, the fission cross section and the radiative cross section. In the radiative capture, a compound nucleus is formed with gamma ray emission (for example ^{236}U). In addition some of the neutrons can suffer inelastic or elastic scattering from the nucleus without being absorbed.

The total cross section for ^{235}U at energies below 0.1 eV is mostly due to the fission cross section (84 %), while the remaining are due to radiative capture (16 %). In this energy range, the total cross section follows the *1/v* law. In the range between 1 eV and 1 keV is the resonance region. In this region the cross section varies very rapidly with the energy. In the third region for energies larger than 1 keV, the cross section for ^{235}U is predominantly due to scattering (inelastic at energies larger than 14 keV). The fission cross section is lower than the total cross section, and much lower than in the first energy range [10].

The cross section for ^{238}U is almost constant in the low energy region, and is predominantly due to elastic scattering and radiative absorption. The fission cross section starts to be significant above 1.4 MeV, although it remains a small fraction of the total cross section.

4.4. Chain reaction

As described above the neutron cross section for ^{235}U fission, is much higher for low energy neutrons. The fission process produces fast neutrons with energies in the interval 0.1 MeV<E<10MeV, and are predominantly emitted with 0.75 MeV. Thus, the neutrons must be slowed down in the reactor to energies at which they can be captured in the ^{235}U with greater probability. The neutrons transfer their energy to the nuclei by multiple scattering, until reaching an approximately thermal distribution, in which they can be captured by the fissionable fuel [13]. The so called thermal neutrons have a range of energies in the interval 0.001 eV<E<1 eV.

The mean free path of a neutron of 2 MeV in the fuel is around 3 cm [10]. However, the probability of inducing a fission in one collision is rather low at that energy (18 %). The probability will increase significantly after six collisions (number of collisions × probability of fission ~1). If the neutron follows a random walk through the fuel, it will move about $\sqrt{6} \times 3\,cm \approx 7\,cm$ from the starting point before inducing fission [10]. If the probability that a neutron induce a new fission is q, then on average each neutron will create (νq-1) additional neutrons, where ν is the mean number of neutrons emitted per fission.

If there are n(t) neutrons at time t, then at time t+δt there will be an additional number of neutrons proportional to the number of neutrons n(t), times the number of neutrons produced by each of these, times the fraction of time they had to react $\delta t/t_p$, where t_p is the time required to produce a fission at the indicated energy, so [10]:

$$n(t+\delta t) = n(t) + (\nu q - 1)\, n(t)\, \delta t / t_p. \tag{15}$$

This equation gives the differential equation for neutron multiplication:

$$\frac{dn(t)}{dt} = (\nu q - 1)\, n(t) / t_p, \tag{16}$$

that has the solution [10]:

$$n(t) = n_0\, e^{(\nu q - 1) t / t_p}. \tag{17}$$

http://dx.doi.org/10.5772/54262

Thus the number of neutrons can grow exponentially if $\nu q > 1$ (supercritical assembly), or decrease exponentially if $\nu q < 1$ (subcritical assembly). The probability q depends on the geometry and size of the sample, and on the energy of the neutrons. The characteristic time for fission is $t_p = 10^{-8}$ s, this means that for a supercritical assembly there will be a huge amount of energy liberated in less than a microsecond. The critical radius for a sphere of ^{235}U is about 8.7 cm, and the critical mass is 52 kgs [10].

The reactors have control rods containing boron, or cadmium, which can absorb neutrons in the thermal range. So, inserting or withdrawing the control rods in the reactor core can modify the reactor power. The reactor must be operated at criticality, but is important that criticality is achieved taking into account the delayed neutrons, and not the prompt neutrons alone. This is because, the prompt neutrons have lifetimes of the order of 10^{-3} s, and this is a very short time to mechanically control the power output of the reactors. The delayed neutrons led to a better timescale to control the reactor. Moreover, the reactors must be designed for thermal stability, i.e., the probability for a new fission is such that [10]:

$$\frac{dq}{dT} < 0, \tag{18}$$

thus an increase in the temperature of the reactor diminishes the reactivity.

4.5. Nuclear fuel of the reactors

Natural uranium has more than 99 % of ^{238}U, and less than 1 % of ^{235}U. When a neutron is emitted in the fission process at 2 MeV, it will more probably interact with the ^{238}U, through inelastic scattering, leaving it in an exited state. After a few scatterings the neutron will lose its energy capable to induce fission in ^{238}U. So, the neutron must find a ^{235}U to induce fission, but as it is so much less abundant it is more probable the neutron is captured in a ^{238}U resonance to form ^{239}U with the emission of a gamma ray. Thus, the proportion of fission neutrons that induce further fission in natural uranium is rather low, and a chain reaction can not be sustained [10].

However, a technology has been developed to produce a chain reaction from natural uranium. The Fukushima Daiichi reactors are thermal reactors. In a thermal reactor the fuel is the ceramic uranium dioxide, and is contained in an array of thin rods containing the fuel pellets. In the reactor the fuel is surrounded by a large volume filled with a material of low mass number, called the moderator. The neutrons can lose their energy in the moderator before encountering a uranium nucleus. Thus the neutron cross section for ^{235}U fission increase and is much larger than for ^{238}U, compensating the low concentration of uranium-235. The neutrons are called thermal because their energy corresponds to the operation temperature of the reactor (0.1 eV=1160 ºK).

The capture of neutrons at thermal energies, lead to the fission of the ^{235}U with large probability, and thus, the chain reaction is made possible with natural uranium using this technology.

The moderator can be ^{12}C, or heavy water (D_2O). But, if the reactors use enriched uranium, the reactor can be moderated using ordinary water.

The reactors at Fukushima have low enriched uranium fuel (LEU), except the unit 3 that has a low percentage (~6 %) of *MOX* (mixed oxide) consisting in a blend of uranium oxide (UO_2) and plutonium oxide (PuO_2) [15]. The mixture has 7 % plutonium and the rest is natural uranium. The LEU is enriched to ~3 % of uranium-235.

4.6. The reactor power

The thermal power generated by a nuclear reactor is approximated by the formula [12]:

$$P = \varphi N\, V \sigma_f w, \tag{19}$$

where $\varphi = n\,v$, is the neutron flux; N is the number of fissile nuclei ^{235}U per unit volume; V is the total fuel volume; σ_f is the neutron cross section for fission; and w is the mean energy liberated per fission.

The efficiency of a nuclear reactor is 1/3 the total thermal power. There is a distinction in the notation between units of megawatt of thermal power -which is the total generated power- indicated as MWt, and the total generated electric power -the energy converted to work- indicated as MWe.

The reactors of Fukushima generated the following energy [18]: i) the unit 1 has a General Electric reactor of 460 MWe; ii) unit 2, has a General Electric reactor of 784 MWe ; iii) unit 3 has a Toshiba reactor of 784 MWe; iv) unit 4, has a Hitachi reactor of 784 MWe; v) unit 5 has a Toshiba reactor of 784 MWe; vi) unit 6 has a General Electric reactor of 1100 MWe.

4.7. Power of the reactor after shutdown

After the scram of the reactors the control rods were inserted into the core, and the uranium-235 fission stopped. However, the fuel rods containing the fission products which are neutron rich, decay emitting beta rays and gamma rays as well. So, the fuel rods continued producing heat after the scram.

The empirical expression for the power of heat decay P is given by the formula [10, 12]:

$$P = 0.07\ P_0[(\tau - \tau_s)^{-0.2} - \tau^{-0.2}], \tag{20}$$

where P_0 is the nominal reactor power, τ is the time since the reactor startup in seconds, and τ_s is the time of reactor shutdown since the startup. This expression has an acceptable error in a certain time interval ranging from 10 seconds and 100 days.

The nominal power for reactor 2 and 3 is 2352 MWt. If the rods have a life of one year, the decay power after one day is: 11.7 MW; after one week is: 6.3 MW; after one month

http://dx.doi.org/10.5772/54262

is: 3.5 MW; and after one year is: 0.7 MW. If the lifetime of the rods is larger the decay power will be higher.

5. The importance of blogging in crisis times

The Ministry of Education, Culture, Sports, Science and Technology (MEXT) of Japan published real time radiation measurement data acquired by the "System for Prediction of Environment Emergency Dose Information" (SPEEDI). However, this data was not very accessible for processing, as was published in html format and in Japanese. Mr. Marian Steinbach, started a "Google Docs Spreadsheet" and called for help to the online world to collect the data [21]. The results are compressed in one file (station_data_1h.csv.gz) that contains the following information:

a. Station identifier: a number identifies each station that performs a measurement.

b. Time UTC (in order to convert to JST just add nine hours).

c. Radiation dose measured by the stations. The measurements are in nGy/h, or equivalently in nSv/h (with a quality factor of one).

d. Rain precipitation in mm.

For example, the file looks like these three sample lines:

1150000004 - 28/02/2011 15:00 - 370

1020000026 - 28/02/2011 15:00 – 21 - 0

1020000004 - 28/02/2011 15:00 – 20 - 0

The measurements are made every ten minutes in more than two hundred stations along Japan, and the file starts in March 1, 2011, until the present. So, the archive has more than 600 Mbytes today, and it requires near 2 Gbytes of free RAM memory to read it with conventional text processors. It's good to emphasize here the great effort made by several bloggers to collect all the measured data (see for example Refs. [20, 21, 22]). The data must be combined with another archive providing information on each station: site name, site identifier, location, city, prefecture, and geographic coordinates (see [24]). The authors of the present chapter programmed a Fortran code to retrieve information from this data base.

Some data on the spreadsheet indicates measured values equal to -999, it is presumed that it is due to a saturation of the detectors that can measure dose rates up to 1000 mSv/h. In addition, several stations close to the nuclear power plant stopped measuring dose rates on March 11, and restarted on September 21, 2011 [21].

The action of bloggers helped to understand the critical situation at Fukushima and bring calm to the general population. It was an alternative and reliable source to retrieve information about the crisis and (in several cases) the information provided was not so confusing like in the main news channels.

6. Study of the radiation distribution

In a simplified model of the radiation emitted from the nuclear power plant, it could be assumed that the particles will travel radially out from a point source, emitting S particles per second isotropically. The particles can survive a certain distance without scattering with other particles, depending on the macroscopic cross section for scattering and the density of target nuclei. If there is a certain amount of radioactive particles being emitted from the site accident, the flux of particles will be attenuated by a geometric factor with the distance and by the scattering with the atoms of the atmosphere.

The flux of particles passing through a sphere of radius r is:

$$\phi(r) = \frac{e^{-\Xi r}}{4\pi r^2} S, \tag{21}$$

where Ξ is the macroscopic cross section (the macroscopic cross section depends on the microscopic cross section σ and the number of air nuclei per unit volume: $\Xi = N\,\sigma$). Thus, it is expected that if the Fukushima nuclear reactors are emitting radioactive particles to the air isotropically, the measured radiation must fall faster than r^{-2} with distance. Of course, the hypothesis of isotropic emission of particles is not true in the case of Fukushima, since there was an important wind carrying the radioactive emission to the sea in the days following the accident. This was a good factor within the disgrace of the accident, since the wind dragged the radiation out from populated zones.

It is difficult to make correct predictions on the nuclear fallout without a detailed numeric simulation taking into account the full meteorological data: including wind direction, wind speed, and rain fall at each date.

The database of dose measurements was surveyed to provide a description of the dependence of the radiation versus time, and radiation versus distance from Fukushima Daiichi nuclear power plant.

Since there are two widely used systems of units to measure the radiation dose, we clarify the definitions and the conversion between the two systems of units.

A dose is the quantity of energy absorbed by a kilogram of exposed tissue to the radiation. In the International System, a dose unit is the gray (*Gy*) equivalent to 1 joule/kg. The other commonly used unit is the *rad*, equivalent to 100 erg/gram, or equivalent to 0.01 Gy.

As the damage produced on the tissue depends on the radiation type, i.e., alpha particles; neutrons; beta particles; or gamma rays, it is used the concept of quality factor *QF*, and it is defined a *dose equivalent* as [12, 25]:

$$H = (QF) \times D. \tag{22}$$

http://dx.doi.org/10.5772/54262

mSv/h	μSv/h	nSv/h	mrem/h
0.0001	0.1	100	10^{-6}
0.001	1	1000	0.00001
0.01	10	10000	0.0001
0.1	100	100000	0.001
1	1000	10^6	0.01
10	10000	10^7	0.1

Table 4.

A larger equivalent dose produces a larger damage to the human health.

If D is given in *Gy*, the units of H are *sieverts*, while if D is measured in *rad*, H are in *rem*.

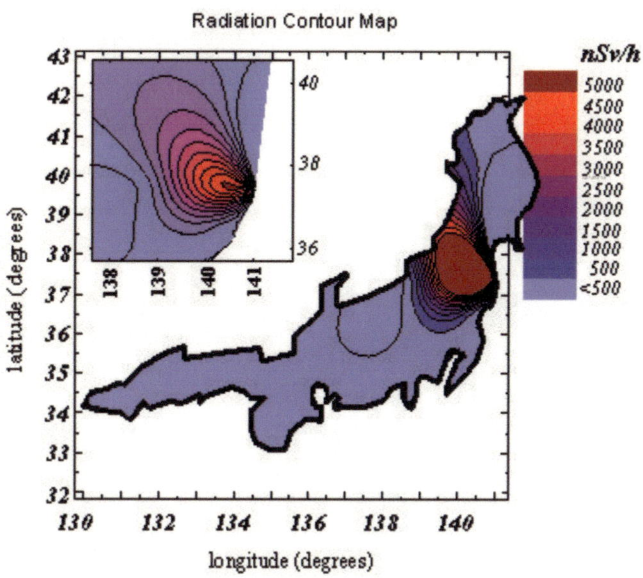

Figure 1. Radiation contour map as November 15, 2011

For fast interpretation and comparison of the results with the data found in the literature, the Table 4 gives the conversion between commonly used units and its multiples for the measured dose equivalent rate, spanning all the relevant range of values for the Fukushima accident. In the analysis of the data is implicitly assumed that the quality factor is one, as is commonly assumed in the literature about Fukushima, i.e., the measurements are expressed as mGy/h or mSv/h, indistinctly, although this is not strictly correct, since the damage depends on the type of radiation and the form of exposure.

The occupational annual dose to individual adults, except for planned special exposures, is a total effective dose equivalent equal to 50 mSv [25]. For the individuals the radiation dose equivalent limit is 5 mSv [25]. This means that in normal conditions a person must not receive a dose equivalent larger than 0.6 μSv/h. In the next set of figures we show the results of the database analysis. The Figure 1 shows the radiation contour map for the Honshu island of Japan, as November 15, 2011, at 15:20 UTC. The silhouette of the island is representative. The color scale increase in units of 500 nSv/h. The darkest red shows the region in which the radiation is above 5000 nSv/h. Note that the background radiation, according to the database, is roughly 30 nSv/h. The limit for non-nuclear workers is in the range 570 nSv/h / 2283 nSv/h (5 mSv/20 mSv annual).

The inset shows the same map but with contour lines in intervals of 1000 nSv/h. The darkest red –in the inset- shows the region in which the radiation is above 10000 nSv/h.

Caveat: the radiation contour map displayed could be scary, but as it will be shown the radiation falls with time due to the half life of the radioactive isotopes and by the action of the wind. There is an important contribution from atmospheric radiation to this map, and not necessarily all the radiation was deposited in the ground. Moreover, this is a contour map and do not mean that a high radiation level was actually measured in all the shadowed areas, but on some point located within it at the date indicated. The map is more representative of the spatial distribution of the radiation at the indicated date.

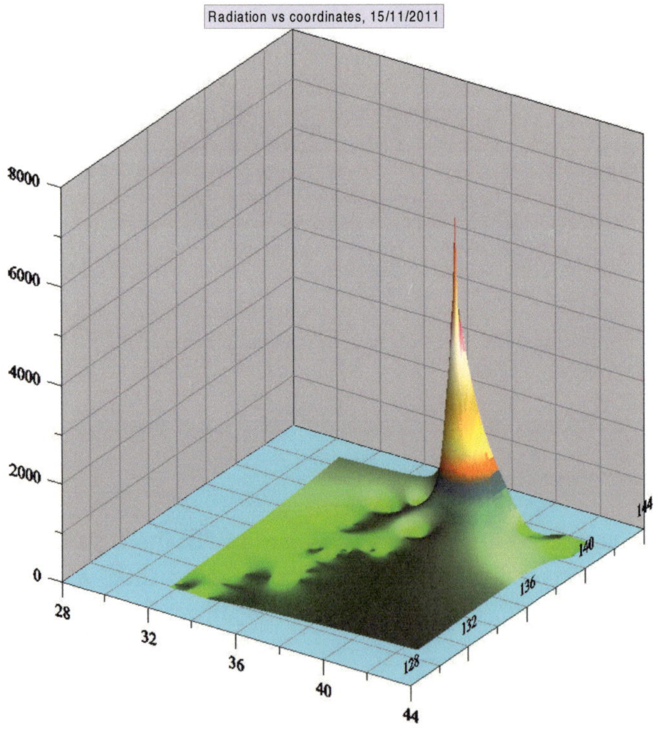

Figure 2. Spatial distribution of the radiation as function of the geographic coordinates.

The release of radioactive isotopes was due to the necessary venting of the reactor pressure vessel and of the dry well of the reactor. The venting was done to reduce the pressure inside

http://dx.doi.org/10.5772/54262

the reactor. The release of radioactive particles was worsened with the explosion of the secondary containment that occurred due to the hydrogen explosions of units 1 and 3, and a breach in the suppression chamber of unit 2 (see the Appendix).

The incident was rated at level 7 according to the International Nuclear Event Scale. However, the total amount of iodine-131 was 73 % of those measured at the 1986 Chernobyl nuclear disaster, and the amount of caesium-137 released from Fukushima is around 60 % of the amount released from Chernobyl [27].

The main isotopes released at Fukushima have been iodine; caesium; strontium; and plutonium. These radioactive elements have been released to the atmosphere and to the Pacific Ocean.

Prefecture	City/Station	Distance (km)	Rad (nGy/h)
Fukushima	Futaba Town – Yamada	0,7	27210
Fukushima	Ono Ookuma	2,5	6417
Fukushima	Futaba Town – Shinzan	3,1	5588
Fukushima	Futaba Town - Kooriyama	4,3	1698
Fukushima	Namie Town - Namie	7,1	1167
Fukushima	Yonomori Tomioka Town	7,3	4461
Fukushima	Namie Town – Kyohashi	8,1	528
Fukushima	Tomioka Tomioka Town	10,1	3612
Fukushima	Kamikooriyama Tomioka Town	10,8	1787
Fukushima	Tomiokacho	11,8	2157
Fukushima	Matsudate Naraha town	13,6	1434
Fukushima	Shigeoka Naraha Town	14,5	1177
Fukushima	Yamadaoka Naraha Town	19,1	352
Fukushima	Futatsunuma Hirono town	22	611

Table 5.

The Figure 2 shows the radiation landscape on November 15, 2011, at 15:20 UTC. The radiation dose rate in nGy/h is a function of the geographic coordinates, i.e., latitude and longitude, in degrees. At this date, the radiation peaked 27210 nGy/h at the nuclear station, which is 50 times more radiation than the maximum limit for non-nuclear workers. The second peak was at Ono Ookuma (distance 2.5 kms), measuring 6417 nGy/h. There was another peak of 5588 nGy/h at Futaba Town, Shinzan (3.1 kms). For clarity, the Table 5 shows the measurements above 100 nGy/h, for the same date as above, specifying the location of each station and its distance to the power plant.

Prefecture	City/Station	Distance (km)	Rad (nGy/h)
Ibaraki	Onuma Hitachi City	104	446
Ibaraki	Mayumi Hitatioota City	106,7	324
Ibaraki	Kuji Hitachi City	107,8	775
Ibaraki	Kume Hitatioota City	108,4	147
Ibaraki	Isobe Hitatioota City	108,5	487
Ibaraki	Ishigami Tokai	110,5	724
Ibaraki	Toyooka Tokai Village	110,5	387
Ibaraki	Nemoto Hitachioomiya City	111	266
Ibaraki	City Nukata Naka	111,7	319
Ibaraki	Funaishikawa Tokai	113	192
Ibaraki	Kadobe Naka City	113,1	735
Ibaraki	City Uridura Naka	113,4	188
Ibaraki	Muramatsu Tokai-mura	113,8	251
Ibaraki	City Yokobori Naka	114,3	309
Ibaraki	Oshinobe Tokai Village	114,5	306
Ibaraki	City Kounosu Naka	115,8	401
Ibaraki	Sawa Hitachinaka City	116,7	734
Ibaraki	Sugaya Naka City	117,1	234
Ibaraki	Mawatari Hitachinaka City	118,5	311
Ibaraki	Hitachinaka Hitachinaka City	119,5	351
Ibaraki	Godai Naka City	120	398
Ibaraki	Ajigaura Hitachinaka City	120,1	199
Ibaraki	Horiguchi, Hitachinaka City	122,7	1045
Ibaraki	Sawa Yanagi Hitachinaka City	124,2	238
Ibaraki	Ishikawa, Mito	125	216
Ibaraki	Isohama Oarai Town	128,3	162

Table 6.

http://dx.doi.org/10.5772/54262

According to these measurements, the region beyond 20 km has an "acceptable" level of radiation at the indicated date, although it is still twice greater than the background up to distances of 140 km (not shown in the table).

The following figures show the decrease of the radiation with the radial distance from the nuclear power plant. The vertical axis displays the radiation dose in nGy/h, in logarithmic scale, while the horizontal axis displays the radial distance to the nuclear power plant in km. The Figure 3 shows the radiation versus distance as March 15, 2011, at 15:20 UTC time. In this date, there are no stations collecting data closer than 100 km from the nuclear plant. There was a blackout of several stations that recovered after September 21, 2011, thus the data is incomplete. There is a large scatter of the data, which may be due to the topography of Japan, the action of the wind, and the rain fall in each location.

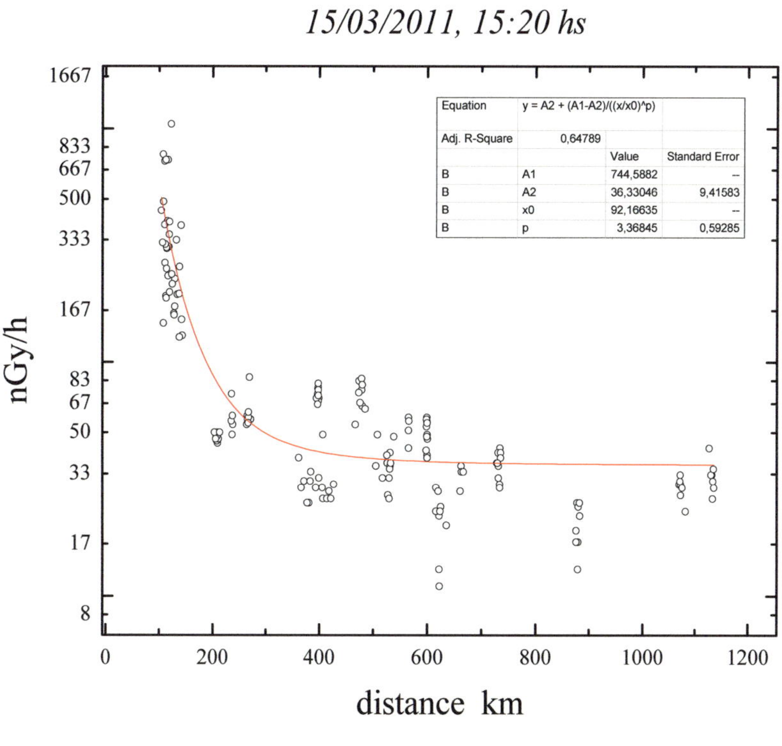

Figure 3. Radiation versus distance as March 15, 2011.

A non-linear fit of the data was performed with Origin 8.0. The result of the fit is shown as the red curve in the figures. It was found that the optimal fit is obtained with a function proportional to $1/r^p$, with $p=3.36$. Thus, the radiation decrease faster than $1/r^2$, as expected.

As shown in the figures and tables the radiation has a large heterogeneous distribution, sometimes with large variations in dose measurements (>200 nGy/h) in very short distances (sometimes of the order one hundred meters).

Prefecture	City/Station	Distance (km)	Rad (nGy/h)
Ibaraki	Onuma Hitachi City	104,1	166
Ibaraki	Kuji Hitachi City	107,8	180
Ibaraki	Toyooka Tokai Village	110,5	120
Ibaraki	Ishigami Tokai	110,5	122
Ibaraki	Muramatsu Tokai-mura	113,8	113
Ibaraki	Sawa Hitachinaka City	116,7	112
Ibaraki	Mawatari Hitachinaka City	118,5	129
Ibaraki	Hitachinaka Hitachinaka City	119,5	177
Ibaraki	Ajigaura Hitachinaka City	120,1	116
Ibaraki	Horiguchi, Hitachinaka City	122,7	136
Ibaraki	Isohama Oarai Town	128,3	117
Ibaraki	Town Ooarai Onuki	129,9	125
Ibaraki	Hiroura town Ibaraki	132,6	147
Ibaraki	Tasaki Hokota City	136,4	107
Ibaraki	Ebisawa town Ibaraki	137,7	110
Ibaraki	Araji Hokota City	138,3	128
Ibaraki	Tsukuriya Hokota City	138,4	158
Ibaraki	Momiyama Hokota City	141,5	189

Table 7.

As March 15, 2011, due to the stations blackout in Fukushima the largest radiation measurements made were in the Ibaraki prefecture as shown in the Table 6.

http://dx.doi.org/10.5772/54262

The Table 6 displays the radiation measurements larger than 100 nGy/h. It is possible to appreciate the large dispersion in the measured radiation with the distance. For example, there is a large difference in the dose measured at Funaishikawa Tokai with what is measured at its neighbor Kadobe Naka city.

The Figure 4 displays the radiation versus the radial distance at July 15, 2011, at 15:20 UTC time. It is seen that the non-linear fit decrease slower with the distance. The best fit gives *p=1.5*, so in this case, the rate of decrease is less than the r^{-2} law. This could be to the action of the wind, not taken into account in the Equation (21). Moreover, the radioactive particulate could act as nucleation centers to form rain drops, and the rain could significantly modify the radiation distribution.

In addition, is observed that the radiation distribution at this date is more scattered.

It is seen that the maximum radiation dose rate at Ibaraki decreased respect to the measurements made in March. The maximum dose rate was 189 nGy/h at Momiyama Hokota City (141.5 km from Fukushima Daiichi I, see Table 7).

The Figure 5 shows the radiation decrease with distance as November 15, 2011. The non-linear fit gives p=1.09, thus, the radiation profile is becoming flatter as time goes by. The stations were recovered near Fukushima, so the readings are higher.

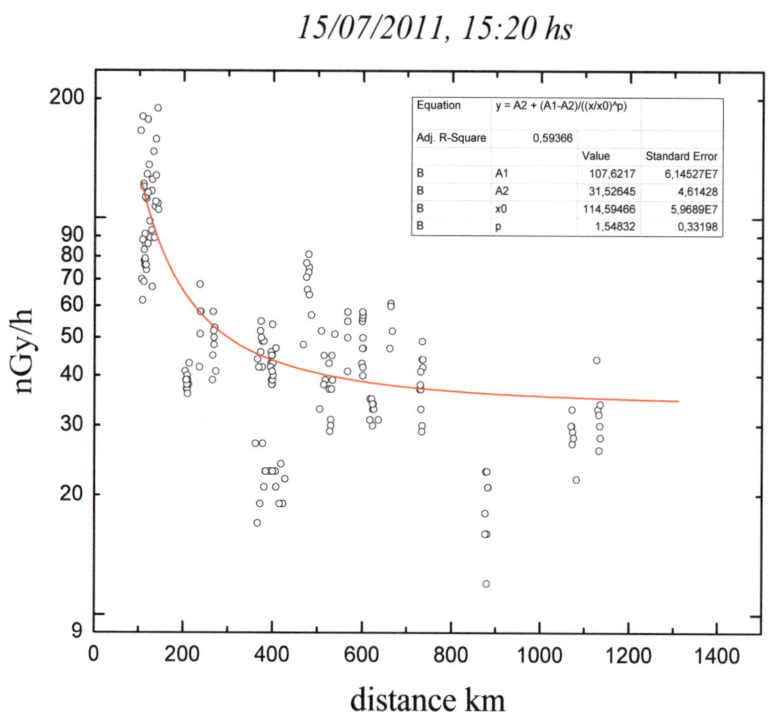

Figure 4. Radiation versus the radial distance at July 15, 2011

There are records of dose rates as high as 27210 nGy/h, near the central (see Table 5).

The following figures show the decrease of radiation with time at two locations. The Figure 6 shows the decrease of radiation with time at the nuclear power station. It is seen that the time evolution can be approximated with a straight line in the indicated period of time. The slope of the line is roughly 31 nGy/day. A decreasing exponential with a half life of 3.4 years can also be adjusted (not shown).

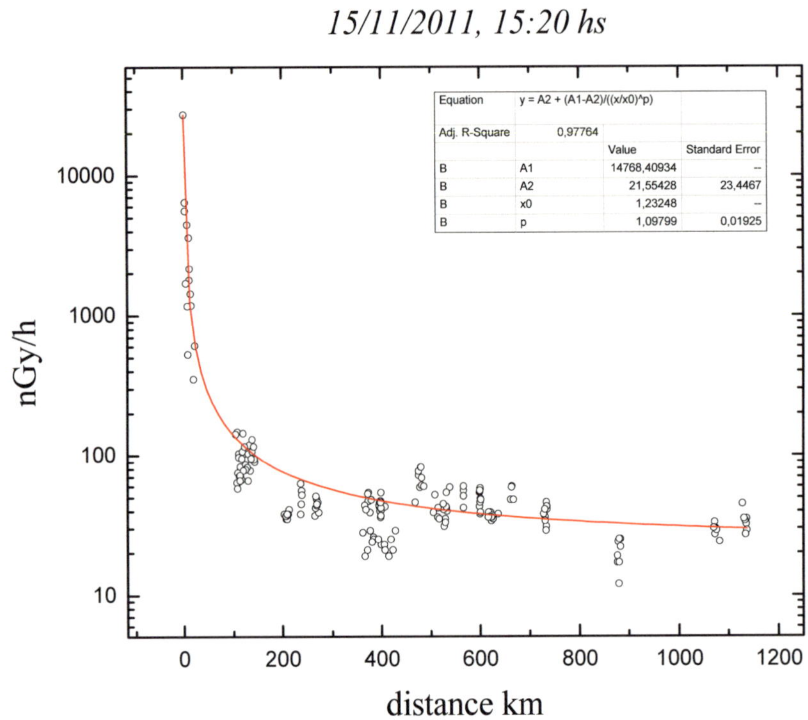

Figure 5. Radiation decrease with distance as November 15, 2011

The Figure 7, shows the radiation decrease with time for the City-Nukata-Naka station, at the Ibaraki prefecture. The decay of the radiation with time is more acute for this station (located at the south of the nuclear power plant). The tail of the distribution can be adjusted with a decreasing exponential with half life 15.6 days (not shown in the graph). The decay of the radiation with time at the nuclear power plant is compatible (although not exactly the same) with the mean life of caesium-137, and strontium-90. The difference between the adjustment and the mean life could disappear with measurements over a larger period of time. Meanwhile, the decrease of radiation at City-Nukata-Naka station has the order of magnitude of the mean life of the iodine-131, and radon-222.

The difference in the decay with time at the two stations can be due to the size of the particulate of each radioactive species, i.e., if the size of the cesium and of the strontium is larger, they could fall closer to the plant. The spectrum of elements in the fallout depends on the

http://dx.doi.org/10.5772/54262

volatility of the isotopes as well, i.e., not all the isotopes produced at the reactor will contaminate a large area. Since the boiling point of each isotope is different, it is expected that the percentage of mass released will be different for each isotope.

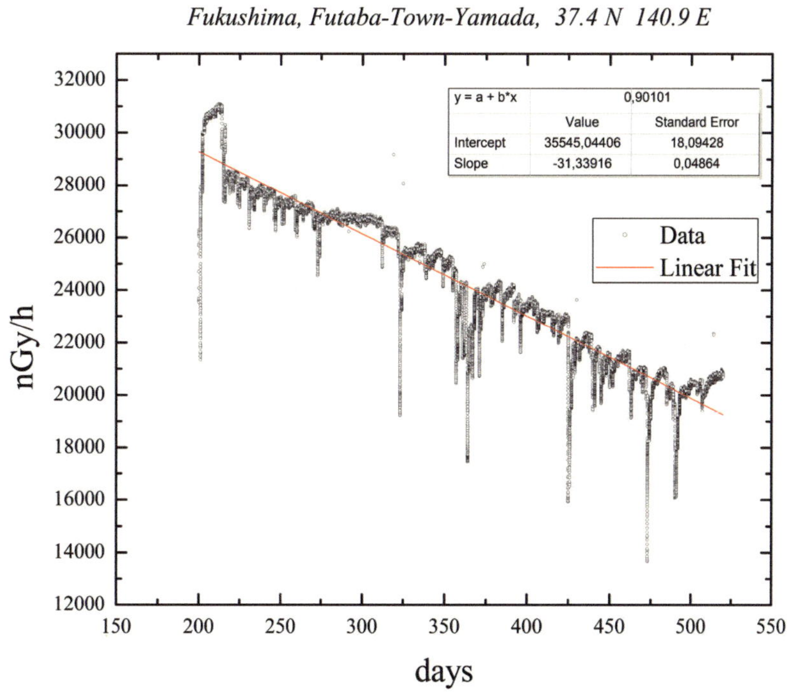

Figure 6. Decrease of the radiation with time at 0.7 km from the nuclear central.

Although there was a stations blackout in the SPEEDi network, TEPCO continued monitoring the radiation at the nuclear power station. These data are not contained within the SPEEDi database analyzed here. The Reference [23] contains data measurements at the nuclear central and presents an artistic representation of the data.

However, is important to note that in the literature is claimed that the TEPCO reports are confusing. Some organizations and news media showed doubts about the published data [30],[31]. As a consequence of this, a global project called Safecast -independently of governments or multinational companies- started to map radiation levels around Japan, using static and moving sensors [30].

On the wake of the disaster, Japan shifted the energetic policy and has plans to drop out nuclear energy by 2030. The decision of the Japan government is accompanied by similar actions of the governments of Germany and Switzerland. On the other hand, Italy has suspended the plans to reinforce the nuclear power in the country [31].

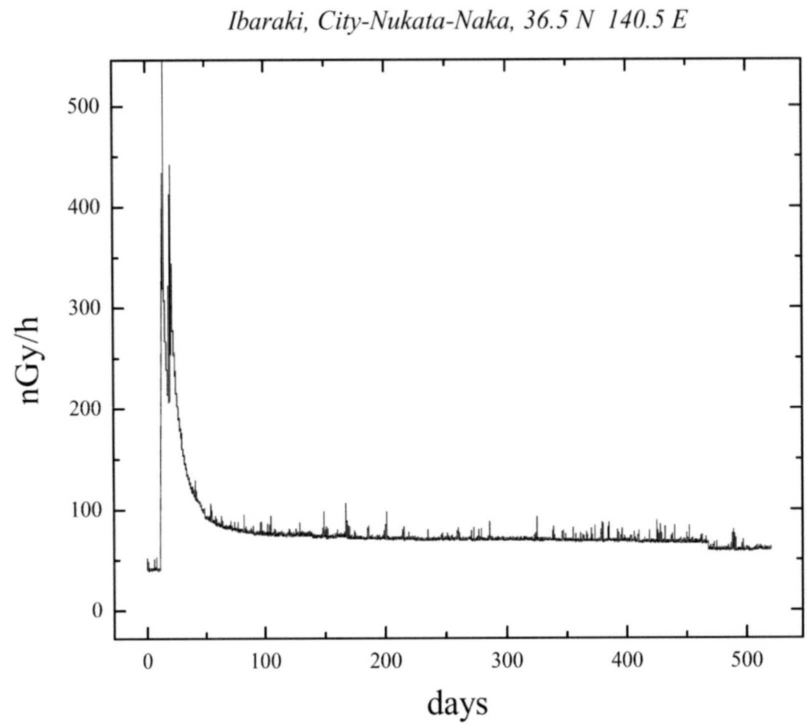

Figure 7. Decrease of the radiation with time at 111 km from the nuclear central.

7. Biological effects of radiation

Biological effects of ionizing radiation depend on dose, dose rate and type of radiation. Dose rates after an accident situation will show large variations in time and position according to meteorological conditions, topography, and whether the area in consideration is urban or rural. Their determination will also be subjected to large uncertainties. Transport characteristics and half life of involved radionuclide will also play a role in determining exposure levels after the accident, as will emergency response and active measures taken regarding the exposed population.

Dose rates after a major accident are dominated in the short time by atmospheric submersion, while most radioactive material is still in suspension. From this source, dose commitment by inhalation is much larger than direct external exposure coming from the radioactive cloud. Given its short half life of 8 days, exposure due to ^{131}I was dominant during the first days after the emissions. Besides, ^{131}I will be absorbed through contaminated food, and will accumulate in the thyroid gland. Its beta particles, with mean energy around 190 keV, have a tissue penetration of 0.6 to 2 mm, and can kill or transform thyroid tissue and affect lung cells if inhaled. Supplying non radioactive iodine in order to saturate thyroid tissue and or swift evacuation are the recommended procedure. Estimation of dose commitment is usually difficult but a crude number can be derived as follows:

Dose commitment from breathing is around 1.5×10^{-8} Sv/Bq for ^{131}I. Activity concentration was measured at the plant by Tepco and was found to be 1×10^{3} Bq/m^3 on April 7th, 2011. Dose rate at that particular point would have been 20 μSv/hr, considering a normal breathing rate of about 1.2 m^3/hr. However, ^{131}I concentrations must have been orders of magnitude larger shortly after pressure relief events. Moreover, other radionuclides like ^{137}Cs and ^{134}Cs were also suspended in air, though in smaller concentrations. Uncertainties will therefore be large for dose commitment from breath [32].

After contamination plume has passed, main source of radiation is from contaminated material deposited on the ground. ^{137}Cs and ^{134}Cs are the dominant radionuclides that provide the ground shine. ^{137}Cs with a half-life of 30 years is also the main long term contamination source.

In order to make a rough estimate of exposure due to ^{137}Cs, an infinite ground plane uniformly contaminated yielding scatterless photons with a semi isotropic emission pattern is the simplest model that can be considered. A simple geometric calculation for the photon fluence in that model, will show that for a given height above the ground, radiation is dominated by photons travelling horizontally [33]. With a mean free path in air of 280 m for absorption, scatterless photons is quite a good an simple approximation for primary photons. Obtained result shows that the infinite ground plane may not be a particularly good approximation, especially for urban areas. If exposure is dominated by horizontally travelling photons, ground roughness as well as structures above ground level will significantly affect the exposure field. In other words, in order to obtain accurate dose rate convertion factor, i.e., equivalent dose rates for a given ground activity, detailed Monte Carlo calculations would be required. For horizontally travelling isotropic primary photons calculations performed with MC code PENELOPE yield an estimated of 1.3×10^{-12} Sv/(Bq/ m^2) 1 m above ground level, which is in agreement with published data for smooth plane surface [33,34].

The Figure 8 shows that the photon fluence from a flat surface at 1 m height is dominated by horizontally travelling photons.

More detailed Monte Carlo calculations show that average dose rate conversion factor yield 2×10^{-12} (Sv/h)/(Bq/ m^2) for a fresh contaminated surface. Consequently, accumulated dose for a whole year exposure is around 1.6×10^{-8} Sv/(Bq/ m^2). Analog calculation were made for ^{134}Cs, yielding an exposure for the first year of about 3.7×10^{-8} Sv/(Bq/m^2)

In order to roughly estimate exposure levels to the areas surrounding Fukushima plant, we can take published values of ^{134}Cs and ^{137}Cs release from the power plant. Different estimates put that quantity in 1×10^{16} Bq for each radionuclide [35]. If we consider the 20 km evacuation zone, and assume half the amount of radioactive material was deposited inside the semicircle with origin in the power plant (with the other half released over the ocean), mean deposition would be 1.6×10^{7} Bq/m^2 for both radionuclides, yielding a 75 mSv/yr average exposure for the first year.

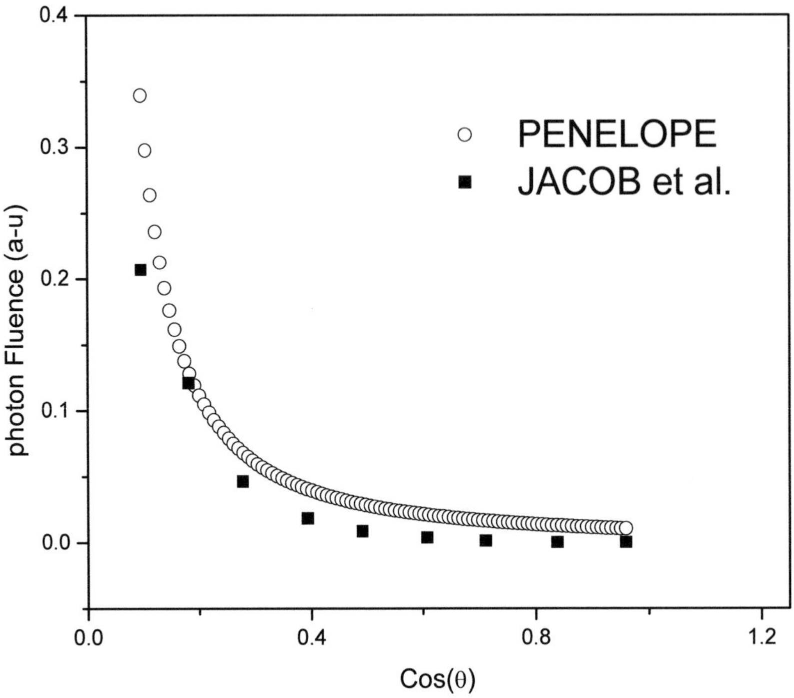

Figure 8. Photon fluence from a flat surface at 1m height

Deterministic radiation effects start at levels above 1000 mSv, but for exposures as low as 100 mSv, statistically significant increases in cancer cases are expected among the exposed population. Assuming that outside the evacuation area, exposure levels will not be higher than the average inside it, which is a conservative assumption, no significant increase in cancer cases should be expected outside the evacuation zone.

Of course, deposition is far from uniform, and a significant amount of Cs-134 and Cs-137 have been carried away by wind. Atmospheric transport models suggest that only 20% of the total release was deposited in Japan land, while 80 % was carried out by wind towards the Pacific Ocean thanks to the prevailing winds during all but one major release events, throughout the emergency [36].

Taking these results into account the average radioactive Cs deposition would fall from 1.7×10^7 Bq/ m^2 to 6.8×10^6 Bq/ m^2. and average exposure for the first year inside the 20 km exclusion zone to approximately 15 mSv/yr. Actual deposition maps from flight measurements carried out by DOE show that deposition in Japan soil took place towards the northwest with maximum values for the combined Cs deposition of 1.8×10^7 Bq/ m^2 close to the plant in the northwest direction. Measurements 20 km away from the plant, in the same direction show numbers around 1×10^6 Bq/ m^2, which will produce roughly the same exposure as average background natural radiation in Japan (3.8 mSv/yr), but still larger than the 1 mSv/yr exposure limit for the public [37].

http://dx.doi.org/10.5772/54262

8. Conclusion

In this chapter it is given a comprehensive introduction to the Fukushima nuclear accident. A large database of dose measurements taken in time intervals of ten minutes over more than 200 stations was studied using a code specifically designed for it. It is seen that the radiation decays with time following an exponential decay compatible with iodine contamination at large distances. The radiation profile decrease slower than the r^{-2} law, and seems to become flatter as time goes by. There is a large scatter of radiation, with large variations in distances of the order of a hundred meters, but it seems to become more homogenous with time. Most affected zone is a strip pointing to the northwest of Fukushima. Numeric simulations are performed with the Monte Carlo code PENELOPE to find the equivalence between the published values of the activity and the dose equivalent. It is found that the exclusion zone of 20 km give an exposure of the same order as the background radiation in Japan or less.

Appendix

Chronology of the disaster of Fukushima

	Unit 1	**Unit2**	**Unit 3**	**Unit 4**	**Unit 5**	**Unit 6**
March 11, 2011 14:46 JST, earthquake	scrammed	scrammed	scrammed	-	-	-
	loss-of-offsite power lines	loss-of-offsite power lines	loss-of-offsite power lines	loss-of-offsite power lines	loss-of-offsite power lines	loss-of-offsite power lines
	Emergency diesel generators started	Emergency diesel generators started	Emergency diesel generators started	Emergency diesel generators started	Emergency diesel generators started	Emergency diesel generators started
14:50 JST		reactor core isolation system (RCIC) started				
14:52 JST	IC (isolation condenser) automatically actuated					
15:03 JST	IC manually shutdown					
15:05 JST			reactor core isolation system (RCIC) started			

	Unit 1	Unit2	Unit 3	Unit 4	Unit 5	Unit 6
15:27 JST, 1st tsunami wave						
15:35 JST, 2nd tsunami wave	Alternate and direct current power lines were lost	Alternate and direct current power lines were lost	Alternate and direct current power lines were lost	Alternate and direct current power lines were lost	electric power sourced by diesel generator at unit 6	Air cooled electric diesel generator survived
	residual heat removal pumps lost	residual heat removal pumps lost	residual heat removal pumps lost	residual heat removal pumps lost	residual heat removal pumps lost	residual heat removal pumps lost
	High pressure coolant injection (HPCI) inoperable					
18:00 JST	water level reach the top of the fuel, temperature rose					
March 12, 2011 02:44 JST			emergency battery for high pressure core-flooder runs out			
05:30 JST	steam and hydrogen vented into secondary containment					
05:50 JST	Fresh water injection started					
10:58 JST		steam and hydrogen vented into secondary containment				
14:50 JST	fresh water injection halted					
15:36 JST	Hydrogen explosion, secondary containment blown up					
19:00 JST	Sea water injection					
March 13, 2011 02:44 JST			high pressure coolant injection stops			

http://dx.doi.org/10.5772/54262

	Unit 1	Unit2	Unit 3	Unit 4	Unit 5	Unit 6
07:00 JST			water level reaches top of the fuel			
13:00 JST	reactor vented, refilled with water and boric acid. Declared level 4 accident.	stable	reactor vented, refilled with water and boric acid			
March 14, 2011 11:01 JST		water supply damaged by explosion in reactor 3	hydrogen explosion, secondary containment blown up.			
13:15 JST		reactor core isolation cooling system stops				
18:00 JST		water level reaches the top of the fuel				
March 15, 2011 6:00 JST		Explosion in the pressure suppression room.				
11:00 JST		temporary cooling systems damaged by explosion at reactor 3.	second explosion. Radiation rates of 400 mSv/h.	fire breaks out		
March 16, 2011 06:00 JST	Workers withdrawn from the plant					
March 17, 2011 07:00 JST			helicopters drop water on the spent fuel pool. Radiation spike of 3.75 Sv/h. Police and fire trucks sprayed water into the reactor with high pressure hoses.	helicopters drop water on the spent fuel pool		
March 24, 2011	electrical power restored					
August 21, 2011	cold shutdown achieved					

Acknowledgements

We acknowledge Mr. Marian Steinbach for sharing the radiation data sheet and for clarifications about it. CG thanks Mr. Rama Hoetzlein for useful references.

Author details

Cristian R. Ghezzi[1], Walter Cravero[1,2] and Nestor Sanchez Fornillo[1]

1 National University of the South, Department of Physics, Bahía Blanca, Provincia de Buenos Aires, Argentina

2 Institute of Physics of the South, Bahía Blanca, Provincia de Buenos Aires, Argentina

References

[1] "Nuclear Testing and Nonproliferation", http://www.iris.iris.edu/HQ/Bluebook/contents.html

[2] "Fukushima faced 14-metre tsunami", World Nuclear News, http://www.world-nuclear-news.org/RS_Fukushima_faced_14-metre_tsunami_2303113.html

[3] TEPCO press release 3, 2011, http://www.tepco.co.jp/en/press/corp-com/release/11031103-e.html

[4] Japan earthquake update (2210 CET), IAEA press release, 2011, http://www.iaea.org/press/?p=1133

[5] United States Nuclear Regulatory Commission Technical Training Center, The Boiling Water Reactor Systems: Reactor Concepts Manual, 2009.

[6] GENERAL ELECTRIC, Boiling Water Reactor GE-BWR4 Technology, Technology Manual, chapter 6: BWR Differences, Editor: United States Nuclear Regulatory Commission Technical Training Center, 2009.

[7] "Insights from review and analysis of the Fukushima Dai-ichi accident", Hirano M., Yonomoto T., Ishigaki M., Watanabe N., Maruyama Y., Sibamoto Y., Watanabe T., and Moriyama K., 2012, Journal of Nuclear Science and Technology, 49, 1.

[8] Wikipedia, "Fukushima Daiichi Nuclear Disaster", http://en.wikipedia.org/wiki/Fukushima_Daiichi_nuclear_disaster#cite_note-tepco11b-75

[9] Gundersen A., http://fairewinds.org/content/gundersen-postulates-unit-3-explosion-may-have-been-prompt-criticality-fuel-pool

[10] Cottingham W. N., & Greenwood D. A., "An introduction to nuclear physics", Cambridge University Press, 1986.

[11] "How Much Spent Nuclear Fuel Does the Fukushima Daiichi Facility Hold?" Scientific American, March 17 2011, http://www.scientificamerican.com/article.cfm?id=nuclear-fuel-fukushima

[12] Murray R. L., "Nuclear energy: An Introduction to the Concepts, Systems, and Applications of the Nuclear Processes", Butterworth-Heinemann, 2001.

[13] Lewis, E. E., "Fundamentals of Nuclear Reactors Physics", Academic Press, 2008.

[14] United States Nuclear Regulatory Commission, http://www.nrc.gov/reading-rm/doc-collections/fact-sheets/

[15] Wikipedia, "MOX fuel", http://en.wikipedia.org/wiki/MOX_fuel

[16] Fukushima Nuclear Accident Analysis Report, Tokyo Electric Power Company (TEPCO), June 20, 2012.

[17] Fukushima Daiichi Status Report, International Atomic Energy Agency, April 27, 2012.

[18] Wikipedia, "Fukushima Daiichi Nuclear Power Plant", http://en.wikipedia.org/wiki/Fukushima_Daiichi_Nuclear_Power_Plant#cite_note-pu-7

[19] United States Nuclear Regulatory Commission, Technical Training Center, Chattanooga, TN, "Boiling Water Reactor Systems: Reactor Concepts Manual", http://www.nrc.gov/ reactors/ power.html.

[20] Japan Radiation Map, 2012, http://jciv.iidj.net/map/.

[21] Steinbach, M., "A Crowd Sourced Japan Radiation Spreadsheet", 2012, http://www.sendung.de/japan-radiation-open-data/

[22] GebWeb, http://gebweb.net/blogpost/2011/03/17/japan-radiation-map/

[23] Hoetzlein, R., Leonardo, "Visual Communication in Times of Crisis: The Fukushima Nuclear Accident", 24, 2, 113, 2012; http://www.rchoetzlein.com.

[24] "Radiation Measuring Station Locations", SPEEDi (System for Prediction of Environmental Emergency Dose Information), http://goo.gl/iDo0N

[25] "Ocupational Dose Limits", United States Nuclear Regulatory Commission, http://www.nrc.gov/reading-rm/doc-collections/cfr/part020/part020-1201.html

[26] Villarreal, E., http://public.tableausoftware.com/views/JapanRadiationLevels/ JapanRadiationLevelsDashboard

[27] "Fukushima radioactive fallout nears Chernobyl levels", New Scientist, March 24, 2011.

[28] "Japanese nuclear firm admits error on radiation reading", The Guardian, 27 March 2011, http://www.guardian.co.uk/world/2011/mar/27/japan-nuclear-error-radiation-reading

[29] "Fukushima radiation higher than first estimated", Reuter, Kevin Krolicki, May 24, 2012.

[30] Safecast, 2012, blog.safecast.org.

[31] "Japan targets phasing out nuclear power in 2030s", The Mainichi, September 15, 2012.

[32] Journal of Environmental Radioactivity, 109, 103, 2012.

[33] Saito and P. Jacob, Radiation Protection Dosimetry, 58, 29, 1995.

[34] F. Salvat, J.M. Fernández-Varea and J. Sempau, "PENELOPE–2008: A Code System for Monte Carlo Simulation of Electron and Photon Transport", OECD Nuclear Energy Agency, Issy-les-Moulineaux, France, 2008.

[35] Chino et al., J. Nucl. Sci. Tec., 48, 1129, 2011

[36] A. Stohl et al, J. Chem. Phys., 12, 2313, 2012

[37] U.S. Department of Energy: http://energy.gov/articles/us-department-energy-releases-radiation-monitoring-data-fukushima-area